MOLECULAR BIOLOGY
INTELLIGENCE
UNIT

Molecular Mechanisms of Muscular Dystrophies

Steve J. Winder, Ph.D.

Department of Biomedical Science
University of Sheffield
Sheffield, U.K.

CRC Press
Taylor & Francis Group
Boca Raton London New York

CRC Press is an imprint of the
Taylor & Francis Group, an **informa** business

MOLECULAR MECHANISMS OF MUSCULAR DYSTROPHIES

Molecular Biology Intelligence Unit

First published 2006 by Landes Bioscience

Published 2018 by CRC Press
Taylor & Francis Group
6000 Broken Sound Parkway NW, Suite 300
Boca Raton, FL 33487-2742

© 2006 by Taylor & Francis Group, LLC
CRC Press is an imprint of Taylor & Francis Group, an Informa business

First issued in paperback 2019

No claim to original U.S. Government works

ISBN 13: 978-0-367-44637-6 (pbk)
ISBN 13: 978-1-58706-264-3 (hbk)

Library of Congress Cataloging-in-Publication Data

Molecular mechanisms of muscular dystrophies / [edited by] Steve J. Winder.
    p. ; cm. -- (Molecular biology intelligence unit)
  Includes index.
  ISBN 1-58706-264-X
  1. Muscular dystrophy--Molecular aspects. I. Winder, Steve J. II. Series: Molecular biology intelligence unit (Unnumbered).
    [DNLM: 1. Dystrophin--genetics. 2. Muscular Dystrophies--genetics. 3. Dystrophin-Associated Proteins--genetics. WE 559 M717 2006]
  RC935.M7M67 2006
  616.7'48042--dc22

                                                    2005028600

To Kathryn

CONTENTS

EDITOR

Steve J. Winder
Department of Biomedical Science
University of Sheffield
Sheffield, U.K.
Chapter 16

CONTRIBUTORS

Simone Abmayr
Department of Neurology
Muscular Dystrophy Cooperative
  Research Center
University of Washington School
  of Medicine
Seattle, Washington, U.S.A.
Chapter 2

Marvin E. Adams
Department of Physiology
  and Biophysics
University of Washington
Seattle, Washington, U.S.A.
Chapter 4

Qing Bai
Department of Human Anatomy
  and Genetics
University of Oxford
Oxford, U.K.
Chapter 3

Derek J. Blake
Department of Pharmacology
University of Oxford
Oxford, U.K.
Chapter 5

Susan C. Brown
Dubowitz Neuromuscular Unit
Imperial College
Hammersmith Campus
London, U.K.
Chapter 7

Edward A. Burton
Department of Clinical Neurology
Radcliffe Infirmary
and
Department of Human Anatomy
  and Genetics
University of Oxford
Oxford, U.K.
Chapter 3

Kate M.D. Bushby
Institute of Human Genetics
International Centre for Life
Newcastle upon Tyne, U.K.
Chapter 8

Olli Carpén
Department of Pathology
University of Helsinki
Helsinki, Finland
Chapter 11

Jeff Chamberlain
Department of Neurology,
  Medicine and Biochemistry
Muscular Dystrophy Cooperative
  Research Center
University of Washington School
  of Medicine
Seattle, Washington, U.S.A.
Chapter 2

Kay E. Davies
MRC Functional Genetics Unit
Department of Human Anatomy
  and Genetics
University of Oxford
Oxford, U.K.
Chapter 3

Alan E.H. Emery
Department of Neurology
Royal Devon and Exeter Hospital
Exeter, U.K.
Foreword

James M. Ervasti
Department of Physiology
University of Wisconsin Medical School
Madison, Wisconin, U.S.A.
Chapter 1

Roland Foisner
Max F. Perutz Laboratories
University Departments
  at the Vienna Biocenter
Department of Medical Biochemistry
University of Vienna
Vienna, Austria
Chapter 12

Stanley C. Froehner
Department of Physiology
  and Biophysics
University of Washington
Seattle, Washington, U.S.A.
Chapter 4

Ferruccio Galbiati
Department of Pharmacology
University of Pittsburgh School
  of Medicine
Pittsburgh, Pennsylvania, U.S.A.
Chapter 9

Josef Gotzmann
Max F. Perutz Laboratories
University Departments
  at the Vienna Biocenter
Department of Medical Biochemistry
University of Vienna
Vienna, Austria
Chapter 12

Rami Jarjour
IBLS
Division of Molecular Genetics
University of Glasgow
Anderson College
Glasgow, Scotland
Chapter 13

Keith Johnson
IBLS
Division of Molecular Genetics
University of Glasgow
Anderson College
Glasgow, Scotland
Chapter 13

Steven H. Laval
Institute of Human Genetics
International Centre for Life
Newcastle upon Tyne, U.K.
Chapter 8

Michael P. Lisanti
Department of Molecular Pharmacology
The Albert Einstein Cancer Center
Albert Einstein College of Medicine
Bronx, New York, U.S.A.
Chapter 9

Elizabeth M. McNally
Department of Medicine
Department of Human Genetics
The University of Chicago
Chicago, Illinois, U.S.A.
Chapter 10

Judith Melki
Molecular Neurogenetics Laboratory
Institut National de la Sant et de la
    Recherche Médicale (INSERM)
Evry, France
Chapter 14

Francesco Muntoni
Dubowitz Neuromuscular Unit
Imperial College
Hammersmith Campus
London, England, U.K.
Chapter 7

Kay Ohlendieck
Department of Biology
National University of Ireland
Maynooth, Ireland
Chapter 15

Robert Olaso
Molecular Neurogenetics Laboratory
Institut National de la Sant et de la
    Recherche Médicale (INSERM)
Evry, France
Chapter 14

Terence A. Partridge
Muscle Cell Biology Group
MRC Clinical Sciences Centre
ICSM Hammersmith Hospital Site
London, England, U.K.
Chapter 18

Justin M. Percival
Department of Physiology
    and Biophysics
University of Washington
Seattle, Washington, U.S.A.
Chapter 4

Markus A. Ruegg
Biozentrum
University of Basel
Basel, Switzerland
Chapter 6

Nouzha Salah
Molecular Neurogenetics Laboratory
Institut National de la Sant et de la
    Recherche Médicale (INSERM)
Evry, France
Chapter 14

Roy V. Sillitoe
Department of Pharmacology
University of Oxford
Oxford, U.K.
Chapter 5

Jérémie Vitte
Molecular Neurogenetics Laboratory
Institut National de la Sant et de la
    Recherche Médicale (INSERM)
Evry, France
Chapter 14

Dominic J. Wells
Gene Targeting Unit
Department of Neuromuscular Diseases
Division of Neuroscience
    and Psychological Medicine
Imperial College London
Charing Cross Hospital
London, England, U.K.
Chapter 17

FOREWORD

The first detailed study of the clinical features, hereditary nature and pathology of muscular.dystrophy is attributed to Edward Meryon, an English physician, who published his findings in 1852. Yet it was more than 130 years later that the responsible gene for the commonest form of dystrophy, Duchenne muscular dystrophy, was identified and characterised.[1] Subsequently the genes for many other forms of dystrophy have been identified, and it is estimated that there are now approaching 40 such disorders, though some of these are very rare. However once the gene for Duchenne dystrophy had been identified and its protein product, dystrophin, found to be localised to the muscle membrane, many thought it would not be very long before the details of the pathogenesis of this disorder would be understood, with the possibility of finding an effective treatment. But clearly there is still a great deal we do not understand.

The structure and function of the dystrophin-glycoprotein complex still exercises the attention of biochemists. It is now clear that many other proteins interact with this complex. Some components, such as syntrophin, may not themselves be associated with any specific form of dystrophy yet may still play a key role in cell signalling. In certain forms of dystrophy the defective proteins have been found to be associated not with the dystrophin-glycoprotein complex but either lie outside the complex (e.g., laminin A2 chain of merosin) or even be associated with the inner nuclear membrane (lamins A/C and emerin). Furthermore in certain dystrophies primary defects of glycosyltransferase activity have been reported. Finally myotonic dystrophy (a multi system disease and therefore some would argue not a 'true' dystrophy), oculopharyngeal dystrophy and facioscapulohumeral dystrophy are each associated in their different ways with tandem repetitive DNA sequences.

In view of the considerable variation in their clinical features, genetics and molecular pathologies, one questions whether the all-embracing term 'dystrophy' continues to be relevant. Perhaps a solution might be to base nomenclature on specific protein defects such as dystrophinopathies, sarcoglycanopathies, and laminopathies.

An unexpected finding in certain dystrophies, but now reported in some other conditions though not to the same extent, is the phenomenon of very different phenotypes being generated by different alleles. For example, different mutations of the LMNA gene have now been shown to account for two different forms of dystrophy as well as for no less than six other clinically distinct disorders. This finding could have important implications for any approach to therapy in future.

A clear understanding of pathogenetic processes could one day lead to the finding of a drug which in some way interferes with these processes and thereby be therapeutically effective. Meanwhile attention has turned to the possibility of various forms of gene therapy and stem-cell therapy. In gene therapy approaches currently being pursued include using a viral vector for gene replacement. However in view of the recently reported cases of 'insertional mutagenesis' (following gene therapy for severe combined immunodeficiency resulting in T-cell leukaemia), several other approaches are being pursued: for example, appropriate antisense oligonucleotides to effect exon skipping and the possibility of the upregulation of a compensatory gene product. With regard to the latter approach, in Duchenne dystrophy upregulation of utrophin (or perhaps ADAM-12 or calcineurin), may compensate for the deficiency of dystrophin and a search is ongoing to find a possible drug to do this in patients. Stem-cell therapy (as opposed to myoblast transplantation) offers an attractive alternative but to be therapeutically effective a means will have to be found to improve the efficiency of vascular delivery.

There is no doubt that the study of the muscular dystrophies in recent years has been exciting and rewarding. It has attracted the attention of many investigators of international repute, and this is reflected in the various contributions to this volume. One is reminded of what Bertolt Brecht expressed so aptly: 'Beauty in nature is a quality which gives the human senses a chance to be skilful'.

Alan E.H. Emery
Department of Neurology
Royal Devon and Exeter Hospital
Exeter, U.K.

1. Emery AEH, Emery MLH. The history of a genetic disease: Duchenne muscular dystrophy or Meryon's disease. RSM Press, London, 1995.

Acknowledgments

This book represents a personal collection of chapters covering a spectrum of muscular dystrophies and one myopathy that to some extent reflects my own interests and expertise rather than any attempt to produce a comprehensive treatise on every muscular dystrophy so far characterised. Having said that, the works included represent most of the major muscular dystrophies and in particular those where a molecular understanding of the underlying mechanisms is most advanced. I would like to thank Ron Landes for the concept and opportunity to make this possible, the many contributors who provided the excellent chapters contained in the book, making my role as an editor so much easier, and not least to my own research group who have put up with my absence from the bench for the last few months.

Steve J. Winder
Sheffield, U.K.

CHAPTER 1

Structure and Function of the Dystrophin-Glycoprotein Complex

James M. Ervasti

Introduction

Duchenne muscular dystrophy (DMD) is the most prevalent and severe form of human muscular dystrophy. While clinical descriptions of DMD date back to the 1850s, over 100 years passed before evidence suggested that the muscle cell plasma membrane, or sarcolemma, is compromised in DMD muscle. The molecular basis for DMD and its associated sarcolemmal instability became more clear with landmark studies published in the mid-to-late 1980s which identified the gene defective in DMD.[1] The DMD locus spans over 2.5 million bases distinguishing it as the largest gene in the human genome. The array of transcripts expressed from the DMD gene is complex due to the presence of multiple promoters and alternative splicing. The largest transcripts encode a four-domain protein with a predicted molecular weight of 427,000, named dystrophin. Dystrophin is the predominant DMD transcript expressed in striated muscle and DMD gene mutations, deletions or duplications most frequently result in a loss of dystrophin expression in muscle of patients afflicted with DMD. Based on its localization to the cytoplasmic face of the sarcolemma and sequence similarity with domains/motifs common to proteins of the actin-based cytoskeleton, dystrophin was hypothesized early on to play a structural role in anchoring the sarcolemma to the underlying cytoskeleton and protect the sarcolemma against stress imposed during muscle contraction or stretch. Biochemical studies aimed at confirming the hypothesized structure and function of dystrophin revealed its tight association with a multi-subunit complex, the so-named dystrophin-glycoprotein complex. Since its description, the dystrophin-glycoprotein complex has emerged as an important structural unit of muscle and also as a critical nexus for understanding muscular dystrophies arising from defects in several distinct genes.

Initial Isolation and Characterization of the Dystrophin-Glycoprotein Complex

Following close behind the identification of dystrophin as the protein missing from DMD muscle, the laboratory of Kevin Campbell demonstrated that dystrophin could be solubilized from the membrane vesicle fraction of skeletal muscle homogenates using the detergent digitonin and dramatically enriched by wheat germ agglutinin chromatography.[2] Given that the primary sequence contained no hydrophobic stretches to directly anchor dystrophin within the sarcolemma, it seemed most likely that dystrophin indirectly bound to wheat germ agglutinin through association with a membrane glycoprotein embedded in the sarcolemma and this idea was confirmed by experiments showing that dystrophin binding to wheat germ agglutinin was disrupted by chaotropic agents.[2] While anion-exchange chromatography further amplified and purified dystrophin over wheat germ agglutinin chromatography alone, numerous proteins remained in the peak dystrophin fractions and lectin blotting minimally revealed

Molecular Mechanisms of Muscular Dystrophies, edited by Steve J. Winder. ©2006 Eurekah.com.

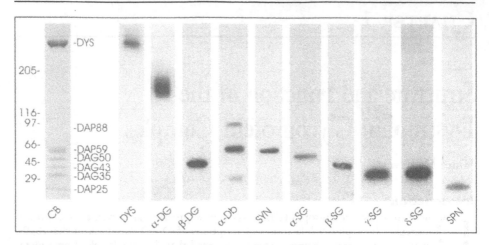

Figure 1. Protein constituents of the dystrophin-glycoprotein complex. Shown on the left is a Coomassie blue-stained SDS-polyacrylamide gel loaded with dystrophin-glycoprotein complex purified from rabbit skeletal muscle. Molecular weight standards are indicated on the left while the molecular weights of each dystrophin-associated protein (DAP) or glycoprotein (DAG) are indicated on the right. Also shown are western blots containing electrophoretically-resolved dystrophin-glycoprotein complex that was enriched from digitonin-solubilized mouse skeletal muscle by wheat germ agglutinin chromatography. Western blots were stained with antibodies specific for dystrophin (DYS), α- and β-dystroglycan (DG), α-dystrobrevin (α-Db), syntrophin (SYN), α-, β-, γ- and δ-sarcoglycans (SG) and sarcospan (SPN).

12 copurifying glycoproteins as potential molecular partners for dystrophin.[2] Sucrose gradient centrifugation further resolved potential candidates to 9 proteins stained by Coomassie blue that were shown to strictly cosediment with dystrophin (Fig. 1): a singlet of 88,000, a triplet of 59,000, a singlet of 50,000, a doublet of 43,000, a singlet of 35,000 (but present at a molar ratio of ~2:1 relative to dystrophin) and a singlet of 25,000 apparent M_r. Lectin blotting identified the 50,000, 43,000 and 35,000 species as glycoproteins and further revealed a broad band with an apparent molecular weight of 156,000.[3] While the 156,000 M_r protein was poorly stained by Coomassie blue, its strong staining by wheat germ agglutinin and strict cosedimentation with dystrophin nonetheless elevated its candidacy as a sarcolemmal glycoprotein receptor for dystrophin.[3] Thus, the list of potential dystrophin-associated proteins was narrowed down to 10 distinct M_r proteins, 5 of which were glycosylated.

Importantly, the Campbell lab was simultaneously pursuing a long-term project aimed at generating new monoclonal antibodies to calcium channels expressed in muscle. Since several calcium channel subunits were known to be glycosylated, wheat germ agglutinin-enriched fractions from detergent solubilized muscle membranes were used to immunize mice and screen hybridomas. Screening positive clones against dystrophin-enriched preparations yielded monoclonal antibodies to dystrophin, the 156,000 and 50,000 M_r dystrophin associated glycoproteins.[3] The new antibodies were instrumental in confirming through coimmunoprecipitation experiments the tight association of proteins that cosedimented with dystrophin as well as their colocalization with dystrophin at the sarcolemma.[3-5] Moreover, the monoclonal antibodies were critical in demonstrating that the abundance of the 156,000 and 50,000 M_r dystrophin-associated glycoproteins was dramatically reduced in DMD muscle.[3,6,7] Since these proteins colocalized with dystrophin in situ, copurified with dystrophin in stoichiometric amounts even after several distinct protein purification methodologies, and were diminished in dystrophin-deficient muscle, it was concluded that dystrophin was part of a large, hetero-oligomeric complex that may serve to stabilize the sarcolemma. Because several constituents were glycosylated and exploitation of this characteristic was so important in their isolation, the assembly of proteins associated with dystrophin was named the dystrophin-glycoprotein complex.

Additional biochemical analyses identified the 156,000 glycoprotein and 59,000 M_r triplet as peripheral membrane proteins[5] while the 50,000, 43,000, 35,000 and 25,000 M_r species behaved as a subcomplex of integral membrane proteins.[5,8] Based on its extensive glycosylation[5,9] and peripheral membrane association, the 156,000 M_r dystrophin-associated glycoprotein was hypothesized to reside on the extracellular face of the sarcolemma and possibly function as a receptor for a component of the extracellular matrix. These hypotheses were born out with the cloning/sequencing of the gene encoding the 156,000 dystrophin-associated glycoprotein, which is expressed from a single transcript along with one of the 43,000 M_r dystrophin-associated glycoproteins.[6] The propeptide is proteolytically processed into a wholly extracellular 156,000 subunit and a 43,000 M_r single-pass transmembrane subunit. Based on its extensive glycosylation and association with dystrophin, the 156,000 and 43,000 M_r subunits were renamed α- and β-dystroglycan, respectively. A screen of known extracellular matrix molecules for skeletal muscle α-dystroglycan binding activity identified laminin as the first extracellular ligand for α-dystroglycan.[6,9] Laminin-Sepharose pull-down of the entire dystrophin complex definitively demonstrated that α-dystroglycan was a stoichiometric component of the complex.[9] These results also led to the hypothesis that the dystrophin-glycoprotein complex may play a role in muscle cell adhesion to the basal lamina.

Working independently, Ozawa and colleagues corroborated[10-12] many of the key findings first reported by the Campbell laboratory and they made some important original contributions in elucidating several of the protein-protein interactions within the complex (discussed below). However, Ozawa and colleagues strongly disputed two important conclusions of Campbell's group. First, they initially dismissed α-dystroglycan as an important component of the dystrophin-glycoprotein complex because it could not be stained by Coomassie blue.[13] Ozawa and colleagues also strongly contested the initial identification of the 59,000 M_r α-dystrobrevin/syntrophin triplet as cytoplasmic peripheral membrane proteins because their experiments led them to conclude that one of the proteins was a transmembrane glycoprotein.[12] In subsequent work,[14] it is clear that Ozawa and colleagues ultimately concurred that α-dystroglycan is an important component of the complex and that syntrophins are nonglycosylated cytoplasmic proteins. Using limited proteolysis, wheat germ agglutinin chromatography and an array of site-specific antibodies, Ozawa and colleagues first demonstrated that the cysteine-rich and first half of the C-terminal domains of dystrophin were important for its binding to the glycoprotein complex.[11] By blot overlay assay, they showed that β-dystroglycan, and the 88,000 and 59,000 M_r dystrophin-associated proteins (α-dystrobrevins and syntrophins) directly bound the cysteine-rich and/or C-terminal domains of dystrophin,[15] despite failing to reconcile their prior crosslinking studies where they concluded that the 50,000 and 35,000 M_r dystrophin-associated glycoproteins were directly associated with dystrophin and most important in anchoring it to the sarcolemma.[10] Ozawa and colleagues also more conclusively showed[16] that the dystrophin-glycoprotein complex could be dissociated into 3 sub-complexes consisting of α- and β-dystroglycan (dystroglycan complex), the 50,000, 43,000, and 35,000 M_r dystrophin-associated glycoproteins (sarcoglycan complex), dystrophin plus the 87,000 and 59,000 Mr dystrophin-associated proteins (dystrophin/dystrobrevin/syntrophin complex). However, it bears noting that several studies predating the work of Ozawa and colleagues reported data suggesting resolution of sub-complexes with similar molecular compositions.[5,8,17,18]

The largely biochemical studies described above suggested that the dystrophin-glycoprotein complex may function to physically couple the sarcolemmal cytoskeleton with the extracellular matrix (Fig. 2) and that loss of this structural linkage may render the sarcolemma more susceptible to damage when exposed to mechanical stress. The purified dystrophin-glycoprotein complex also provided a substrate for peptide sequencing and antibody production which yielded new probes important in the identification of genes encoding all dystrophin associated proteins and elucidation of their respective roles in Duchenne and other forms of muscular dystrophy. Notably, the genes encoding several dystrophin-associated proteins cause forms of muscular dystrophy when mutated in humans or when knocked out in mice. Since dystrophin

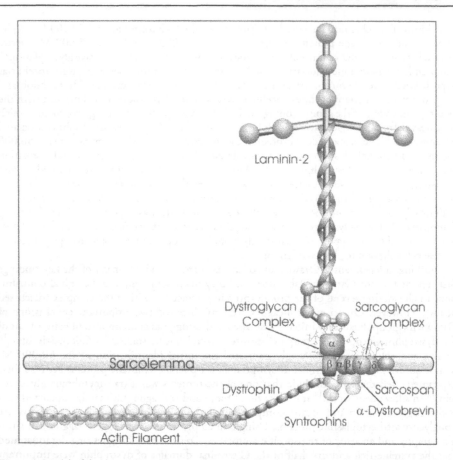

Figure 2. Model of the dystrophin-glycoprotein complex. Dystrophin is thought to physically couple the sarcolemma with the costameric cytoskeleton (see Fig. 3) through lateral association of its N-terminal and rod domains with cytoplasmic γ-actin filaments and through direct binding of its cysteine-rich domain to the β-subunit of the dystroglycan complex. Oligosaccharides present on α-dystroglycan are important for its binding to laminin-2 of the extracellular matrix. The sarcoglycan-sarcospan complex is thought to strengthen the noncovalent binding of dystrophin and α-dystroglycan with β-dystroglycan. α-Dystrobrevin appears to couple dystrophin with the sarcoglycan-sarcospan complex and also link the dystrophin-glycoprotein complex with the intermediate filament cytoskeleton (see Fig. 4). Two syntrophin molecules are anchored through homologous binding sites present in the C-terminal domains of dystrophin and α-dystrobrevin. Syntrophins are thought to serve as adaptor molecules that localize signaling molecules near the sarcolemma (see Fig. 4). Redrawn from ref. 95.

and its associated proteins are each a story in and of themselves, I will leave their detailed discussions to be elaborated in the specific chapters that follow. Below I will summarize the large body of evidence supporting an important mechanical role for the dystrophin-glycoprotein complex and then discuss its function within the larger protein network of skeletal muscle. Finally, I will propose an engineering design analogy that I believe best fits existing data.

In Support of a Mechanical Function for the Dystrophin-Glycoprotein Complex

Within skeletal myofibers, dystrophin is enriched in a discrete, rib-like lattice termed costameres.[19,20] Costameres are protein assemblies that circumferentially align in register with

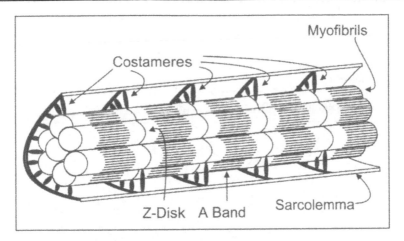

Figure 3. The costameric cytoskeleton of striated muscle. Dystrophin is enriched at costameres, protein assemblies that circumferentially align in register with the Z disk of peripheral myofibrils and physically couple force-generating sarcomeres with the sarcolemma. Redrawn from ref. 21.

the Z disk of peripheral myofibrils and physically couple force-generating sarcomeres with the sarcolemma in striated muscle cells (Fig. 3). A variety of data indicate that costameres are a striated muscle-specific elaboration of the focal adhesions expressed by nonmuscle cells.[21] Classical experiments by Street[22] and the Sangers[23] suggest that costameres function to laterally transmit contractile forces from sarcomeres across the sarcolemma to the extracellular matrix and ultimately to neighboring muscle cells. Lateral transmission of contractile force would be useful for maintaining uniform sarcomere length between adjacent actively contracting and resting muscle cells comprising different motor units within a skeletal muscle. It is also logical that the sites of lateral force transmission across the sarcolemma would be mechanically fortified to minimize stress imposed on the relatively labile lipid bilayer. Other results suggest that costameres may also coordinate an organized folding, or "festooning" of the sarcolemma,[22,24] which again may minimize stress experienced by the sarcolemmal bilayer during forceful muscle contraction or stretch. Thus, in support of its hypothesized mechanical function, dystrophin is enriched within a structure (the costamere) that likely transmits mechanical force to and through the sarcolemma.

Consistent with its enrichment at costameres in normal muscle, the absence of dystrophin in humans and mice leads to a disorganized costameric lattice.[19,25-28] Extensive data consistently report that the dystrophin-deficient sarcolemma is exceedingly fragile[29-31] resulting in dramatically increased movement of membrane impermeant molecules across the sarcolemma of dystrophin-deficient muscle.[30-42] Both necrosis and sarcolemmal permeability of dystrophin-deficient muscle are exacerbated by physical exercise and improved by muscle immobilization.[33,35,39,42-45] Physiological studies have demonstrated that force production by dystrophin-deficient muscle is significantly decreased when normalized against muscle cross-sectional area.[40,41,46-59] Interestingly, force output by dystrophin-deficient muscle is hypersensitive to lengthening, or eccentric contraction[60,61] and the force decrement exhibited by dystrophin-deficient muscle undergoing eccentric contraction positively correlates with acutely increased sarcolemmal permeability.[40,41,53-55,59-63] Immunofluorescence analysis of mechanically peeled sarcolemma has demonstrated that dystrophin at costameres is tightly attached to the sarcolemma[20] and its presence is necessary for strong coupling between the sarcolemma and costameric actin filaments comprised of cytoplasmic γ-actin.[64] Transgenic overexpression of the dystrophin homologue utrophin, or a dystrophin construct retaining the β-dystroglycan binding site and one actin binding domain is sufficient to restore coupling

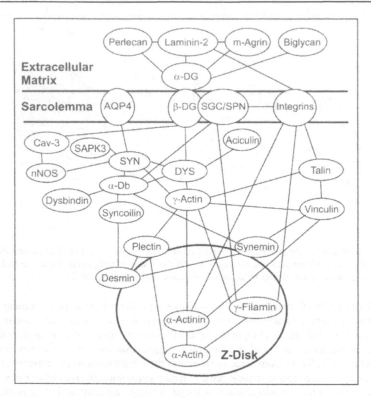

Figure 4. The protein interacting network of the dystrophin-glycoprotein complex. AQP4: aquaporin 4; Cav-3: caveolin-3; nNOS: neuronal nitric oxide synthase; SAPK3: stress-activated protein kinase 3 For a full list of supporting references, refer to the supplementary information in ref. 21 from which this figure was redrawn.

between the sarcolemma and costameric actin and rescue the sarcolemmal permeability defects accompanying dystrophin deficiency.[65,66] Dystrophin is also enriched in costameres of cardiac muscle.[67] Like skeletal muscle, dystrophin-deficient cardiac myocytes are abnormally vulnerable to mechanical stress-induced contractile failure and injury.[68,69] Finally, knockout of the dystroglycan or sarcoglycan complexes also causes muscular dystrophy that is accompanied by defects in sarcolemmal integrity.[70-79] When taken together, the above studies provide compelling evidence that the dystrophin-glycoprotein complex mainly functions to anchor the sarcolemma to costameres and stabilize it against the mechanical forces transduced through costameres during muscle contraction or stretch.

Expanding beyond the Dystrophin-Glycoprotein Complex

Since its initial description in 1990, many additional proteins have been shown to interact with different dystrophin-glycoprotein complex components (Fig. 4). Once the laminin α-chain G-domain was identified as the binding site for α-dystroglycan,[80] several other proteins containing homologous G-domain modules were interrogated and shown to bind α-dystroglycan with high affinity. The current list of such proteins includes agrins,[81-84] neurexins[85] and perlecan.[86,87] Like laminins, these proteins all bind to α-dystroglycan in a manner dependent on its oligosaccharide modifications.[88] In contrast, the chondroitin sulfate chains of the proteoglycan biglycan have been shown to mediate its binding to the core protein of α-dystroglycan.[89] While the functional significance of α-dystroglycan binding to several different extracellular matrix molecules is not fully clear, the results suggest that the dystroglycan complex may serve multiple roles that vary with the extracellular ligand to which it is bound. That agrins, neurexins and perlecan have all been implicated in various aspects of synapse formation or function[90] further suggests participation by the dystroglycan complex in this process.

Based on its sequence similarity with the calponin homology actin binding domains of β-spectrin and α-actinin, the N-terminal domain of dystrophin was hypothesized to bind actin filaments. Indeed, recombinant proteins encoding the dystrophin N-terminal domain do bind actin filaments,[91] but with relatively low affinity.[92] In contrast, purified dystrophin-glycoprotein complex bound actin filaments with substantially higher affinity compared to the isolated amino-terminal domain. The stoichiometry of binding however, suggested a more extensive lateral association between dystrophin and actin filaments than could be explained by actin binding solely through the N-terminal domain alone.[93] Mapping of the actin binding sites in dystrophin through F-actin cosedimentation of fragments after limited proteolysis led to the identification of a second actin binding site situated in the middle third of the dystrophin rod domain.[93] The spectrin-like repeats in this novel actin binding site were found to carry an excess of basic amino acid residues and were shown to bind acidic actin filaments largely through electrostatic interaction.[94] Because the middle rod actin binding site of dystrophin was separated from the amino-terminal actin binding domain by ~1200 amino acids, the two sites were proposed to act in concert to effect an extended lateral association that could account for the measured stoichiometry of binding. In addition, the dystrophin-glycoprotein complex was shown to slow depolymerization of actin filaments in vitro.[93,95] These data supported a novel actin filament side-binding model for dystrophin and are entirely consistent with a role for the dystrophin-glycoprotein complex in mechanically coupling the sarcolemma with actin filaments of the costameric cytoskeleton.[64]

The development of two-hybrid methodologies for the identification of protein-protein interactions has led to the discovery of several proteins that interact with the dystrophin-glycoprotein complex. Two-hybrid screens using α-dystrobrevin as bait identified several novel proteins.[96-98] Two of these proteins, synemin/desmuslin[98] and syncoilin,[99] are structurally related to intermediate filament proteins and interact with the classical intermediate filament protein desmin. Interestingly, mice knocked out for either α-dystrobrevin[100] or desmin[101,102] exhibit skeletal and cardiomyopathy, which suggests that mechanical coupling of the dystrophin-glycoprotein complex to the intermediate filament cytoskeleton is necessary for normal muscle function (see Chapter 5). Two hybrid screens using the cytoplasmic domains of sarcoglycans identified a skeletal muscle-specific form of filamin (γ-filamin) as a sarcoglycan interacting protein.[103] Like dystrophin, filamin contains an N-terminal calponin homology actin binding domain and a large number of repeated motifs, although the repeats in filamin differ in structure from those in dystrophin. Interestingly, filamin A is recruited to focal adhesions of nonmuscle cells in response to local mechanical stress applied via collagen-coated magnetic beads.[104] Since γ-filamin is upregulated and recruited to the sarcolemma in dystrophin-deficient muscle,[103] it is tempting to speculate that the dystrophin-deficient costamere may "sense" increased mechanical stress and attempt to compensate by recruitment of filamin. In addition to γ-filamin, several more actin binding proteins of costameres are upregulated in dystrophin-deficient muscle including the cytolinker plectin,[105] the integrin-associated proteins talin and vinculin[106] and the laminin receptor α7β1 integrin.[107,108] While not normally present at costameres, the dystrophin homologue utrophin is upregulated and recruited to costameres in dystrophin-deficient muscle.[64,65] Based on the protein interaction network illustrated in Figure 4, it seems most reasonable that all of these structural proteins are upregulated by the dystrophin-deficient muscle cell in an attempt to compensate for the absence of dystrophin by fortifying the weakened costamere through the recruitment of parallel mechanical linkages. Because dystrophy persists, these parallel linkages are either not completely redundant with the dystrophin-glycoprotein complex, or the compensatory upregulation/recruitment is incomplete. In support of the latter possibility, transgenic overexpression of utrophin[37,53,54] or $\alpha 7$ integrin[109] has been shown to further compensate for dystrophin deficiency. Thus, many of the proteins found to interact with the dystrophin-glycoprotein complex, or upregulated in its absence, appear to couple the complex with other structural elements of muscle, or form parallel mechanical links between the sarcolemma and myofibrillar apparatus. As such, these findings further reinforce an important mechanical function for the dystrophin-glycoprotein complex.

An Engineering Design Analogy

In many respects, the bulk of experimental data indicate that the dystrophin-glycoprotein complex functions in a manner analogous to the two-by-four (~2 inch x 4 inch) timbers used to frame the typical American stick house. The architect utilizes two-by-fours as one structural element of a sturdy support matrix (i.e., the costamere) intended to securely hold in place weatherproof siding and shingles (i.e., the sarcolemma) as well as doors and windows (i.e., ion channels and pumps) that allow for controlled movement of occupants, air and light both into and out of the house. When built to the architect's specifications, the house (or normal muscle cell) can withstand the stress imposed on it by extremes in weather such as high winds or heavy snowfall (or muscle contraction). If, however, the house is built during a shortage of two-by-fours (i.e., the dystrophin-deficient muscle cell), the carpenter may be forced to substitute two-by-two timbers instead (i.e., compensatory upregulation of partially redundant structural proteins). Such an alteration from original design may indeed allow the carpenter to construct a house that stands in calm weather. Conversely, the compromised structure may distort sufficiently under the force of gravity to cause doors and windows to stick or not close tightly. Moreover, the house built with substandard structural elements is certainly less likely to remain intact when more severe weather strikes.

Up to this point, the dystrophin-glycoprotein complex as two-by-four analogy has not taken into account that several interacting proteins suggest additional roles for the dystrophin-glycoprotein complex in organizing molecules involved in cellular signaling (Fig. 4). For example, α-syntrophin anchors neuronal nitric oxide synthase to the sarcolemma,[110] which is necessary to properly regulate vascular perfusion in active muscle.[111,112] Other data indicate that MAP kinase signaling pathways are perturbed in dystrophin-deficient muscle.[113-115] Because the putative role(s) for the dystrophin-glycoprotein complex in cell signaling will no doubt be elaborated in subsequent chapters, the reader may be left wondering whether and how a role in signaling fits with its well-supported mechanical function. However, I believe the dystrophin-glycoprotein complex as two-by-four analogy fits well with its role in anchoring signaling molecules and also can explain the signaling perturbations observed in dystrophin-deficient muscle. While the architect clearly intended the two-by-fours as structural support for the weather-proofing and controlled entry components of the house, this primary function does not prevent the electrician, plumber, telephone and cable television installers from subsequently utilizing the two-by-four framework as a support to route and organize additional regulatory and communication systems (i.e., cell signaling pathways) that further enhance functionality of the house. These additional systems would very reasonably be expected to malfunction in a house constructed of substandard structural components (i.e., two-by-twos instead of two-by-fours) as a secondary consequence of its distortion under gravity or when challenged by more stringent weather conditions. Alternatively, one could as reasonably argue that mechanical distortion of the structurally weak two-by-two framework house may cause a short in the electrical system (i.e., altered cell signaling) that in turn destroys the house by catastrophic fire (i.e., apoptosis). In either case, I believe the architect could successfully argue that compromised mechanical integrity precipitated destruction of the house (and that was not his, but the carpenter's fault!). The mere association of signaling molecules with the dystrophin-glycoprotein complex and perturbations of signal transduction pathways in dystrophin-deficient muscle are not sufficient evidence to refute the compelling data supporting a primary mechanical function for the complex. Moreover, such data fail to provide compelling support that the dystrophin-glycoprotein complex actively regulates cellular signaling, or that altered signaling initiates the pathologies observed in dystrophin-deficient muscle. While these hypotheses certainly remain attractive (especially with respect to development of treatments for muscular dystrophy), the current challenge to the field is to design and perform experiments that rigorously test their validity.

In at least one respect, the dystrophin-glycoprotein complex as two-by-four analogy fails. When a house becomes too small for its occupants, the two-by-fours supporting static walls are demolished in order to expand rooms. In the case of muscle however, cells simultaneously grow under the influence of muscle contraction so the "two-by-four" framework of muscle cells must be sufficiently dynamic to expand with growth while simultaneously protecting against stress-induced membrane damage. Several studies have demonstrated that the dystrophin-glycoprotein complex and costameres are indeed dynamic structures capable of remodeling in vivo.[21] However, a remaining challenge is to understand how such fascinating and opposing functions are effected at the molecular level, or perhaps at the level of interacting protein networks.

Acknowledgements

I thank Ariana Combs and Inna Rybakova for the blot images used in Figure 1 and Kevin Sonnemann for helpful discussions. The author is supported by grants from the Muscular Dystrophy Association and the National Institutes of Health (AR42423).

References

1. O'Brien KF, Kunkel LM. Dystrophin and muscular dystrophy: Past, present, and future. Mol Genet Metab 2001; 74:75-88.
2. Campbell KP, Kahl SD. Association of dystrophin and an integral membrane glycoprotein. Nature 1989; 338:259-262.
3. Ervasti JM, Ohlendieck K, Kahl SD et al. Deficiency of a glycoprotein component of the dystrophin complex in dystrophic muscle. Nature 1990; 345:315-319.
4. Ohlendieck K, Ervasti JM, Snook JB et al. Dystrophin-glycoprotein complex is highly enriched in isolated skeletal muscle sarcolemma. J Cell Biol 1991; 112:135-148.
5. Ervasti JM, Campbell KP. Membrane organization of the dystrophin-glycoprotein complex. Cell 1991; 66:1121-1131.
6. Ibraghimov-Beskrovnaya O, Ervasti JM, Leveille CJ et al. Primary structure of dystrophin-associated glycoproteins linking dystrophin to the extracellular matrix. Nature 1992; 355:696-702.
7. Ohlendieck K, Matsumura K, Ionasescu VV et al. Duchenne muscular dystrophy: Deficiency of dystrophin-associated proteins in the sarcolemma. Neurology 1993; 43:795-800.
8. Ervasti JM, Kahl SD, Campbell KP. Purification of dystrophin from skeletal muscle. J Biol Chem 1991; 266:9161-9165.
9. Ervasti JM, Campbell KP. A role for the dystrophin-glycoprotein complex as a transmembrane linker between laminin and actin. J Cell Biol 1993; 122:809-823.
10. Yoshida M, Ozawa E. Glycoprotein complex anchoring dystrophin to sarcolemma. J Biochem 1990; 108:748-752.
11. Suzuki A, Yoshida M, Yamamoto H et al. Glycoprotein-binding site of dystrophin is confined to the cysteine-rich domain and the first half of the carboxy-terminal domain. FEBS Lett 1992; 308:154-160.
12. Yamamoto H, Hagiwara Y, Mizuno Y et al. Heterogeneity of dystrophin-associated proteins. J Biochem (Tokyo) 1993; 114:132-139.
13. Ozawa E, Yoshida M, Hagiwara Y et al. Formation of dystrophin lining on muscle cell membrane in myogenesis in vivo. In: Ozawa E, Masaki T, Nabeshima Y, eds. Frontiers in Muscle Research. Amsterdam: Elsevier, 1991:305-319.
14. Suzuki A, Yoshida M, Ozawa E. Mammalian α1- and β1-syntrophin bind to the alternative splice-prone region of the dystrophin COOH terminus. J Cell Biol 1995; 128:373-381.
15. Suzuki A, Yoshida M, Hayashi K et al. Molecular organization at the glycoprotein-complex-binding site of dystrophin—Three dystrophin-associated proteins bind directly to the carboxy-terminal portion of dystrophin. Eur J Biochem 1994; 220:283-292.
16. Yoshida M, Suzuki A, Yamamoto H et al. Dissociation of the complex of dystrophin and its associated proteins into several unique groups by n-octyl β-D-glucoside. Eur J Biochem 1994; 222:1055-1061.

17. Butler MH, Douville K, Murname AA et al. Association of the Mr 58,000 postsynaptic protein of electric tissue with Torpedo dystrophin and the Mr 87,000 postsynaptic protein. J Biol Chem 1992; 267:6213-6218.

18. Kramarcy NR, Vidal A, Froehner SC et al. Association of utrophin and multiple dystrophin short forms with the mammalian M$_r$ 58,000 dystrophin-associated protein (syntrophin). J Biol Chem 1994; 269:2870-2876.

19. Porter GA, Dmytrenko GM, Winkelmann JC et al. Dystrophin colocalizes with β-spectrin in distinct subsarcolemmal domains in mammalian skeletal muscle. J Cell Biol 1992; 117:997-1005.

20. Straub V, Bittner RE, Léger JJ et al. Direct visualization of the dystrophin network on skeletal muscle fiber membrane. J Cell Biol 1992; 119:1183-1191.

21. Ervasti JM. Costameres: The Achilles' heel of Herculean muscle. J Biol Chem 2003; 278:13591-13594.

22. Street SF. Lateral transmission of tension in frog myofibers: A myofibrillar network and transverse cytoskeletal connections are possible transmitters. J Cell Physiol 1983; 114:346-364.

23. Danowski BA, Imanaka-Yoshida K, Sanger JM et al. Costameres are sites of force transmission to the substratum in adult rat cardiomyocytes. J Cell Biol 1992; 118:1411-1420.

24. Shear CR, Bloch RJ. Vinculin in subsarcolemmal densities in chicken skeletal muscle: Localization and relationship to intracellular and extracellular structures. J Cell Biol 1985; 101:240-256.

25. Minetti C, Tanji K, Bonilla E. Immunologic study of vinculin in Duchenne muscular dystrophy. Neurology 1992; 42:1751-1754.

26. Minetti C, Tanji K, Rippa PG et al. Abnormalities in the expression of β-spectrin in Duchenne muscular dystrophy. Neurology 1994; 44:1149-1153.

27. Ehmer S, Herrmann R, Bittner R et al. Spatial distribution of β-spectrin in normal and dystrophic human skeletal muscle. Acta Neuropathol (Berl) 1997; 94:240-246.

28. Williams MW, Bloch RJ. Extensive but coordinated reorganization of the membrane skeleton in myofibers of dystrophic (mdx) mice. J Cell Biol 1999; 144:1259-1270.

29. Mokri B, Engel AG. Duchenne dystrophy: Electron microscopic findings pointing to a basic or early abnormality in the plasma membrane of the muscle fiber. Neurology 1975; 25:1111-1120.

30. Menke A, Jockusch H. Decreased osmotic stability of dystrophin-less muscle cells from the mdx mouse. Nature 1991; 349:69-71.

31. Menke A, Jockusch H. Extent of shock-induced membrane leakage in human and mouse myotubes depends on dystrophin. J Cell Sci 1995; 108:727-733.

32. Engel AG. Duchenne Dystrophy. In: Engel AG, Banker BQ, eds. Myology: Basic and Clinical. New York: McGraw-Hill, 1986:1185-1240.

33. Weller B, Karpati G, Carpenter S. Dystrophin-deficient mdx muscle fibers are preferentially vulnerable to necrosis induced by experimental lengthening contractions. J Neurol Sci 1990; 100:9-13.

34. Cox GA, Cole NM, Matsumura K et al. Overexpression of dystrophin in transgenic mdx mice eliminates dystrophic symptoms without toxicity. Nature 1993; 364:725-729.

35. Clarke MSF, Khakee R, McNeil PL. Loss of cytoplasmic basic fibroblast growth factor from physiologically wounded myofibers of normal and dystrophic muscle. J Cell Sci 1993; 106:121-133.

36. Matsuda R, Nishikawa A, Tanaka H. Visualization of dystrophic muscle fibers in mdx mouse by vital staining with evans blue: Evidence of apoptosis in dystrophin-deficient muscle. J Biochem (Tokyo) 1995; 118:959-964.

37. Tinsley JM, Potter AC, Phelps SR et al. Amelioration of the dystrophic phenotype of mdx mice using a truncated utrophin transgene. Nature 1996; 384:349-353.

38. Straub V, Rafael JA, Chamberlain JS et al. Animal models for muscular dystrophy show different patterns of sarcolemmal disruption. J Cell Biol 1997; 139:375-385.

39. Vilquin JT, Brussee V, Asselin I et al. Evidence of mdx mouse skeletal muscle fragility in vivo by eccentric running exercise. Muscle Nerve 1998; 21:567-576.

40. Harper SQ, Hauser MA, DelloRusso C et al. Modular flexibility of dystrophin: Implications for gene therapy of Duchenne muscular dystrophy. Nat Med 2002; 8:253-261.

41. Barton ER, Morris L, Musaro A et al. Muscle-specific expression of insulin-like growth factor I counters muscle decline in mdx mice. J Cell Biol 2002; 157:137-148.

42. Bansal D, Miyake K, Vogel SS et al. Defective membrane repair in dysferlin-deficient muscular dystrophy. Nature 2003; 423:168-172.

43. Karpati G, Carpenter S. Small-caliber skeletal muscle fibers do not suffer deleterious consequences of dystrophic gene expression. Am J Med Genet 1986; 25:653-658.

44. Mizuno Y. Prevention of myonecrosis in mdx mice: Effect of immobilization by the local tetanus method. Brain Dev 1992; 14:319-322.

45. Mokhtarian A, Lefaucheur JP, Even PC et al. Hindlimb immobilization applied to 21-day-old mdx mice prevents the occurrence of muscle degeneration. J Appl Physiol 1999; 86:924-931.

46. Coulton GR, Curtin NA, Morgan JE et al. The mdx mouse skeletal muscle myopathy: II. Contractile properties. Neuropathol Appl Neurobiol 1988; 14:299-314.
47. Kometani K, Tsugeno H, Yamada K. Mechanical and energetic properties of dystrophic (mdx) mouse muscle. Jpn J Physiol 1990; 40:541-549.
48. Stedman HH, Sweeney HL, Shrager JB et al. The mdx mouse diaphragm reproduces the degenerative changes of Duchenne muscular dystrophy. Nature 1991; 352:536-539.
49. Sacco P, Jones DA, Dick JR et al. Contractile properties and susceptibility to exercise-induced damage of normal and mdx mouse tibialis anterior muscle. Clin Sci (Lond) 1992; 82(2):227-236.
50. Quinlan JG, Johnson SR, McKee MK et al. Twitch and tetanus in mdx mouse muscle. Muscle Nerve 1992; 15:837-842.
51. Cox GA, Cole NM, Matsumura K et al. Overexpression of dystrophin in transgenic mdx mice eliminates dystrophic symptoms without toxicity. Nature 1993; 364:725-729.
52. Pastoret C, Sebille A. Time course study of the isometric contractile properties of mdx mouse striated muscles. J Muscle Res Cell Motil 1993; 14:423-431.
53. Deconinck N, Tinsley J, De Backer F et al. Expression of truncated utrophin leads to major functional improvements in dystrophin-deficient muscles of mice. Nature Med 1997; 3:1216-1221.
54. Tinsley J, Deconinck N, Fisher R et al. Expression of full-length utrophin prevents muscular dystrophy in mdx mice. Nature Med 1998; 4:1441-1444.
55. Deconinck N, Rafael JA, Beckers-Bleukx G et al. Consequences of the combined deficiency in dystrophin and utrophin on the mechanical properties and myosin composition of some limb and respiratory muscles of the mouse. Neuromuscular Disorders 1998; 8:362-370.
56. Bobet J, Mooney RF, Gordon T. Force and stiffness of old dystrophic (mdx) mouse skeletal muscles. Muscle Nerve 1998; 21:536-539.
57. Stevens ED, Faulkner JA. The capacity of mdx mouse diaphragm muscle to do oscillatory work. J Physiol 2000; 522:457-466.
58. Lynch GS, Hinkle RT, Chamberlain JS et al. Force and power output of fast and slow skeletal muscles from mdx mice 6-28 months old. J Physiol 2001; 535:591-600.
59. DelloRusso C, Crawford RW, Chamberlain JS et al. Tibialis anterior muscles in mdx mice are highly susceptible to contraction-induced injury. J Muscle Res Cell Motil 2001; 22:467-475.
60. Petrof BJ, Shrager JB, Stedman HH et al. Dystrophin protects the sarcolemma from stresses developed during muscle contraction. Proc Natl Acad Sci USA 1993; 90:3710-3714.
61. Moens P, Baatsen PH, Marechal G. Increased susceptibility of EDL muscles from mdx mice to damage induced by contractions with stretch. J Muscle Res Cell Motil 1993; 14:446-451.
62. Deconinck N, Ragot T, Maréchal G et al. Functional protection of dystrophic mouse (mdx) muscles after adenovirus-mediated transfer of a dystrophin minigene. Proc Natl Acad Sci USA 1996; 93:3570-3574.
63. Brooks SV. Rapid recovery following contraction-induced injury to in situ skeletal muscles in mdx mice. J Muscle Res Cell Motil 1998; 19:179-187.
64. Rybakova IN, Patel JR, Ervasti JM. The dystrophin complex forms a mechanically strong link between the sarcolemma and costameric actin. J Cell Biol 2000; 150:1209-1214.
65. Rybakova IN, Patel JR, Davies KE et al. Utrophin binds laterally along actin filaments and can couple costameric actin with the sarcolemma when overexpressed in dystrophin-deficient muscle. Mol Biol Cell 2002; 13:1512-1521.
66. Warner LE, DelloRusso C, Crawford RW et al. Expression of Dp260 in muscle tethers the actin cytoskeleton to the dystrophin-glycoprotein complex. Hum Mol Genet 2002; 11:1095-1105.
67. Kaprielian RR, Stevenson S, Rothery SM et al. Distinct patterns of dystrophin organization in myocyte sarcolemma and transverse tubules of normal and diseased human myocardium. Circulation 2000; 101:2586-2594.
68. Danialou G, Comtois AS, Dudley R et al. Dystrophin-deficient cardiomyocytes are abnormally vulnerable to mechanical stress-induced contractile failure and injury. FASEB J 2001; 15:1655-1658.
69. Kamogawa Y, Biro S, Maeda M et al. Dystrophin-deficient myocardium is vulnerable to pressure overload in vivo. Cardiovasc Res 2001; 50:509-515.
70. Hack AA, Ly CT, Jiang F et al. γ-Sarcoglycan deficiency leads to muscle membrane defects and apoptosis independent of dystrophin. J Cell Biol 1998; 142:1279-1287.
71. Duclos F, Straub V, Moore SA et al. Progressive muscular dystrophy in α-sarcoglycan-deficient mice. J Cell Biol 1998; 142:1461-1471.
72. Araishi K, Sasaoka T, Imamura M et al. Loss of the sarcoglycan complex and sarcospan leads to muscular dystrophy in β-sarcoglycan-deficient mice. Hum Mol Genet 1999; 8:1589-1598.
73. Coral-Vazquez R, Cohn RD, Moore SA et al. Disruption of the sarcoglycan-sarcospan complex in vascular smooth muscle: A novel mechanism for cardiomyopathy and muscular dystrophy. Cell 1999; 98:465-474.

74. Cote PD, Moukhles H, Lindenbaum M et al. Chimaeric mice deficient in dystroglycans develop muscular dystrophy and have disrupted myoneural synapses. Nat Genet 1999; 23:338-342.
75. Hack AA, Lam MY, Cordier L et al. Differential requirement for individual sarcoglycans and dystrophin in the assembly and function of the dystrophin-glycoprotein complex. J Cell Sci 2000; 113:2535-2544.
76. Durbeej M, Cohn RD, Hrstka RF et al. Disruption of the β-sarcoglycan gene reveals pathogenetic complexity of limb-girdle muscular dystrophy type 2E. Mol Cell 2000; 5:141-151.
77. Straub V, Donahue KM, Allamand V et al. Contrast agent-enhanced magnetic resonance imaging of skeletal muscle damage in animal models of muscular dystrophy. Magn Reson Med 2000; 44:655-659.
78. Cohn R, Henry M, Michele D et al. Disruption of Dag1 in differentiated skeletal muscle reveals a role for dystroglycan in muscle regeneration. Cell 2002; 110:639-648.
79. Sasaoka T, Imamura M, Araishi K et al. Pathological analysis of muscle hypertrophy and degeneration in muscular dystrophy in γ-sarcoglycan-deficient mice. Neuromuscul Disord 2003; 13:193-206.
80. Gee SH, Blacher RW, Douville PJ et al. Laminin-binding protein 120 from brain is closely related to the dystrophin-associated glycoprotein, dystroglycan, and binds with high affinity to the major heparin binding domain of laminin. J Biol Chem 1993; 268:14972-14980.
81. Bowe MA, Deyst KA, Leszyk JD et al. Identification and purification of an agrin receptor from Torpedo postsynaptic membranes: A heteromeric complex related to the dystroglycans. Neuron 1994; 12:1173-1180.
82. Campanelli JT, Roberds SL, Campbell KP et al. A role for dystrophin-associated glycoproteins and utrophin in agrin-induced AChR clustering. Cell 1994; 77:663-674.
83. Gee SH, Montanaro F, Lindenbaum MH et al. Dystroglycan-α, a dystrophin-associated glycoprotein, is a functional agrin receptor. Cell 1994; 77:675-686.
84. Sugiyama J, Bowen DC, Hall ZW. Dystroglycan binds nerve and muscle agrin. Neuron 1994; 13:103-115.
85. Sugita S, Saito F, Tang J et al. A stoichiometric complex of neurexins and dystroglycan in brain. J Cell Biol 2001; 154:435-445.
86. Peng HB, Ali AA, Daggett DF et al. The relationship between perlecan and dystroglycan and its implication in the formation of the neuromuscular junction. Cell Adhes Commun 1998; 5:475-489.
87. Peng HB, Xie HB, Rossi SG et al. Acetylcholinesterase clustering at the neuromuscular junction involves perlecan and dystroglycan. J Cell Biol 1999; 145:911-921.
88. Michele DE, Campbell KP. Dystrophin-Glycoprotein Complex: Post-translational Processing and Dystroglycan Function. J Biol Chem 2003; 278:15457-15460.
89. Bowe MA, Mendis DB, Fallon JR. The small leucine-rich repeat proteoglycan biglycan binds to α-dystroglycan and is upregulated in dystrophic muscle. J Cell Biol 2000; 148:801-810.
90. Montanaro F, Carbonetto S. Targeting dystroglycan in the brain. Neuron 2003; 37:193-196.
91. Hemmings L, Kuhlman PA, Critchley DR. Analysis of the actin-binding domain of α-actinin by mutagenesis and demonstration that dystrophin contains a functionally homologous domain. J Cell Biol 1992; 116:1369-1380.
92. Way M, Pope B, Cross RA et al. Expression of the N-terminal domain of dystrophin in E. coli and demonstration of binding to F-actin. FEBS Lett 1992; 301:243-245.
93. Rybakova IN, Amann KJ, Ervasti JM. A new model for the interaction of dystrophin with F-actin. J Cell Biol 1996; 135:661-672.
94. Amann KJ, Renley BA, Ervasti JM. A cluster of basic repeats in the dystrophin rod domain binds F- actin through an electrostatic interaction. J Biol Chem 1998; 273:28419-28423.
95. Rybakova IN, Ervasti JM. Dystrophin-glycoprotein complex is monomeric and stabilizes actin filaments in vitro through a lateral association. J Biol Chem 1997; 272:28771-28778.
96. Newey SE, Howman EV, Ponting CP et al. Syncoilin, a novel member of the intermediate filament superfamily that interacts with α-dystrobrevin in skeletal muscle. J Biol Chem 2001; 276:6645-6655.
97. Benson MA, Newey SE, Martin-Rendon E et al. Dysbindin, a novel coiled-coil-containing protein that interacts with the dystrobrevins in muscle and brain. J Biol Chem 2001; 276:24232-24241.
98. Mizuno Y, Thompson TG, Guyon JR et al. Desmuslin, an intermediate filament protein that interacts with α-dystrobrevin and desmin. Proc Natl Acad Sci USA 2001; 98:6156-6161.
99. Poon E, Howman EV, Newey SE et al. Association of syncoilin and desmin: linking intermediate filament proteins to the dystrophin-associated protein complex. J Biol Chem 2002; 277:3433-3439.
100. Grady RM, Grange RW, Lau KS et al. Role for α-dystrobrevin in the pathogenesis of dystrophin-dependent muscular dystrophies. Nature Cell Biol 1999; 1:215-220.
101. Milner DJ, Weitzer G, Tran D et al. Disruption of muscle arhitecture and myocardial degeneration in mice lacking desmin. J Cell Biol 1996; 134:1255-1270.

102. Li Z, Mericskay M, Agbulut O et al. Desmin is essential for the tensile strength and integrity of myofibrils but not for myogenic commitment, differentiation, and fusion of skeletal muscle. J Cell Biol 1997; 139:129-144.
103. Thompson TG, Chan YM, Hack AA et al. Filamin 2 (FLN2): A muscle-specific sarcoglycan interacting protein. J Cell Biol 2000; 148:115-126.
104. Glogauer M, Arora P, Chou D et al. The role of actin-binding protein 280 in integrin-dependent mechanoprotection. J Biol Chem 1998; 273:1689-1698.
105. Schroder R, Mundegar RR, Treusch M et al. Altered distribution of plectin/HD1 in dystrophinopathies. Eur J Cell Biol 1997; 74:165-171.
106. Law DJ, Allen DL, Tidball JG. Talin, vinculin and DRP (utrophin) concentrations are increased at mdx myotendinous junctions following onset of necrosis. J Cell Sci 1994; 107:1477-1483.
107. Vachon PH, Xu H, Liu L et al. Integrins ($\alpha7\beta1$) in muscle function and survival - Disrupted expression in merosin-deficient congenital muscular dystrophy. J Clin Invest 1997; 100:1870-1881.
108. Hodges BL, Hayashi YK, Nonaka I et al. Altered expression of the $\alpha7\beta1$ integrin in human and murine muscular dystrophies. J Cell Sci 1997; 110:2873-2881.
109. Burkin DJ, Wallace GQ, Nicol KJ et al. Enhanced expression of the $\alpha7\beta1$ integrin reduces muscular dystrophy and restores viability in dystrophic mice. J Cell Biol 2001; 152:1207-1218.
110. Adams ME, Mueller HA, Froehner SC. In vivo requirement of the α-syntrophin PDZ domain for the sarcolemmal localization of nNOS and aquaporin-4. J Cell Biol 2001; 155:2-10.
111. Thomas GD, Sander M, Lau KS et al. Impaired metabolic modulation of α-adrenergic vasoconstriction in dystrophin-deficient skeletal muscle. Proc Natl Acad Sci USA 1998; 95:15090-15095.
112. Sander M, Chavoshan B, Harris SA et al. Functional muscle ischemia in neuronal nitric oxide synthase-deficient skeletal muscle of children with Duchenne muscular dystrophy. Proc Natl Acad Sci USA 2000; 97:13818-13823.
113. Kolodziejczyk SM, Walsh GS, Balazsi K et al. Activation of JNK1 contributes to dystrophic muscle pathogenesis. Curr Biol 2001; 11:1278-1282.
114. Nakamura A, Harrod GV, Davies KE. Activation of calcineurin and stress activated protein kinase/p38-mitogen activated protein kinase in hearts of utrophin-dystrophin knockout mice. Neuromusc Disord 2001; 11:251-259.
115. Nakamura A, Yoshida K, Takeda S et al. Progression of dystrophic features and activation of mitogen-activated protein kinases and calcineurin by physical exercise, in hearts of mdx mice. FEBS Lett 2002; 520:18-24.

CHAPTER 2

The Structure and Function of Dystrophin

Simone Abmayr and Jeff Chamberlain

Abstract

Duchenne muscular dystrophy (DMD) and the allelic Becker muscular dystrophy (BMD) are X-linked recessive disorders caused by mutations in the dystrophin gene. The dystrophin gene spans 2.4 MB on the human X-chromosome and contains seven promoters, each with a unique first exon, and 78 additional coding exons. Despite years of study, the function of this protein is not completely understood, and it appears to serve slightly different roles in muscle and nonmuscle tissues. This chapter will focus on the role of dystrophin in muscle, where it is thought to be a structural protein providing a flexible and elastic link between the cortical actin cytoskeleton and the extracellular matrix via interactions with the dystrophin-glycoprotein complex. This link helps to dissipate muscle contractile force from the intracellular cytoskeleton to the extracellular matrix. The absence of dystrophin disrupts this link, resulting in membrane fragility and rendering the sarcolemma susceptible to mechanical injury during contraction. The absence of dystrophin also results in destabilization of the DGC. Loss of the DGC, combined with transient disruptions of sarcolemma integrity, leads to myofiber necrosis by mechanisms that are poorly understood. Some of the proteins that interact with dystrophin, particularly certain components of the DGC, appear to be involved in signaling cascades. Consequently, a complete understanding of the function of dystrophin requires consideration of its direct structural role as well as its indirect role in localizing signaling molecules.

Introduction

Characterization of a variety of separate mutations in different DMD and BMD patients, coupled with studies of genotype/phenotype correlations, provided the first evidence for the function of dystrophin. DMD and BMD are progressive, degenerative muscle diseases that result in cycles of skeletal muscle fiber necrosis and regeneration, accompanied by the gradual replacement of myofibers with fibrotic connective tissue and adipocytes.[1,2] In addition to skeletal muscle degeneration, most of the patients develop cardiomyopathy and one-third display cognitive deficits.[2-5] The major difference between DMD and BMD is that BMD has a later age of onset and a slower progression. While DMD patients generally succumb to respiratory or cardiac failure in their late second to third decade, BMD patients often survive past age thirty, and one extremely mild case of BMD has been reported where the patient was still walking in his seventies.[1,6] What is it about dystrophin mutations that lead to these phenotypic differences?

The earliest genotype studies of the dystrophin gene revealed that almost two-thirds of DMD and BMD cases resulted from various genomic deletions.[7-9] DMD was associated with the absence or near complete absence of dystrophin, or more rarely, from expression of a nonfunctional protein. In contrast, most BMD patients expressed a partially functional dystrophin protein, or lower than normal levels of dystrophin.[10,11] DMD was observed to result when a gene deletion removed a combination of exons such that the normal open reading frame of the dystrophin mRNA was disrupted. Frameshifted transcripts are typically unstable, and the encoded truncated proteins rarely accumulate to a detectable level. BMD resulted if the deleted exon(s) maintained the normal mRNA open reading frame, thus encoding an internally truncated protein.[12-15] Importantly, any deletion that removed exons near the C-terminus of dystrophin, regardless of the impact on the open reading frame, invariably led to DMD.[16-18] Further studies revealed that the critical C-terminal sequences are responsible for assembling and stabilizing the DGC, while much of the N-terminal two-thirds of dystrophin maintains a lateral association with cortical actin. Since the DGC and DMD pathophysiology are explored elsewhere in this book, this chapter will focus on structural features of dystrophin that are important for its function.

Structural Domains of Dystrophin Overview

The muscle isoform of dystrophin is a 427 kDa filamentous protein that localizes to the sarcolemma, and which has been historically described as being composed of four domains (Fig. 1).[19,20] While this description is an oversimplification, it provides a useful basis from which to consider its function. The amino-terminal portion of the molecule is an actin-binding domain (ABD) that displays a high degree of sequence conservation with the ABDs of other members of the spectrin super family of proteins, including α-actinin.[21] This N-terminal ABD provides a link between dystrophin and the cytoskeleton via direct binding to short, F- (filamentous) actin filaments. The N-terminal ABD is followed by a rod domain composed of 24 moderately conserved repeating units that display conservation with the repeat units found in spectrin, hence their name "spectrin-like repeats".[22-24] The rod domain forms a flexible and elastic linker between the N- and C-terminal portions of dystrophin, and may help localize dystrophin to the sarcolemma.[25] Many mildly affected BMD patients express dystrophins deleted for significant portions of this domain (Fig. 2).[6,11,26] The rod domain contains a second actin-binding domain, located over an extended region of basic repeats.[27] The N-terminal and the rod ABDs appear to collaborate in forming a lateral association with actin filaments.[28] Four short, proline-rich spacer sequences located before and after the rod domain, between repeats 3 and 4, and between repeats 19 and 20 have been referred to as 'hinges' as they may confer flexibility upon the protein (Fig. 1).[23] Hinge 4, which is located at the end of the rod domain, contains a WW domain, a protein-protein interaction motif (see below).[29] Following the rod-domain is the cysteine-rich (CR) domain, composed of EF-hands and a "ZZ" zinc-finger motif.[20,30,31] The WW, cysteine-rich and ZZ regions fold together into a unique structure required for binding dystroglycan, the portion of the DGC that spans the sarcolemma (Figs. 1 and 2).[32] The carboxy-terminal (CT) domain binds to members of the syntrophin and dystrobrevin families via two coiled-coil regions, and displays a variety of sequence alterations in different tissues that arise from alternative RNA splicing.[33-39] A detailed atomic structure of all the subdomains in dystrophin was recently presented by Winder and colleagues.[31]

The N-Terminal Actin-Binding Domain (ABD1)

The N-terminal ABD (ABD1) is the region of dystrophin upstream of the first spectrin-like repeat and is encoded on exons 1 through the middle of exon 8 (amino acids 1-252; Fig. 1).[21] Dystrophin does not appear capable of cross-linking actin filaments, although this may depend on the fine structure of the filaments.[40-42] Fragments spanning amino acids 1-233 or 1-246 bind F-actin, but not nonpolymerized G-actin, in vitro.[43-45] This region contains two calponin-homology (CH) domains, a motif originally identified in smooth muscle calponin.[46,47] Linked pairs of the ~100 amino acid CH motif sequence have been shown to fold into an actin binding site within a number of actin cross-linking proteins.[47,48] NMR spectroscopy studies performed in vitro between F-actin and a variety of dystrophin peptides have identified two specific contact sites within the CH domains named ABS1 and 3 (amino acids 18-27 and 131-147, respectively).[49-51] A third contact site, ABS2 (amino acids 88-116), was inferred from binding studies between actin and α-actinin.[52] Deletion studies that attempted to narrow

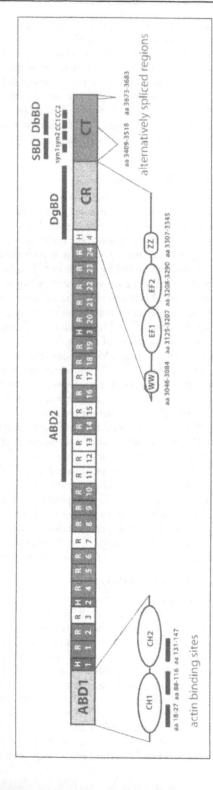

Figure 1. Domain structure of the human dystrophin gene (scale is approximate). Details are discussed in the main text. ABD: actin-binding domains; H: hinge regions; R: spectrin-like repeats; CR: cysteine-rich domain; CT: C-terminal domain; CH: calponin homology motifs; WW: WW domain; EF: EF-hand motifs; ZZ: ZZ domain; DgBD: dystroglycan-binding domain;[20,32,106] SBD: syntrophin-binding domain; DbBD: dystrobrevin-binding domain; syn: location of the syntrophin contact sites (syn1 spans amino acids 3427-3461; syn2 spans amino acids 3462-3483);[38] CC: coiled coil motifs (CC1 spans amino acids 3506-3593; CC2 spans amino acids 3558-3593 as determined from the primary cDNA sequence). Also shown at the bottom left are the locations and extents of ABS1-3. Hinge 4 spans amino acids 3041-3112.[23] Basic repeats are shown in white, the others in blue. The location of the alternatively spliced exons is from Feener et al.[37] The locations of other features are described in the main text.

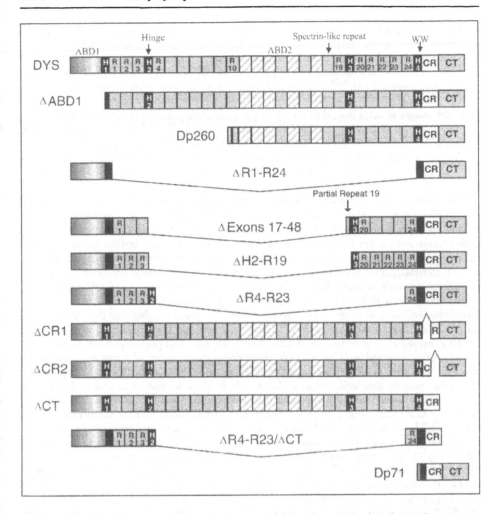

Figure 2. Schematic illustration of the domain structure of dystrophin proteins described in the main text. At the top is the structure of full-length dystrophin, abbreviations are as in Figure 1. Note that the ΔExons 17-48 encoded protein retains a partial repeat 19 (see also Fig. 3). The exon 13-48 deletions referred to in the text are predicted to encode a protein similar to the exon 17-48 structure, except that repeats 2 and 3 are also deleted. The micro-dystrophin construct deleted for exons 3-9, 12-48 and 71-78 is similar to the exon 13-48 deletion, but lacks all of ABD1 except for the first 31 amino acids, most of hinge 1, repeats 2-18 and the CT domain. Dp260 and Dp71 are naturally occurring isoforms of dystrophin.[150,177]

down the major sequences required for binding to actin indicated that dystrophin fragments spanning amino acids 1-90 or 1-68 were able to bind F-actin in vitro.[44,53] However, these isolated fragments of dystrophin appear to bind actin with reduced affinity and higher capacity than when dystrophin is complexed with the DGC.[54] Studies that measured the interaction between the entire complex and actin showed that dystrophin binds actin with a K_d of 0.5 μM and a stoichiometry of one dystrophin protein to 24 actin monomers.[27,41] Dystrophin is also able to protect actin filaments from depolymerization.[27,41] The N-terminal ABD appears to bind actin primarily via hydrophobic interactions that are largely insensitive to ionic strength.[28,43] Studies of β-spectrin suggest that its first repeat, which has a basic charge, contributes to actin binding, and a utrophin ABD linked to the first 2.5 spectrin-like repeats displays enhanced actin binding relative to fragments lacking the repeats.[55,56] The third repeat of dystrophin is also highly basic, raising the possibility that the first several repeats in dystrophin and related family members might facilitate binding to actin.[28,55-58]

A high-resolution atomic model for the binding of ABD1 to actin was recently derived by electron microscopy (EM).[40,59] The dystrophin CH domains were found to be globular structures formed by four α-helices, with the two CH domains separated by a central helix.[59] The cryo-EM images of actin bound to the dystrophin ABD suggested that the ABD bound as a monomer, similar to that observed for utrophin and fimbrin.[40,59] Most reports indicate that dystrophin and the DGC do not dimerize, although a recent study suggests that this may depend on the fine structure of individual F-actin filaments.[40-42,60,61] Finally, the EM images also provided support for the suggestion that the first several spectrin repeats might stabilize actin binding.

Several aspects of the interaction between dystrophin and actin have been confirmed by in vivo studies. Ervasti et al developed a technique whereby intact, isolated myofibers could be mechanically peeled to remove the sarcolemma from the underlying cytoskeleton.[62] When peeled fibers from normal mice are immunolabeled with antibodies against actin, the actin staining is clearly observed on both the sarcolemma and the cytoskeleton in a costameric pattern. In contrast, peeled fibers from dystrophin-deficient *mdx* mice revealed actin localization only in the cytoskeletal layer. These studies revealed that dystrophin is required to maintain a mechanically strong link between actin filaments and the sarcolemma. Furthermore, antisera specific for individual isoforms of actin demonstrated that γ-actin, rather than β- or α- (muscle) actin was bound to dystrophin in vivo.[62] Thus, dystrophin binds to cortical actin below the sarcolemma, rather than contractile α-actin. Evidence that small portions of ABD1 can functionally tether dystrophin to actin in vivo have come from studies of transgenic *mdx* mice, which do not express endogenous dystrophin.[63] One series of transgenic *mdx* mice expressed a dystrophin protein that lacked most of ABD1 (ΔABD1; a deletion of amino acids 45-273, corresponding approximately to a genomic deletion of exons 3-8; Fig. 2). Despite this deletion, muscles from the transgenic mice displayed only a mild dystrophy and a moderate reduction in strength.[64] These results suggested that dystrophin either had a second actin-binding domain, or that the N-terminal 44 amino acids were capable of actin binding in vivo, perhaps through ABS1 (Fig. 1). More recent studies indicate that both possibilities are likely true. An internal ABD (ABD2) has been found in the rod domain (Fig. 1; see below). Also, a transgenic *mdx* mouse expressing a dystrophin construct deleted for exons 3-9, 12-48 and 71-78 also displayed a mild phenotype.[65] This latter deletion removed all but the first 31 amino acids of ABD1, all of ABD2, rod domain repeats 2-18 and the CT domain. Thus, the first 31 amino acids of dystrophin, perhaps in collaboration with repeat 1, can weakly bind F-actin in vivo.

These in vitro and in vivo mouse studies are in agreement with studies of patients with in-frame deletions in ABD1. These patients typically display a moderate to severe BMD phenotype.[15,66-68] Muscle biopsies from these patients contained very low levels of dystrophin, suggesting that an intact ABD1 helps stabilize dystrophin in muscle.[67] A similar lack of stability was inferred from the transgenic *mdx* mice expressing the ΔABD1 deletion (Fig. 2). In that study, only 1 of 12 independent lines of mice expressed normal dystrophin levels, despite the use of a strong promoter to drive transgene expression.[64] A number of BMD patients have been described who carry a deletion of exons 3-7.[69-72] This deletion results in a shift of the mRNA open reading frame and was predicted to lead to DMD. The relatively mild phenotype has been explained by translational reinitiation occurring in exon 8 at a match with the Kozak consensus sequence, leading to production of low levels of an N-terminally truncated dystrophin that lacks most of ABD1.[71] Also, a few DMD patients have been reported with missense mutations in first CH domain of dystrophin, suggesting that small alterations can profoundly disrupt either dystrophin stability or its binding to actin.[73-75] Collectively, these studies indicate that ABD1 binds actin in vivo and is required for stabile dystrophin expression. However, ABD1 can partially function with only a small portion of its sequence intact.

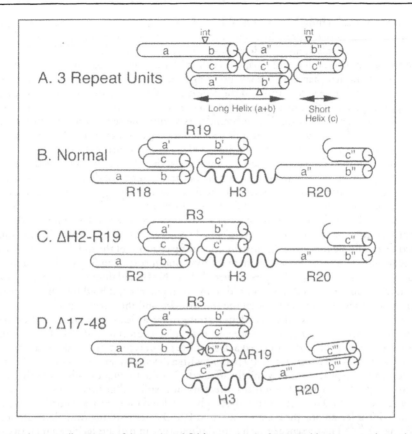

Figure 3. Schematic illustration of the predicted folding patterns of spectrin-like repeats in dystrophin. A) An example of the interdigitated folding pattern for a stretch of 3 repeats. a and b form the long section, while c forms the short section. a', etc, denotes consecutive repeats. Arrowheads show the location of the conserved exon/intron boundary within the b-helix (for simplicity, only 2 are shown in A). Adapted from the manuscripts by Kahana, Cross and colleagues.[25,77] B) Schematic illustration of the predicted folding patterns for 3 spectrin-like repeats flanking hinge 3 in dystrophin. Note how the hinge domain disrupts the interdigitated repeat array. C) Structure of the repeat array flanking hinge 3 in the ΔH2-R19 protein. The overall structure is identical to that shown in B. D) Structure of the repeat array flanking hinge 3 in the ΔExon 17-48 encoded protein. This deletion results in the last half of the b helix (b") and the full c helix (c") from R19 being retained in the protein, creating an extra helical region that might disrupt the folding pattern of the protein. Note that the fragment of the b helix from repeat 19 begins at the exon/intron border immediately upstream of exon 49 (arrowhead).

The Central Rod Domain

The central rod domain is encoded on exons 8-61.[20,23,76] This region spans ~77% of the dystrophin molecule and consists of 24 'spectrin-like' repeats and four proline-rich regions (hinges 1-4; Fig. 1). The rod domain folds into a triple-helical coiled-coil of heptad repeats with alternating long and short helical sections.[23-25,77-79] Each repeat is approximately 110 amino acids in length and is encoded on between 2 and 3 exons. The triple helix is made up of the short helix folded with the N-and C-terminal halves of the adjacent long helices.[25,77,80] In this arrangement, adjacent repeats are nested within one another, allowing for formation of a flexible rod domain with properties similar to a strong spring (Fig. 3A).[24] The internal hinges, 2 and 3, disrupt this nested folding pattern and may provide regions of additional flexibility within the rod domain.[23] Structural and sequence differences within the repeats adjacent to these hinges, as well as those at each end of the rod domain, may distinguish them functionally from the internal, nested repeats in dystrophin and other spectrin family members.[57,81]

The 24 repeats display only a moderate level of sequence conservation.[23] However, most of the conservation is by amino acid type, rather than identity, and repeats in dystrophin are less similar to one another than are the corresponding repeats in β-spectrin.[51] In an unusual relationship between the gene structure and the repeat domains, one exon/intron boundary splits each repeat at precisely the same place in the long section (helix b; Fig. 3A).[23] In contrast, the locations of the remaining exon/intron boundaries within the repeat unit structures are not well conserved.[20,23,76] Various observations have led to suggestions that the 24 repeat structure of dystrophin may have evolved from a primordial 4 repeat ancestor.[23,82]

The organization of the triple helical repeats suggest that the rod domain serves as a flexible and elastic spacer or linker domain between the N- and C-terminal domains of dystrophin.[83] This region could also serve as a molecular shock absorber, helping to buffer the mechanical stresses associated with contraction and force development in muscle.[23,51,54,84] An additional role for the rod domain may be to facilitate an association between dystrophin and the sarcolemmal membrane. Expression of several small subfragments of dystrophin in *mdx* muscles showed that multiple regions of the protein, including portions of the rod domain, localized to the sarcolemma.[85] A recombinant fragment of dystrophin corresponding to the second repeat interacts specifically with lipid bilayers in vitro.[86] Also, NMR and fluorescence spectra analyses of the interactions between dystrophin rod domain fragments and lipid bilayers has provided evidence for a tight interaction between the rod domain and the sarcolemma.[87]

A variety of mutation studies has demonstrated that at least part of the rod domain is required for dystrophin function. Transgenic *mdx* mice expressing a dystrophin lacking the entire rod domain (ΔR1-R24) displayed a dystrophic phenotype at least as severe as that of the *mdx* mouse (Fig. 2).[57] However, a BMD patient with a deletion of exons 17-48, which removes ~16 spectrin-like repeats, was only mildly affected. Despite lacking two-thirds of the rod domain and 46% of the dystrophin protein, this patient is still ambulatory in his late seventies.[6] Numerous other BMD patients express dystrophins deleted for significant portions of the rod domain.[11] Included among these are several mild BMD cases resulting from deletion of exons 13-48 (Fig. 2).[26,88] The presence of such a mild phenotype in patients lacking a majority of the dystrophin coding sequence is remarkable; however, equally perplexing is that each of these large deletions removes the only portion of the rod domain with an obvious function: the internal actin-binding domain (ABD2).

ABD2 was identified and characterized through in vitro binding studies.[27,28] These studies revealed that a broad region of the rod domain that carries a preponderance of basic amino acids binds F-actin. The primary regions involved in binding are repeats 11-13, 15 and 17. Repeats 3 and 7 are also basic and could possibly contribute to binding. Studies to examine the effect of ionic strength on the binding revealed significant differences in the way in which ABD1 and 2 interact with actin filaments. ABD2 interacts primarily by an electrostatic mechanism, in contrast to the largely hydrophobic interactions of ABD1.[28,43] This latter association might be expected to form a tighter bond and may explain a seemingly greater functional importance of ABD1 relative to ABD2 (below). Intriguingly, the dystrophin homologue utrophin lacks the preponderance of basic amino acids that characterize the repeats in ABD2.[28] Compared with dystrophin, utrophin appears to have a somewhat less stable interaction with actin that is dispersed across the mostly acidic repeats 1-10 and its own ABD1.[28,58]

Evidence that ABD2 interacts with F-actin in vivo was derived from studies of mechanically peeled myofibers. When this approach was applied to transgenic *mdx* mice expressing Dp260, which lacks ABD1 (Fig. 2), the Dp260 protein was able to rescue the association of costameric γ-actin filaments with the sarcolemma.[89] Studies examining interactions between actin and an intact dystrophin complexed with the DGC suggest that actin binding is much tighter, and occurs with reduced stoichiometry, than is observed with isolated ABD1 or 2 fragments.[27,28,41] These authors provided compelling evidence that both ABDs collaborate to form a high affinity, yet flexible lateral association between dystrophin and actin at a ratio of 1 dystrophin per 24 actin monomers. Studies of the various actin-binding regions in dystrophin, utrophin and spectrin are collectively leading to a model where the N-terminal 60% of dystrophin, from the N-terminus through repeat 17, may function as a single, broadly dispersed ABD.[28,40,41,55,56,58] The presence of multiple, lower affinity actin binding sites may lead to a fluid or elastic link between the DGC and actin, depending on the type of mechanical forces being applied to the muscle sarcolemma.[41]

Although the normal dystrophin protein appears to have a large dispersed ABD, it is clear that relatively small portions of this region can be highly functional in vivo. Transgenic mice that expressed a dystrophin molecule lacking either (1) ABD1 (ΔABD1) or (2) ABD1 and a significant portion of the rod domain, but which retained the internal ABD2 (i.e., the Dp260 isoform of dystrophin), each displayed a dystrophic phenotype significantly milder than in *mdx* mice (Fig. 2).[44,64,89] Both proteins supported normal expression and assembly of the dystrophin-glycoprotein complex and protected muscles from contraction-induced injury, but a low level on ongoing myofibers degeneration was apparent. These data are in contrast to the markedly milder phenotype observed in the exon 17-48 deletion patient and in transgenic *mdx* mice expressing a dystrophin cDNA deleted for exons 17-48, which lacks ABD2 (see below).[6,57,90] Thus, while actin can bind to multiple regions along dystrophin, deletion of the rod domain contact sites is associated with significantly healthier muscle tissue than is deletion of the N-terminal actin-binding regions.

The mild BMD phenotypes associated with exon 13-48 and 17-48 deletions[6,26,88] have spurred the development of a variety of optimized mini- and micro-dystrophin constructs that are being tested for use in gene therapy applications.[91] Muscles from transgenic *mdx* animals that expressed the dystrophin construct deleted for exon 17-48 sequences displayed correct expression and localization of the DGC (with the possible exception of nNOS);[92,93] the muscles had little evidence of ongoing necrosis and regeneration; and they generated 95% of the specific force as did control muscles.[90] Nonetheless, we hypothesized that protein engineering could be used to design a mini-dystrophin with significantly greater functional capacity.[94] Part of this hypothesis was based on the observation that the location of exon/intron boundaries within the spectrin repeat coding region is variable, except for the intron that always splits helix b.[23,24] Consequently, the exon 17-48 deletion removes 4 amino acids from repeat 3, all of hinge 2 and spectrin-like repeats 4 through 18, but only half of repeat 19 (Fig. 2). We speculated that this residual partial repeat could destabilize the folding of dystrophin and compromise its function (Fig. 3B-D).[57] A modified exon 17-48 deletion construct was designed, designated ΔH2-R19, that contained a precise deletion of the sequences between spectrin-like repeats 3 and hinge 3, rather than a deletion whose ends corresponded to exon/intron boundaries (Fig. 3C). When tested in transgenic *mdx* mice, the ΔH2-R19 mini-dystrophin compensated for the lack of normal dystrophin. The ΔH2-R19 transgenic *mdx* muscles displayed a morphological appearance and had strength, contractile properties and resistance to contraction-induced injury that were not different from wild-type controls.[57] The ΔH2-R19 construct is smaller and more functional than the Δexon 17-48 construct, and displays functional properties in *mdx* mice not different than full-length dystrophin. We refer to a deletion such as ΔH2-R19 as having preserved the normal phasing of the repeat units.

In order to identify the minimal portion of the rod domain needed to maintain normal muscle physiology, additional deletion constructs with between zero and 7 spectrin-like repeats were tested.[57,95,96] Constructs with less than four repeats were poorly functional, defining a minimum size requirement for the rod domain.[57,97] Expression in *mdx* muscles of a construct with zero spectrin repeats (ΔR1-R24) resulted in a muscular dystrophy more severe than in dystrophin-deficient *mdx* muscles, perhaps because this nonfunctional dystrophin is able to displace utrophin from the sarcolemma[57,98] Elevated utrophin levels in *mdx* muscles have previously been shown to partially compensate for the absence of dystrophin, and indeed dystrophin/utrophin double knockout mice display a significantly worse phenotype than do *mdx* mice.[99-101]

Dystrophin constructs with 4 spectrin-like repeats have displayed a remarkably high degree of functional activity, although the relative functional activity depended critically on which repeats were included in the construct, as well as whether or not hinge domains were included.[57,96] The best micro-dystrophin (ΔR4-R23) retained spectrin-repeats 1-3 and hinges 1 and 2, and thus displays a normal structure from the N-terminus through hinge 2, which is then joined to repeat 24.[57] These data suggest that retention of repeats 1-3 and the first hinge may be especially important for dystrophin function. These studies also suggest that the different spectrin-like repeats are not functionally interchangeable and that simply having a hinge in the middle of the repeat domain does not necessarily increase the functionality of the protein. Further evidence for this noninterchangeability came from a study where the 4 repeats from the highly functional ΔR4-R23 micro-dystrophin were replaced by the natural four repeat unit from the homologous protein α-actinin. That protein was nonfunctional in *mdx* mouse muscles, despite being expressed well and supporting assembly of the DGC.[81] Additional mini-dystrophins with 5 and 6 repeats have also been shown to display considerable activity after being delivered to muscles of adult *mdx* mice using AAV.[95,102] A 7 repeat micro-dystrophin was poorly functional.[57]

An intriguing aspect of the micro-dystrophins is that several display an extraordinary degree of functional activity in vivo despite lacking ABD2. Could the rod domain bind other proteins beside actin? Each of the mini and micro-dystrophins tested support assembly of the entire DGC, with the notable exception of nNOS and the DGC-associated protein aquaporin-4.[92,103] Nonetheless, no direct or compelling evidence has been presented to suggest that any proteins bind the rod domain of dystrophin other than actin. Also the truncated dystrophins that fail to restore nNOS localization, including ones expressed in revertant *mdx* myofibers, do not share a common region of overlapping deletion.[92] Thus, the available evidence points to three functions for the rod domain in muscle: (1) a mechanical role linking the cortical actin cytoskeleton and DGC binding domains, where it appears to act as an elastic shock absorber and/or force transducer; (2) a trafficking role, helping to target dystrophin to the sarcolemma; and (3) a structural role facilitating and stabilizing a lateral association between dystrophin and actin.

The Dystroglycan-Binding Domain (Cysteine-Rich (CR) Domain)

The cysteine-rich (CR) domain forms part of a tightly packed binding site for β-dystroglycan that has recently been named the dystroglycan-binding domain (DgBD).[32] The DgBD is composed of both the WW domain and the CR domain (Fig. 1).[104] The WW domain is encoded by exons 62-63 and is located at the C-terminal side of the fourth 'hinge' domain, which separates the final (24th) spectrin-like repeat from the CR domain.[105] The CR domain is encoded on exons 64-70 and has two distinguishing sequence motifs. The N-terminal portion folds into 2 "EF-hand" structures of ~83 amino acids that display homology with the calcium binding EF-hands from calmodulin.[20,31,106] The C-terminal portion of the CR domain contains a zinc-finger "ZZ" domain (~39 amino acids).[107] These WW, EF-hand, and ZZ modules fold into a single dystroglycan-binding site.[31,32,106]

Each subportion of the DgBD contributes to binding in a unique way. The WW domain is ~39 amino acids in length and is distinguished by the presence of two tryptophan (W) residues separated by 21 amino acids. Sudol and colleagues have shown that a WW domain is a protein-protein interaction motif found in numerous proteins, most of which serve to localize protein complexes to specialized regions of plasma membranes.[29,108,109] WW domains bind proline-rich sequences containing a core PPxY motif.[110] In the case of dystrophin, the ligand is a 15 amino acid stretch near the C-terminus of β-dystroglycan that contains the sequence PPPY.[106,111] Recent evidence suggests that the C-terminal PPxY motif on β-dystroglycan can be tyrosine-phosphorylated in an adhesion-dependent manner, and that this phosphorylation event can weaken or destabilize binding to the dystrophin and/or utrophin WW domains.[112,113] Phosphorylation of the PPxY motif also enables recruitment and binding of caveolin, which has a WW domain, and a variety of SH2 domain containing signaling molecules, such as, C-Src, Fyn, CsK and Grb-2.[113-115] These observations imply that β-dystroglycan exists in a fluid state, acting alternately as a ligand for dystrophin or as a scaffold for various signaling molecules.[29,113] This dynamic structure could enable the DGC to adapt quickly to altered mechanical stress, inflammation or regenerative processes.

A variety of studies shows that the WW domain can form a tight bond with β-dystroglycan only when linked to the CR domain.[32,111,116-118] The proximal portion of the CR domain is composed of two helix-loop-helix EF-hand structures that are connected to the WW domain by two short α-helices (Fig. 1). These EF-hand motifs are not highly conserved with functional EF-hand motifs in calcium binding proteins, and they do not appear capable of binding calcium.[106] The ZZ domain contains 5 cysteine residues and is located distal to the EF-hand region in the CR domain (Fig. 1). This region is a putative zinc-finger domain with two $Cys-X_2-Cys$ motifs that characterize a variety of members of metal-binding zinc-finger families.[30]

Cosedimentation and blot overlay assays in vitro show that the interaction between dystrophin and β-dystroglycan fragments is severely compromised when any portion of the DgBD region is deleted. Removal of the ZZ motif greatly weakens binding to β-dystroglycan, while further removal of the WW or EF-hand region eliminates binding.[32,116,117] These studies suggest that the primary binding pocket for the C-terminus of β-dystroglycan is formed by an interaction between the WW and EF-hand regions of dystrophin. Indeed, the crystal structure of a fragment including the WW and EF-hands showed that the C-terminal 15 amino acids of β-dystroglycan fit within a tightly packed structure created by the WW domain overlaid upon the adjacent helical structures of the EF hands.[106] However, that crystal structure lacked the ZZ domain, known to be required for full binding. More recently, Ozawa and colleagues showed that full binding in blot overlay assays required the ZZ domain, and further, that mutation of 3 of the 5 cysteine residues within the ZZ domain significantly decreased binding between dystrophin and β-dystroglycan.[32] These authors proposed a model whereby the ZZ domain folds back across the interface between the two EF-hand domains to bind EF1, completing the binding fold for the C-terminus of β-dystroglycan. Full binding between dystrophin and β-dystroglycan in vitro thus requires an intact and contiguous DgBD formed by close association between the WW, EF hand and ZZ domains.[32]

This requirement for all three subdomains of the DbBD is dramatically manifested by in vivo studies in humans and mice. An important difference between DMD and BMD is that the milder BMD is almost always associated with expression of C-terminal regions of dystrophin. For example, a DMD patient with an unusually severe phenotype has been described with a deletion of exons encoding the WW and CR domains.[16] This patient was unique not only for his severe phenotype, but also for the observed stable expression of a mutant dystrophin on the sarcolemma. The severe phenotype likely resulted from this unusually stable expression of a mutant dystrophin that is: (1) not able to bind β-dystroglycan; and (2) able to displace utrophin from the sarcolemma. Other reports have reached similar conclusions.[18,119,120] Evidence that deletion of the DgBD was responsible, rather than the downstream CT domain, comes from the report of a relatively mild BMD phenotype in a patient with a deletion of exons encoding the CT domain.[121] Also, a severely affected DMD patient was described whose only identified mutation was an amino acid substitution of Cys(3340)Tyr in the ZZ domain.[122,123] That same amino acid substitution abrogated binding of dystrophin fragments to dystroglycan in vitro.[32] Surprisingly, a dystrophin transcript lacking exon 68 was identified in human aorta.[37] The predicted protein encoded by this transcript would end after EF-hand 2 and before the ZZ domain. Since this protein should be unable to bind β-dystroglycan, its significance is unclear.

Studies in mice provide direct evidence for the functional importance of the CR domain. Transgenic *mdx* animals that expressed a dystrophin lacking the EF-hand region (ΔCys1; encoded by exons 64-67), displayed a severe dystrophic phenotype (Fig. 2).[118] Analysis of DGC members in ΔCys1 muscles showed a nearly complete absence of α- and β-dystroglycan and all four sarcoglycans (e.g., Fig. 4). This ΔCys1 truncated dystrophin protein also displaced utrophin, resulting in significantly lower DGC levels than in *mdx* mice and a more severe phenotype.[118] Similar results were observed in transgenic *mdx* mice that expressed a dystrophin deleted for the ZZ domain (ΔCys2, Fig. 2; encoded by exons 68-70).[118] A third deletion of the CR domain was identified in *mdx*[3cv] mice. These animals have a mutation of the splice acceptor sequence in dystrophin intron 65, which leads to a complex pattern of alternative mRNA splicing.[124] The only transcript identified that carried an open reading frame was deleted for exons 65-66, which

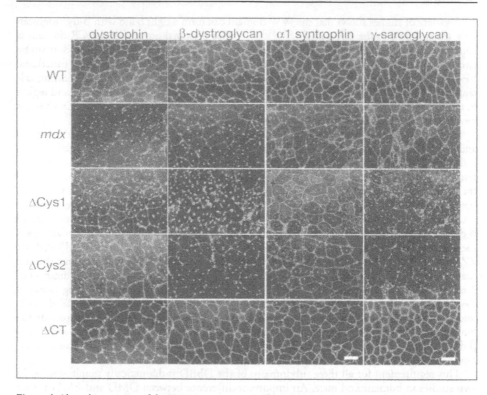

Figure 4. Altered expression of the DGC in transgenic *mdx* mice expressing truncated dystrophins. Shown are skeletal muscle cryosections from various lines of mice immunostained with antibodies against various DGC members. WT: normal C57Bl/10 mice; *mdx*: dystrophin-deficient mice; ΔCys1, ΔCys2 and ΔCT are transgenic *mdx* mice expressing the constructs illustrated in Figure 2. Note that each of the DGC proteins shown is present at very low levels in *mdx* muscles. Syntrophin is expressed at normal levels in skeletal muscles of each of the 3 transgenic strains.[118,125] In contrast, β-dystroglycan and γ-sarcoglycan are expressed at levels lower than in *mdx* muscles when the transgenic dystrophin lacks any portion of the cysteine-rich domain.[118] Mice lacking the CT domain display normal expression of the entire DGC.[125] Immunostained proteins are shown in green, while nuclei were counterstained with DAPI (blue in online version). Adapted from Rafael et al.[118] A color version of this figure is available online at http://www.eurekah.com/chapter.php?chapid=2280&bookid=133&catid=20.

encodes the first EF-hand region of dystrophin. Although this mutant dystrophin is expressed on the sarcolemma, *mdx*[3cv] mice also display a more severe phenotype than do *mdx* mice (unpublished observations).[124] Expression of the Δ65-66 dystrophin in transgenic *mdx* mice also led to assembly of a DGC lacking dystroglycan and the sarcoglycans, and a severe phenotype.[118] These mouse studies together support the idea that the WW and CR domain serve to bind directly to β-dystroglycan, but not to other members of the DGC. Any mutation that eliminates binding to β-dystroglycan also results in the loss from the sarcolemma of all four sarcoglycans and α-dystroglycan.[118,125] In contrast, no mutations have been found in dystrophin, either in patients or in mouse models, that primarily affect sarcoglycan expression. These observations support the model that the sarcoglycans require β-dystroglycan for stable accumulation and assembly into the DGC, and that the sarcoglycans do not directly contact dystrophin in vivo.[116-118,125,126]

In summary, the WW domain and both parts of the cysteine-rich domain of dystrophin (the EF-hand region and the ZZ domain) are essential for an interaction with β-dystroglycan in vivo, and are indispensable for normal muscle structure and function. Importantly, the tripartite DgBD is the only small portion of dystrophin absolutely required for the mechanical role of dystrophin in vivo. The requirement for three different sequence motifs distinguishes dystrophin from other WW domain-containing proteins, where an autonomous WW domain typically supports ligand binding. This larger, and presumably higher affinity, binding interface may be needed to buttress the dystrophin-dystroglycan link such that it can protect myofibers from mechanical injury when experiencing the extreme tensile forces that are developed during contraction.[106] Although utrophin has a similar β-dystroglycan binding domain structure as does dystrophin, the intracellular localization of utrophin at neuromuscular and myotendinous junctions implies that it is not required to withstand stresses as great as those that pull on dystrophin. In this context, utrophin does not appear to bind β-dystroglycan as tightly as does dystrophin.[32]

The Carboxy-Terminal (CT) Domain

The CT domain displays a number of unique features in dystrophin. Located adjacent to the CR domain, the CT domain is encoded on exons 71-79, although exon 79 contains only 2 and 1/3 codons (Figs. 1 and 2). At least one naturally occurring, but rare, isoform of dystrophin is encoded on mRNAs that lack exons 71-79. These transcripts carry intron 70 as the 3' untranslated region and the encoded proteins lack the CT domain.[37,127] Exons 71-74 and 78 display a complex pattern of alternative splicing that is both developmentally regulated and tissue-specific.[34,37] The splicing of exons 71-74 appears to modulate the stoichiometry and ratio of binding between dystrophin and two DGC protein family members, the syntrophins and the dystrobrevins (see below). Exon 78 contains only 31 nucleotides, and its removal from the dystrophin transcript results in a frameshift that encodes an alternate C-terminal extension of dystrophin.[34,37] The function of this alternative C-terminus is unclear. It is composed predominately of hydrophobic residues, whereas the normal C-terminus is hydrophilic. In muscle, the alternative C-terminus is found in embryonic and neonatal tissues, which express predominately the Dp71 isoform of dystrophin (Fig. 2).[34,128,129] The alternate C-terminus may only be found on Dp71, where it might serve to localize this isoform to structures other than the sarcolemma, such as the actin cytoskeleton.[130]

A variety of in vitro binding assays has established the CT domain as the region that binds to the syntrophin and dystrobrevin family members (see Chapters 4 and 5). Two regions appear capable of binding to syntrophin. A region encoded by exon 72-74 was implicated by several labs, and different binding assays have suggested that there may be two adjacent binding motifs in this syntrophin-binding domain (SBD; Fig. 1).[33,36,131-134] The entire CR and CT domains of dystrophin display considerable homology with α-dystrobrevin, and a yeast two hybrid screen with regions of dystrobrevin led to the clear identification of two adjacent syntrophin binding sites on both dystrophin and α-dystrobrevin (see Chapter 5). These sites are indicated as syn1 and syn2 in Figure 1. Syn1 spans exons 72-73, while syn2 spans exons 73-74.[38] These observations indicate that dystrophin and α-dystrobrevin might each bind up to 2 syntrophin molecules, and suggest that their alternative splicing could enable modulation of not only the number of syntrophins present in the DGC, but also the choice of which syntrophin family members localize with the DGC.[38,39,125]

A somewhat similar series of studies has shown that the dystrobrevin-binding domain (DbBD) is formed by two coiled-coil regions located at the exon 74-75 border and in the middle of exon 75.[39,135] Coiled coils are α-helical structures with a heptad repeat similar to leucine zippers. CC1 spans 4 heptad repeats, while CC2 spans 5. These motifs are protein-protein interaction domains, and their presence on both dystrophin and α-dystrobrevin mediates their binding.[39] As noted for the SBD, exon 74 is alternatively spliced in dystrophin, which might allow for modulation of which isoforms and/or family members of the dystrobrevin family associate with dystrophin.[38,125]

Despite the fact that the dystrophin CT mediates binding to the syntrophins and dystrobrevins, the overall importance of this domain in vivo is unclear. Relatively few mutations that only disrupt the CT domain have been observed in patients. One report described a patient with a deletion of exons 70-79, which removes the entire CT domain.[121] This individual displayed a relatively mild BMD phenotype. Since dystrophin expression levels were not examined, it remains unclear whether this phenotype was due to lack of the CT domain *per se*, or due to lower than normal levels of dystrophin. Experiments in mice have addressed this issue more clearly. Transgenic *mdx* animals that expressed a truncated dystrophin molecule deleted for either the SBD (a deletion of the regions encoded by exon 71-74), or the DbBD (a deletion of the regions encoded by exon 75-78) or both, (ΔCT, Fig. 2; exon 71-78 deletion) displayed essentially no signs of dystrophic pathology.[118,125] The only exceptions were in very old animals, which demonstrated a slightly higher level of regeneration.[125] While α-dystrobrevin and α1-syntrophin were nearly absent from the sarcolemma in *mdx* muscles, they were retained at the sarcolemma in the presence of the truncated dystrophins that lacked their binding sites (Fig. 4).[125] The ΔCT transgenic mice displayed normal muscle structure and function despite the lack of a direct association between dystrophin and either syntrophin or dystrobrevin. These data strongly suggest that the latter two proteins are not likely to participate in a mechanical role with dystrophin, since they can function fully without binding to dystrophin. Since syntrophin and dystrobrevin are required for normal muscle function, these data might suggest a subtle signaling role.[125,136-138]

Since syntrophin and dystrobrevin can localize to the sarcolemma in the ΔCT muscles, there must be an alternative interaction within the DGC complex. A potential candidate is the sarcoglycan complex, which has been shown to interact with dystrobrevin using in vitro binding assays.[139] A link between α-dystrobrevin and the sarcoglycan complex would provide a direct connection between core DGC members implicated in cell signaling (dystrobrevin, syntrophin, nNOS and the sarcoglycans). Finally, the ΔCT mice displayed an altered expression ratio of the syntrophin and dystrobrevin isoforms.[125] These observations further suggest that alternative splicing of dystrophin in different tissues may modulate assembly of DGC signaling components.

The Function of Dystrophin

Mechanical Roles

The studies summarized above indicate that dystrophin primarily serves a mechanical role in muscle cells by linking the cortical actin cytoskeleton to the extracellular matrix. This role is carried out by the extended actin-binding domains spread across the N-terminal and rod domains, and the dystroglycan-binding domain (Fig. 1). Muscle fibers lacking dystrophin are less stiff than normal myofibers, suggesting that dystrophin and the DGC may also reinforce the integrity of the membrane by imparting stiffness to the sarcolemma.[140] This role is emphasized by the observed localization pattern of dystrophin in skeletal muscle. Dystrophin has been shown to localize on the sarcolemma in a costameric pattern at the M- and Z-lines.[19,141,142] It is also enriched at sites of cell-cell contact: the neuromuscular and the myotendinous junctions.[143-145] The costameric localization of dystrophin helps to dissipate the forces of contraction laterally from the internal cytoskeleton to the extracellular matrix. The enhanced expression at the myotendinous junction likely provides additional mechanical reinforcement at a site expected to display high levels of stress during muscle contraction.[144,146,147]

Evidence has been presented that the DGC displays little activity in the absence of a direct connection to the actin cytoskeleton. For example, transgenic *mdx* mice that expressed the Dp71 dystrophin isoform developed a severe dystrophic phenotype.[148,149] Dp71 is the major dystrophin isoform in nonmuscle tissues and lacks the N-terminal and rod domain (Fig. 2).[150] Although transgenic Dp71 localized to the sarcolemma and assembled the DGC, the muscles displayed extensive sarcolemmal damage, which is comparable or worse than in *mdx* muscles. Mechanical failure of the sarcolemma eventually leads to myofiber necrosis by mechanisms that are unclear. These may include altered calcium homeostasis, activation of apoptotic pathways, and altered energy metabolism, and are described in Chapters 15 and 16.

Signaling Roles

Dystrophin also facilitates the localization of signaling molecules to the sarcolemma.[151] DGC members that directly bind signaling proteins include β-dystroglycan and syntrophin. As discussed earlier, tyrosine phosphorylation of the C-terminal tail of β-dystroglycan enables binding of caveolin-3, and a number of SH2 domain containing signaling molecules. The syntrophins, which bind both dystrophin and α-dystrobrevin, have been shown to bind directly to nNOS, serine/threonine kinases, calmodulin, aquaporin-4, and a voltage-gated sodium channel.[115,152-160]

The most direct evidence for signaling by the dystrophin complex comes from studies of the interaction between syntrophin and nNOS.[137,161,162] Dystrophin-deficient muscles in mice and humans lack nNOS on the sarcolemma and generate insufficient amounts of NO, resulting in impaired metabolic modulation of α-adrenergic vasoconstriction and functional ischemia.[156,163] Mice with mutant α-1 syntrophin genes also display highly reduced expression of nNOS.[136,138] Interestingly, neither α-1 syntrophin nor nNOS knock-out mice develop a dystrophic phenotype, however both these mutants display abnormal NMJs.[136,164-166] Mice with mutant α-dystrobrevin genes also display abnormalities suggestive of impaired nNOS signaling. Abnormal maturation of postsynaptic membranes was also observed, as has been seen in the dystrophin/utrophin double knockout (dko) and the α1-syntrophin knockout mice.[167-169] Transgenic dko mice expressing a ΔCR truncated dystrophin construct showed an amelioration of the post-synaptic membrane and fiber-type abnormalities, despite continuing to display a dystrophic pathology.[170] Since ΔCR is not able to rescue mechanical abnormalities in dystrophic muscle, the amelioration of the post-synaptic membrane and fiber-type abnormalities may be the result of restoring the localization of signaling molecules.[118,170] The common theme here is that the entire DGC, and hence all these additional accessory proteins, require normal expression of dystrophin for their proper localization to the sarcolemma.

Several DGC members, including dystrophin, are phosphorylated in vivo; however, little is known about the kinases involved or the purpose of phosphorylation.[112,148,157,171-173] The only clear example is that of SAPK3, which has been shown to phosphorylate α1-syntrophin. SAPKs are activated by cellular stress and are connected to the SAPK/c-Jun pathway.[157] The c-Jun NH2-terminal kinase JNK1 displays elevated activity in dystrophic muscle, and this activation may occur through a syntrophin-linked pathway.[174,175] An additional example is the aforementioned adhesion-dependent phosphorylation of β-dystroglycan.[112] SAPK and other protein kinases may therefore directly phosphorylate components of the DGC to enable binding of signaling molecules, and/or to adapt their properties to respond to changes in muscle metabolism, contractile status or stress.[176]

Conclusions

Dystrophin was the first gene isolated by positional cloning methods, establishing a precedence for the human genome project.[22] Sixteen years later, a great deal of information has been learned about how dystrophin works, which proteins it interacts with, and how mutations in this enormous gene lead to the devastating features of DMD and BMD. Unfortunately, there is still no cure, and therapeutic options remain limited. However, the knowledge gained has led to an enormous number of new ideas on how to potentially treat this disorder, from direct gene repair or replacement to treating secondary manifestations of the pathology. More work is needed, but as will be explored later in this book, increasing optimism is spreading that methods can and will be developed to halt the destructive events resulting from the mechanical and signaling abnormalities associated with dystrophin gene mutation.

Acknowledgments

The authors thank Paul Gregorevic, Mike Blankinship, Sheng Li, Luke Judge, Scott Harper, Laura Warner, Gregory Crawford, Jill Rafael-Fortney, Carey Lumeng, Stephanie Phelps, Kathleen Corrado and Greg Cox for helpful discussions. Supported by grants from the National Institutes of Health and the Muscular Dystrophy Association (USA). We are also grateful for the generous support from Chip and Betsy Erwin, Harmonize for Hope, and the Bruce and Jolene McCaw Foundation.

References

1. Emery AE. The muscular dystrophies. Lancet 2002; 359(9307):687-695.
2. Emery AEH. Duchenne muscular dystrophy. Oxford: Oxford Medical Publications, 1993:24.
3. Bresolin N, Castelli E, Comi GP et al. Cognitive impairment in Duchenne muscular dystrophy. Neuromuscul Disord 1994; 4:359-369.
4. North KN, Miller G, Iannaccone ST et al. Cognitive dysfunction as the major presenting feature of Becker's muscular dystrophy. Neurology 1996; 46:461-465.
5. Polakoff RJ, Morton AA, Koch KD et al. The psychosocial and cognitive impact of Duchenne's muscular dystrophy. Semin Pediatr Neurol 1998; 5(2):116-123.
6. England SB, Nicholson LV, Johnson MA et al. Very mild muscular dystrophy associated with the deletion of 46% of dystrophin. Nature 1990; 343:180-182.
7. Kunkel LM et.al. Analysis of deletions in DNA from patients with Becker and Duchenne muscular dystrophy. Nature 1986; 322:73-77.
8. Monaco AP, Neve RL, Coletti-Feener C et al. Isolation of candidate cDNA clones for portions of the Duchenne muscular dystrophy gene. Nature 1986; 323:646-650.
9. Chamberlain JS, Gibbs RA, Ranier JE et al. Deletion screening of the Duchenne muscular dystrophy locus via multiplex DNA amplification. Nucleic Acids Res 1988; 16:11141-11156.
10. Hoffman EP, Brown Jr RH, Kunkel LM. Dystrophin: The protein product of the Duchenne muscular dystrophy locus. Cell 1987; 51:919-928.
11. Koenig M, Beggs AH, Moyer M et al. The molecular basis for Duchenne versus becker muscular dystrophy: Correlation of severity with type of deletion. Am J Hum Genet 1989; 45:498-506.
12. Monaco AP, Bertelson CJ, Liechti-Gallati S et al. An explanation for the phenotypic differences between patients bearing partial deletions of the DMD locus. Genomics 1988; 2:90-95.
13. Arahata K, Beggs AH, Honda H et al. Preservation of the C-terminus of dystrophin molecule in the skeletal muscle from becker muscular dystrophy. J Neurol Sci 1991; 101:148-156.
14. Bulman DE, Murphy EG, Zubrzycka-Gaarn EE et al. Differentiation of Duchenne and becker muscular dystrophy phenotypes with amino- and carboxy-terminal antisera specific for dystrophin. Am J Hum Genet 1991; 48:295-304.
15. Malhotra SB, Hart KA, Klamut HJ et al. Frame-shift deletions in patients with Duchenne and Becker muscular dystrophy. Science 1988; 242:755-759.
16. Hoffman EP, Garcia CA, Chamberlain JS et al. Is the carboxyl-terminus of dystrophin required for membrane association? A novel, severe case of Duchenne muscular dystrophy. Ann Neurol 1991; 30:605-610.
17. Vainzof M, Takata RI, Passos-Bueno MR et al. Is the maintainance of the C-terminus domain of dystrophin enough to ensure a milder Becker muscular dystrophy phenotype? Hum Mol Genet 1993; 2:39-42.
18. Bies RD, Caskey CT, Fenwick R. An intact cysteine-rich domain is required for dystrophin function. J Clin Invest 1992; 90:666-672.
19. Zubrzycka-Gaarn EE, Bulman DE, Karpati G et al. The Duchenne muscular dystrophy gene product is localized in sarcolemma of human skeletal muscle. Nature 1988; 333:466-469.
20. Koenig M, Monaco AP, Kunkel LM. The complete sequence of dystrophin predicts a rod-shaped cytoskeletal protein. Cell 1988; 53:219-226.
21. Davison MD, Critchley DR. Alpha-actinins and the DMD protein contain spectrin-like repeats. Cell 1988; 52(2):159-160.
22. Koenig M, Hoffman EP, Bertelson CJ et al. Complete cloning of the Duchenne muscular dystrophy (DMD) cDNA and preliminary genomic organization of the DMD gene in normal and affected individuals. Cell 1987; 50:509-517.
23. Koenig M, Kunkel LM. Detailed analysis of the repeat domain of dystrophin reveals four potential hinge segments that may confer flexibility. J Biol Chem 1990; 265:4560-4566.
24. Winder SJ, Gibson TJ, Kendrick-Jones J. Dystrophin and utrophin: The missing links! FEBS Lett 1995; 369:27-33.

25. Cross RA, Stewart M, Kendrick-Jones J. Structural predictions for the central domain of dystrophin. FEBS Lett 1990; 262:87-92.

26. Passos-Bueno MR, Vainzof M, Marie SK et al. Half the dystrophin gene is apparently enough for a mild clinical course: Confirmation of its potential use for gene therapy. Hum Mol Genet 1994; 3:919-922.

27. Rybakova IN, Ervasti JM. The dystrophin-glycoprotein complex binds but does not cross-link F-actin. Mol Biol Cell 1996; 6:152a.

28. Amann KJ, Renley BA, Ervasti JM. A cluster of basic repeats in the dystrophin rod domain binds F-actin through an electrostatic interaction. J Biol Chem 1998; 273(43):28419-28423.

29. Ilsley JL, Sudol M, Winder SJ. The WW domain: Linking cell signalling to the membrane cytoskeleton. Cell Signal 2002; 14(3):183-189.

30. Ponting CP, Blake DJ, Davies KE et al. New putative zinc fingers in dystrophin and other proteins. Trends Biochem Sci 1996; 21(1):11-13.

31. Broderick MJ, Winder SJ. Towards a complete atomic structure of spectrin family proteins. J Struct Biol 2002; 137(1-2):184-193.

32. Ishikawa-Sakurai M, Yoshida M, Imamura M et al. ZZ domain is essentially required for the physiological binding of dystrophin and utrophin to β-dystroglycan. Hum Mol Genet 2004; 13:693-702.

33. Yang B, Jung D, Rafael JA et al. Identification of α-syntrophin binding to syntrophin triplet, dystrophin, and utrophin. J Biol Chem 1995; 270:4975-4978.

34. Bies RD, Phelps SF, Cortez MD et al. Human and murine dystrophin mRNA transcripts are differentially expressed during skeletal muscle, heart, and brain development. Nucleic Acids Res 1992; 20:1725-1731.

35. Ahn AH, Freener CA, Gussoni E et al. The three human syntrophin genes are expressed in diverse tissues, have distinct chromosomal locations, and each bind to dystrophin and its relatives. J Biol Chem 1996; 271:2724-2730.

36. Suzuki A, Yoshida M, Ozawa E. Mammalian α1- and β1-syntrophin bind to the alternative splice-prone region of the dystrophin COOH terminus. J Cell Biol 1995; 128:373-381.

37. Feener CA, Koenig M, Kunkel LM. Alternative splicing of human dystrophin mRNA generates isoforms at the carboxy terminus. Nature 1989; 338:509-511.

38. Newey SE, Benson MA, Ponting CP et al. Alternative splicing of dystrobrevin regulates the stoichiometry of syntrophin binding to the dystrophin protein complex. Curr Biol 2000; 10(20):1295-1298.

39. Sadoulet-Puccio HM, Rajala M, Kunkel LM. Dystrobrevin and dystrophin: An interaction through coiled-coil motifs. Proc Natl Acad Sci USA 1997; 94:12413-12418.

40. Sutherland-Smith AJ, Moores CA, Norwood FL et al. An atomic model for actin binding by the CH domains and spectrin-repeat modules of utrophin and dystrophin. J Mol Biol 2003; 329(1):15-33.

41. Rybakova IN, Ervasti JM. Dystrophin-glycoprotein complex is monomeric and stabilizes actin filaments in vitro through a lateral association. J Biol Chem 1997; 272:28771-28778.

42. Orlova A, Rybakova IN, Prochniewicz E et al. Binding of dystrophin's tandem calponin homology domain to F-actin is modulated by actin's structure. Biophys J 2001; 80(4):1926-1931.

43. Hemmings L, Kuhlman PA, Critchley DR. Analysis of the actin-binding domain of alpha-actinin by mutagenesis and demonstration that dystrophin contains a functionally homologous domain. J Cell Biol 1992; 116:1369-1380.

44. Corrado K, Mills PL, Chamberlain JS. Deletion analysis of the dystrophin-actin binding domain. FEBS Lett 1994; 344:255-260.

45. Way M, Pope B, Cross RA et al. Expression of the N-terminal domain of dystrophin in E. coli and demonstration of binding to F-actin. FEBS Lett 1992; 301:243-245.

46. Puius YA, Mahoney NM, Almo SC. The modular structure of actin-regulatory proteins. Curr Opin Cell Biol 1998; 10(1):23-34.

47. Bramham J, Hodgkinson JL, Smith BO et al. Solution structure of the calponin CH domain and fitting to the 3D-helical reconstruction of F-actin:calponin. Structure (Camb) 2002; 10(2):249-258.

48. Stradal T, Kranewitter W, Winder SJ et al. CH domains revisited. FEBS Lett 1998; 431(2):134-137.

49. Levine BA, Moir AJ, Patchell VB et al. The interaction of actin with dystrophin. FEBS Lett 1990; 263:159-162.

50. Levine BA, Moir AJ, Patchell VB et al. Binding sites involved in the interaction of actin with the N- terminal region of dystrophin. FEBS Lett 1992; 298:44-48.

51. Winder SJ, Hemmings L, Maciver SK et al. Utrophin actin binding domain: Analysis of actin binding and cellular targeting. J Cell Sci 1995; 108:63-71.

52. Bresnick AR, Warren V, Condeelis J. Identification of a short sequence essential for actin binding by Dictyostelium ABP-120. J Biol Chem 1990; 265(16):9236-9240.
53. Fabbrizio E, Bonet-Kerrache A, Leger JJ et al. Actin-dystrophin interface. Biochemistry 1993; 32:10457-10463.
54. Ervasti JM, Campbell KP. A role for the dystrophin-glycoprotein complex as a transmembrane linker between laminin and actin. J Cell Biol 1993; 122:809-823.
55. Li X, Bennett V. Identification of the spectrin subunit and domains required for formation of spectrin/adducin/actin complexes. J Biol Chem 1996; 271(26):15695-15702.
56. Zuellig RA, Bornhauser BC, Knuesel I et al. Identification and characterisation of transcript and protein of a new short N-terminal utrophin isoform. J Cell Biochem 2000; 77(3):418-431.
57. Harper SQ, Hauser MA, DelloRusso C et al. Modular flexibility of dystrophin: Implications for gene therapy of Duchenne muscular dystrophy. Nat Med 2002; 8(3):253-261.
58. Rybakova IN, Patel JR, Davies KE et al. Utrophin binds laterally along actin filaments and can couple costameric actin with sarcolemma when overexpressed in dystrophin-deficient muscle. Mol Biol Cell 2002; 13(5):1512-1521.
59. Norwood FL, Sutherland-Smith AJ, Keep NH et al. The structure of the N-terminal actin-binding domain of human dystrophin and how mutations in this domain may cause Duchenne or Becker muscular dystrophy. Structure Fold Des 2000; 8(5):481-491.
60. Chan YM, Kunkel LM. In vitro expressed dystrophin fragments do not associate with each other. FEBS Lett 1997; 410:153-159.
61. Kahana E, Flood G, Gratzer WB. Physical properties of dystrophin rod domain. Cell Motil Cytoskeleton 1997; 36:246-252.
62. Rybakova IN, Patel JR, Ervasti JM. The dystrophin complex forms a mechanically strong link between the sarcolemma and costameric actin. J Cell Biol 2000; 150(5):1209-1214.
63. Sicinski P, Geng Y, Ryder-Cook AS et al. The molecular basis of muscular dystrophy in the mdx mouse: A point mutation. Science 1989; 244:1578-1580.
64. Corrado K, Rafael JA, Mills PL et al. Transgenic mdx mice expressing dystrophin with a deletion in the actin-binding domain display a "mild Becker" phenotype. J Cell Biol 1996; 134:873-884.
65. Fabb SA, Wells DJ, Serpente P et al. Adeno-associated virus vector gene transfer and sarcolemmal expression of a 144 kDa micro-dystrophin effectively restores the dystrophin-associated protein complex and inhibits myofibre degeneration in nude/mdx mice. Hum Mol Genet 2002; 11(7):733-741.
66. Baumbach LL, Chamberlain JS, Ward PA et al. Molecular and clinical correlations of deletions leading to Duchenne and Becker muscular dystrophies. Neurology 1989; 39:465-474.
67. Beggs AH, Hoffman EP, Snyder JR et al. Exploring the molecular basis for variability among patients with Becker muscular dystrophy: Dystrophin gene and protein studies. Am J Hum Genet 1991; 49:54-67.
68. Muntoni F, Gobbi P, Sewry C et al. Deletions in the 5' region of dystrophin and resulting phenotypes. J Med Genet 1994; 31:843-847.
69. Gangopadhyay SB, Sherratt TG, Heckmatt JZ et al. Dystrophin in frameshift deletion patients with Becker muscular dystrophy. Am J Hum Genet 1992; 51:562-570.
70. Winnard AV, Klein CJ, Coovert DD et al. Characterization of translational frame exception patients in Duchenne/Becker muscular dystrophy. Hum Mol Genet 1993; 2:737-744.
71. Winnard AV, Mendell JR, Prior TW et al. Frameshift deletions of exons 3-7 and revertant fibers in Duchenne muscular dystrophy: Mechanisms of dystrophin production. Am J Hum Genet 1995; 56:158-166.
72. Thanh LT, Man NT, Hori S et al. Characterization of genetic deletions in Becker muscular dystrophy using monoclonal antibodies against a deletion-prone region of dystrophin. Am J Med Genet 1995; 58:177-186.
73. Prior TW, Papp AC, Snyder PJ et al. A missense mutation in the Dystrophin gene in a Duchenne muscular dystrophy patient. Nature Genet 1993; 4:357-360.
74. Prior TW, Bartolo C, Pearl DK et al. Spectrum of small mutations in the dystrophin coding region. Am J Hum Genet 1995; 57:22-33.
75. Roberts RG, Gardner RJ, Bobrow M. Searching for the 1 in 2,400,000: A review of Dystrophin gene point mutations. Hum Mutat 1994; 4:1-11.
76. Roberts RG, Coffey AJ, Bobrow M et al. Exon structure of the human Dystrophin gene. Genomics 1993; 16:536-538.
77. Kahana E, Marsh PJ, Henry AJ et al. Conformation and phasing of dystrophin structural repeats. J Mol Biol 1994; 235:1271-1277.
78. Kahana E, Gratzer WB. Minimum folding unit of dystrophin rod domain. Biochemistry 1995; 34:8110-8114.

79. Pascual J, Pfuhl M, Walther D et al. Solution structure of the spectrin repeat: A left-handed anti-parallel triple-helical coiled-coil. J Mol Biol 1997; 273(3):740-751.
80. Yan Y, Winograd E, Viel A et al. Crystal structure of the repetitive segments of spectrin. Science 1993; 262(5142):2027-2030.
81. Harper SQ, Crawford RW, DelloRusso C et al. Spectrin-like repeats from dystrophin and alpha-actinin-2 are not functionally interchangeable. Hum Mol Genet 2002; 11(16):1807-1815.
82. Thomas GH, Newbern EC, Korte CC et al. Intragenic duplication and divergence in the spectrin superfamily of proteins. Mol Biol Evol 1997; 14(12):1285-1295.
83. Grum VL, Li D, MacDonald RI et al. Structures of two repeats of spectrin suggest models of flexibility. Cell 1999; 98(4):523-535.
84. Petrof BJ, Shrager JB, Stedman HH et al. Dystrophin protects the sarcolemma from stresses developed during muscle contraction. Proc Natl Acad Sci USA 1993; 90:3710-3714.
85. Fritz JD, Danko I, Roberds SL et al. Expression of deletion-containing dystrophins in mdx muscle: Implications for gene therapy and dystrophin function. Pediatr Res 1995; 37(6):693-700.
86. DeWolf C, McCauley P, Sikorski AF et al. Interaction of dystrophin fragments with model membranes. Biophys J 1997; 72:2599-2604.
87. Le Rumeur E, Fichou Y, Pottier S et al. Interaction of dystrophin rod domain with membrane phospholipids: Evidence of a close proximity between tryptophan residues and lipids. J Biol Chem 2002: M207321200.
88. Love DR, Flint TJ, Genet SA et al. Becker muscular dystrophy patient with a large intragenic dystrophin deletion: Implications for functional minigenes and gene therapy. J Med Genet 1991; 28:860-864.
89. Warner LE, DelloRusso C, Crawford RW et al. Expression of Dp260 in muscle tethers the actin cytoskeleton to the dystrophin-glycoprotein complex and partially prevents dystrophy. Hum Mol Genet 2002; 11(9):1095-1105.
90. Phelps SF, Hauser MA, Cole NM et al. Expression of full-length and truncated Dystrophin mini-genes in transgenic mdx mice. Hum Mol Genet 1995; 4:1251-1258.
91. Chamberlain JS. Gene therapy of muscular dystrophy. Hum Mol Genet 2002; 11(20):2355-2362.
92. Wells KE, Torelli S, Lu Q et al. Relocalization of neuronal nitric oxide synthase (nNOS) as a marker for complete restoration of the dystrophin associated protein complex in skeletal muscle. Neuromuscul Disord 2003; 13(1):21-31.
93. Chao DS, Gorospe JR, Brenman JE et al. Selective loss of sarcolemmal nitric oxide synthase in Becker muscular dystrophy. J Exp Med 1996; 184(2):609-618.
94. Hauser MA, Chamberlain JS. Progress towards gene therapy for Duchenne muscular dystrophy. J Endocrinol 1996; 149(3):373-378.
95. Wang B, Li J, Xiao X. Adeno-associated virus vector carrying human minidystrophin genes effectively ameliorates muscular dystrophy in mdx mouse model. Proc Natl Acad Sci USA 2000; 97(25):13714-13719.
96. Sakamoto M, Yuasa K, Yoshimura M et al. Micro-dystrophin cDNA ameliorates dystrophic phenotypes when introduced into mdx mice as a transgene. Biochem Biophys Res Commun 2002; 293(4):1265-1272.
97. Yuasa K, Miyagoe Y, Yamamoto K et al. Effective restoration of dystrophin-associated proteins in vivo by adenovirus-mediated transfer of truncated dystrophin cDNAs. FEBS Lett 1998; 425(2):329-336.
98. Crawford GE, Lu QL, Partridge TA et al. Suppression of revertant fibers in mdx mice by expression of a functional dystrophin. Hum Mol Genet 2001; 10(24):2745-2750.
99. Deconinck AE, Rafael JA, Skinner JA et al. Utrophin-dystrophin-deficient mice as a model for Duchenne muscular dystrophy. Cell 1997; 90:717-727.
100. Grady RM, Teng HB, Nichol MC et al. Skeletal and cardiac myopathies in mice lacking utrophin and dystrophin: A model for Duchenne muscular dystrophy. Cell 1997; 90:729-738.
101. Tinsley J, Deconinck N, Fisher R et al. Expression of full-length utrophin prevents muscular dystrophy in mdx mice. Nat Med 1998; 4(12):1441-1444.
102. Watchko J, O'Day T, Wang B et al. Adeno-associated virus vector-mediated minidystrophin gene therapy improves dystrophic muscle contractile function in mdx mice. Hum Gene Ther 2002; 13(12):1451-1460.
103. Crosbie RH, Dovico SA, Flanagan JD et al. Characterization of aquaporin-4 in muscle and muscular dystrophy. FASEB J 2002; 16(9):943-949.
104. Suzuki A, Yoshida M, Yamamoto H et al. Glycoprotein-binding site of dystrophin is confined to the cysteine- rich domain and the first half of the carboxy-terminal domain. FEBS Lett 1992; 308:154-160.

105. Bork P, Sudol M. The WW domain: A signalling site in dystrophin? Trends Biochem Sci 1994; 19:531-533.
106. Huang X, Poy F, Zhang R et al. Structure of a WW domain containing fragment of dystrophin in complex with beta-dystroglycan. Nat Struct Biol 2000; 7(8):634-638.
107. Ponting CP, Blake DJ, Davies KE et al. ZZ and TAZ: New putative zinc fingers in dystrophin and other proteins. Trends Biochem Sci 1996; 21:11-13.
108. Sudol M, Chen HI, Bougeret C et al. Characterization of a novel protein-binding module-the WW domain. FEBS Lett 1995; 369(1):67-71.
109. Sudol M. Structure and function of the WW domain. Prog Biophys Mol Biol 1996; 65:113-132.
110. Chen HI, Sudol M. The WW domain of yes-associated protein binds a proline-rich ligand that differs from the consensus established for src homology 3-binding modules. Proc Natl Acad Sci USA 1995; 92(17):7819-7823.
111. Rentschler S, Linn H, Deininger K et al. The WW domain of dystrophin requires EF-hands region to interact with beta-dystroglycan. J Biol Chem 1999; 380(4):431-442.
112. James M, Nuttall A, Ilsley JL et al. Adhesion-dependent tyrosine phosphorylation of (beta)-dystroglycan regulates its interaction with utrophin. J Cell Sci 2000; 113:1717-1726.
113. Sotgia F, Lee H, Bedford MT et al. Tyrosine phosphorylation of beta-dystroglycan at its WW domain binding motif, PPxY, recruits SH2 domain containing proteins. Biochemistry 2001; 40(48):14585-14592.
114. Sotgia F, Lee JK, Das K et al. Caveolin-3 directly interacts with the C-terminal tail of beta -dystroglycan. Identification of a central WW-like domain within caveolin family members. J Biol Chem 2000; 275(48):38048-38058.
115. Yang B, Jung D, Motto D et al. SH3 domain-mediated interaction of dystroglycan and Grb2. J Biol Chem 1995; 270:11711-11714.
116. Suzuki A, Yoshida M, Hayashi K et al. Molecular organization at the glycoprotein-complex-binding site of dystrophin-three dystrophin-associated proteins bind directly to the carboxy-terminal portion of dystrophin. Eur J Biochem 1994; 220:283-292.
117. Jung D, Yang B, Meyer J et al. Identification and characterization of the dystrophin anchoring site on β-dystroglycan. J Biol Chem 1995; 270:27305-27310.
118. Rafael JA, Cox GA, Corrado K et al. Forced expression of dystrophin deletion constructs reveals structurefunction correlations. J Cell Biol 1996; 134:93-102.
119. Helliwell TR, Ellis JM, Mountford RC et al. A truncated dystrophin lacking the C-terminal domains is localized at the muscle membrane. Am J Hum Genet 1992; 50:508-514.
120. Recan D, Chafey P, Leturcq F et al. Are cysteine-rich and COOH-terminal domains of dystrophin critical for sarcolemmal localization? J Clin Invest 1992; 89:712-716.
121. McCabe ER, Towbin J, Chamberlain J et al. Complementary DNA probes for the Duchenne muscular dystrophy locus demonstrate a previously undetectable deletion in a patient with dystrophic myopathy, glycerol kinase deficiency, and congenital adrenal hypoplasia. J Clin Invest 1989; 83:95-99.
122. Lenk U, Hanke R, Thiele H et al. Point mutations at the carboxy terminus of the human Dystrophin gene: Implications for an association with mental retardation in DMD patients. Hum Mol Genet 1993; 2(11):1877-1881.
123. Lenk U, Oexle K, Voit T et al. A cysteine 3340 substitution in the dystroglycan-binding domain of dystrophin associated with Duchenne muscular dystrophy, mental retardation and absence of the ERG b-wave. Hum Mol Genet 1996; 5(7):973-975.
124. Cox GA, Phelps SF, Chapman VM et al. New mdx mutation disrupts expression of muscle and nonmuscle isoforms of dystrophin. Nat Genet 1993; 4:87-93.
125. Crawford GE, Faulkner JA, Crosbie RH et al. Assembly of the dystrophin-associated protein complex does not require the dystrophin COOH-terminal domain. J Cell Biol 2000; 150(6):1399-1410.
126. Rosa G, Ceccarini M, Cavaldesi M et al. Localization of the dystrophin binding site at the carboxyl terminus of β-dystroglycan. Biochem Biophys Res Commun 1996; 223:272-277.
127. Tinsley JM, Blake DJ, Davies KE. Apo-dystrophin-3: A 2.2kb transcript from the DMD locus encoding the dystrophin glycoprotein binding site. Hum Mol Genet 1993; 2:521-524.
128. Austin RC, Howard PL, D'Souza VN et al. Cloning and characterization of alternatively spliced isoforms of Dp71. Hum Mol Genet 1995; 4(9):1475-1483.
129. Tennyson CN, Dally GY, Ray PN et al. Expression of the dystrophin isoform Dp71 in differentiating human fetal myogenic cultures. Hum Mol Genet 1996; 5(10):1559-1566.
130. Howard PL, Dally GY, Ditta SD et al. Dystrophin isoforms DP71 and DP427 have distinct roles in myogenic cells. Muscle Nerve 1999; 22(1):16-27.
131. Ahn AH, Kunkel LM. Syntrophin binds to an alternatively spliced exon of dystrophin. J Cell Biol 1995; 128:363-371.

132. Dwyer TM, Froehner SC. Direct binding of Torpedo syntrophin to dystrophin and the 87 kDa dystrophin homologue. FEBS Lett 1995; 375:91-94.

133. Cox GF, Kunkel LM. Direct binding of gamma2 syntrophin to dystrophin and its relatives. Am J Hum Gen 1997; 61(4):A168.

134. Castello A, Brocheriou V, Chafey P et al. Characterization of the dystrophin-syntrophin interaction using the two-hybrid system in yeast. FEBS Lett 1996; 383(1-2):124-128.

135. Peters MF, O'Brien KF, Sadoulet-Puccio HM et al. β-dystrobrevin, a new member of the dystrophin family - identification, cloning, and protein associations. J Biol Chem 1997; 272:31561-31569.

136. Adams ME, Kramarcy N, Krall SP et al. Absence of alpha-syntrophin leads to structurally aberrant neuromuscular synapses deficient in utrophin. J Cell Biol 2000; 150(6):1385-1398.

137. Grady RM, Grange RW, Lau KS et al. Role for alpha-dystrobrevin in the pathogenesis of dystrophin-dependent muscular dystrophies. Nat Cell Biol 1999; 1(4):215-220.

138. Hosaka Y, Yokota T, Miyagoe-Suzuki Y et al. Alpha1-syntrophin-deficient skeletal muscle exhibits hypertrophy and aberrant formation of neuromuscular junctions during regeneration. J Cell Biol 2002; 158(6):1097-1107.

139. Yoshida M, Hama H, Ishikawa-Sakurai M et al. Biochemical evidence for association of dystrobrevin with the sarcoglycan-sarcospan complex as a basis for understanding sarcoglycanopathy. Hum Mol Genet 2000; 9(7):1033-1040.

140. Pasternak C, Wong S, Elson EL. Mechanical function of dystrophin in muscle cells. J Cell Biol 1995; 128:355-361.

141. Porter GA, Dmytrenko GM, Winkelmann JC et al. Dystrophin colocalizes with beta-spectrin in distinct subsarcolemmal domains in mammalian skeletal muscle. J Cell Biol 1992; 117:997-1005.

142. Straub V, Bittner RE, Leger JJ et al. Direct visualization of the dystrophin network on skeletal muscle fiber membrane. J Cell Biol 1992; 119(5):1183-1191.

143. Sealock R, Butler MH, Kramarcy NR et al. Localization of dystrophin relative to acetylcholine receptor domains in electric tissue and adult and cultured skeletal muscle. J Cell Biol 1991; 113:1133-1144.

144. Law DJ, Tidball JG. Dystrophin deficiency is associated with myotendinous junction defects in prenecrotic and fully regenerated skeletal muscle. Am J Pathol 1993; 142:1513-1523.

145. Ervasti JM, Ohlendieck K, Kahl SD et al. Deficiency of a glycoprotein component of the dystrophin complex in dystrophic muscle. Nature 1990; 345:315-319.

146. Tidball JG, Law DJ. Dystrophin is required for normal thin filament-membrane associations at myotendinous junctions. Am J Pathol 1991; 138:17-21.

147. Law DJ, Caputo A, Tidball JG. Site and mechanics of failure in normal and dystrophin-deficient skeletal muscle. Muscle Nerve 1995; 18:216-223.

148. Cox GA, Sunada Y, Campbell KP et al. Dp71 can restore the dystrophin-associated glycoprotein complex in muscle but fails to prevent dystrophy [see comments]. Nat Genet 1994; 8(4):333-339.

149. Greenberg DS, Sunada Y, Campbell KP et al. Exogenous Dp71 restores the levels of dystrophin associated proteins but does not alleviate muscle damage in mdx mice. Nature Genet 1994; 8:340-344.

150. Lederfein D, Levy Z, Augier N et al. A 71-kilodalton protein is a major product of the Duchenne muscular Dystrophy gene in brain and other nonmuscle tissues. Proc Natl Acad Sci USA 1992; 89:5346-5350.

151. Rando TA. The dystrophin-glycoprotein complex, cellular signaling, and the regulation of cell survival in the muscular dystrophies. Muscle Nerve 2001; 24(12):1575-1594.

152. Brenman JE, Chao DS, Gee SH et al. Interaction of nitric oxide synthase with the postsynaptic density protein PSD-95 and alpha1-syntrophin mediated by PDZ domains. Cell 1996; 84(5):757-767.

153. Lumeng C, Phelps S, Crawford GE et al. Interactions between beta 2-syntrophin and a family of microtubule-associated serine/threonine kinases. Nat Neurosci 1999; 2(7):611-617.

154. Gee SH, Madhavan R, Levinson SR et al. Interaction of muscle and brain sodium channels with multiple members of the syntrophin family of dystrophin-associated proteins. J Neurosci 1998; 18:128-137.

155. Adams ME, Mueller HE, Froehner SC. In vivo requirement of the α-syntrophin PDZ domain for the sarcolemmal localization of nNOS and aquaporin-4. J Cell Biol 2001; 155:113-122.

156. Thomas GD, Sander M, Lau KS et al. Impaired metabolic modulation of alpha-adrenergic vasoconstriction in dystrophin-deficient skeletal muscle [see comments]. Proc Natl Acad Sci USA 1998; 95(25):15090-15095.

157. Hasegawa M, Cuenda A, Spillantini MG et al. Stress-activated protein kinase-3 interacts with the PDZ domain of alpha 1-syntrophin—a mechanism for specific substrate recognition. J Biol Chem 1999; 274(18):12626-12631.

158. Song KS, Scherer PE, Tang ZL et al. Expression of caveolin-3 in skeletal, cardiac, and smooth muscle cells - caveolin-3 is a component of the sarcolemma and cofractionates with dystrophin and dystrophin-associated glycoproteins. J Biol Chem 1996; 271:15160-15165.
159. Crosbie RH, Yamada H, Venzke DP et al. Caveolin-3 is not an integral component of the dystrophin glycoprotein complex. FEBS Lett 1998; 427(2):279-282.
160. Schultz J, Hoffmuller U, Krause G et al. Specific interactions between the syntrophin PDZ domain and voltage- gated sodium channels. Nat Struct Biol 1998; 5(1):19-24.
161. Brenman JE, Chao DS, Xia HH et al. Nitric oxide synthase complexed with dystrophin and absent from skeletal muscle sarcolemma in Duchenne muscular dystrophy. Cell 1995; 82:743-752.
162. Rando TA. Role of nitric oxide in the pathogenesis of muscular dystrophies: A "two hit" hypothesis of the cause of muscle necrosis. Microsc Res Tech 2001; 55(4):223-235.
163. Sander M, Chavoshan B, Harris SA et al. Functional muscle ischemia in neuronal nitric oxide synthase-deficient skeletal muscle of children with Duchenne muscular dystrophy. Proc Natl Acad Sci USA 2000; 97(25):13818-13823.
164. Kameya S, Miyagoe Y, Nonaka I et al. alpha1-syntrophin gene disruption results in the absence of neuronal-type nitric-oxide synthase at the sarcolemma but does not induce muscle degeneration. J Biol Chem 1999; 274(4):2193-2200.
165. Chao DS, Silvagno F, Bredt DS. Muscular dystrophy in mdx mice despite lack of neuronal nitric oxide synthase. J Neurochem 1998; 71(2):784-789.
166. Crosbie RH, Straub V, Yun HY et al. mdx muscle pathology is independent of nNOS perturbation. Hum Mol Genet 1998; 7(5):823-829.
167. Deconinck N, Rafael JA, Beckers-Bleukx G et al. Consequences of the combined deficiency in dystrophin and utrophin on the mechanical properties and myosin composition of some limb and respiratory muscles of the mouse. Neuromusc Disord 1998; 8(6):362-370.
168. Grady RM, Merlie JP, Sanes JR. Subtle neuromuscular defects in utrophin-deficient mice. J Cell Biol 1997; 136:871-882.
169. Grady RM, Zhou H, Cunningham JM et al. Maturation and maintenance of the neuromuscular synapse: Genetic evidence for roles of the dystrophin-glycoprotein complex. Neuron 2000; 25(2):279-293.
170. Rafael JA, Townsend ER, Squire SE et al. Dystrophin and utrophin influence fiber type composition and post-synaptic membrane structure. Hum Mol Genet 2000; 9(9):1357-1367.
171. Campbell KP. Three muscular dystrophies: Loss of cytoskeleton-extracellular matrix linkage. Cell 1995; 80:675-679.
172. Ozawa E, Yoshida M, Suzuki A et al. Dystrophin-associated proteins in muscular dystrophy. Hum Mol Genet 1995; 4:1711-1716.
173. Madhavan R, Jarrett HW. Calmodulin-activated phosphorylation of dystrophin. Biochemistry 1994; 33:5797-5804.
174. Oak SA, Zhou YW, Jarrett HW. Skeletal muscle signaling pathway through the dystrophin glycoprotein complex and Rac1. J Biol Chem 2003; 278(41):39287-39295.
175. Kolodziejczyk SM, Walsh GS, Balazsi K et al. Activation of JNK1 contributes to dystrophic muscle pathogenesis. Curr Biol 2001; 11(16):1278-1282.
176. Kumar A, Khandelwal N, Malya R et al. Loss of dystrophin causes aberrant mechanotransduction in skeletal muscle fibers. FASEB J 2004; 18(1):102-113.
177. D'Souza VN, Man NT, Morris GE et al. A novel dystrophin isoform is required for normal retinal electrophysiology. Hum Mol Genet 1995; 4:837-842.

CHAPTER 3

Utrophin in the Therapy of Duchenne Muscular Dystrophy

Qing Bai, Edward A. Burton and Kay E. Davies

Abstract

Duchenne muscular dystrophy (DMD) is a devastating muscle wasting disease caused by the lack of the cytoskeletal protein dystrophin in muscle. The dystrophin-related protein, utrophin, shows a high degree of sequence similarity to dystrophin and it has been postulated that utrophin replacement could be a possible therapeutic strategy for the disease. We review the evidence that utrophin can functionally replace utrophin and the work being carried out on the transcriptional regulation of the gene.

Structure and Protein Binding Partners of Utrophin

Utrophin is the autosomal paralogue of dystrophin; it was first identified as a 13kb fetal muscle transcript by low-stringency screening of a human fetal muscle library using a probe to the 3' end of dystrophin[1] The gene was localised to human chromosome 6[1] and mouse chromosome 10.[2] The 13kb transcript encodes a 395kDa protein[3] that is ubiquitously expressed.[4-9]

Examination of the sequence of the utrophin transcript revealed marked similarity to the dystrophin transcript, implying that both genes encode proteins with similar functional domains[3] (see Chapters 1 and 2) (Fig. 1).

N-Terminal Domain

The N-terminal domain of utrophin, encompassing the first 250 amino acids, forms a pair of calponin homology (CH) domains and has been shown to function as an F-actin-binding domain that may be regulated by Ca^{2+}/calmodulin.[10-13] The structure of this domain has been solved crystallographically, and it appears that the actin-binding domains of utrophin and dystrophin are similar.[14-16] More recent atomic modeling suggests that the dystrophin CH domain-actin interaction may be more complex than that of utrophin,[17] although this has been debated.[18] The actin-binding domain of utrophin binds actin with a greater affinity than that of dystrophin, an effect that is attributed to a short N-terminal extension in the utrophin sequence.[11,13]

Rod Domain

The rod domain of utrophin consists of a number of spectrin-like repeats, with proline-rich hinge regions.[3] This part of the utrophin molecule is less well conserved with respect to dystrophin; parts of the alignment show as little as 35% identity between the two proteins,[19] implying looser evolutionary constraints or different functions in this domain of the protein.[3] The original description of this structure in dystrophin suggested that the spectrin-repeat/proline-hinge arrangement could give the molecule a flexible rod structure that would participate in its cytoskeletal functions. More recent work has shown that the dystrophin rod domain

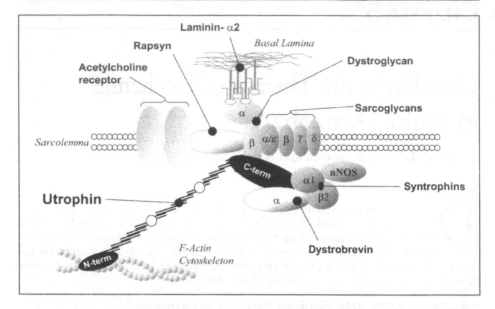

Figure 1. A schematic depiction is shown of utrophin, its functional domains and its protein binding partners at the post-synaptic membrane of the neuromuscular junction. Utrophin is shown in black and the other components of the utrophin-associated glycoprotein complex are shown in grey.

is involved in lateral association with actin filaments, where it functions to stabilize the filaments and prevent de-polymerization. This interaction occurs through clusters of basic residues in the dystrophin rod domain[20] It was initially thought that utrophin did not participate in rod domain-actin association,[21] because recombinant fragments of the utrophin rod domain did not bind to actin filaments under the same conditions in which homologous dystrophin rod domain fragments did. Recent work, however, shows that utrophin is involved in lateral association with actin filaments and prevention of de-polymerization, although intriguingly, this appears to occur via a different mechanism to the association between the dystrophin rod domain and actin. Thus, acidic spectrin-like repeats from the utrophin rod domain are involved in this process, rather than the basic repeats involved in the dystrophin rod domain-actin interaction.[22]

C-Terminal Domain

The cysteine-rich and C-terminal domains of dystrophin and utrophin show 80% sequence identity.[3] Utrophin has similar C-terminal protein binding partners to dystrophin,[23] including β-dystroglycan[23] and syntrophins.[24] It associates with a similar complex of transmembrane components as dystrophin,[25] although unlike dystrophin it preferentially associates with β2-syntrophin.[24] Utrophin associates with components of this complex in tissues other than muscle and the composition of these complexes varies.[26]

Structure of the Utrophin Gene

The utrophin gene spans approximately 900kb of genomic DNA and consists of at least 39 exons that give rise to a 13kb transcript.[19] There are similarities between the utrophin and dystrophin genes, suggesting that they were derived from duplication of a common ancestral gene. Both genes contain multiple short exons, but have a single long exon encoding the 3'UTR. In addition, both have large introns separating the first and second coding exons. Utrophin differs from dystrophin in that there are two full length isoforms which differ by 31 amino acids at their N-terminal end (Fig. 2). These proteins, termed A-utrophin and B-utrophin are

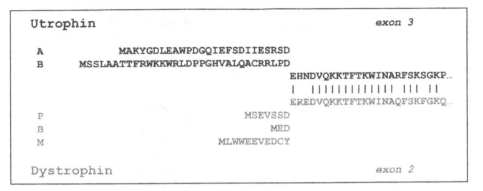

Figure 2. The amino acid sequences of the N-termini of A-utrophin, B-utrophin and the three full-length dystrophin isoforms (P= Purkinje cell; M= muscle; B= brain) are shown aligned with one another. The splice boundary into the first common exon shared by all of the full-length isoforms (utrophin exon 3, dystrophin exon 2) is shown. The sequence encoded by the first common exon is highly conserved between dystrophin and utrophin, whereas there is little similarity between the N-termini encoded by the alternative 5' exons. Whereas the unique N-terminal sequence of each of the dystrophin full-length isoforms is short, A-utrophin and B-utrophin contain 25 and 31 N-terminal amino acids, which share little similarity to one another. It is likely that these proteins interact with actin in subtly different ways.

expressed from different promoters[27] (see later) and have different localization patterns. B-utrophin is mainly found in endothelia whereas A-utrophin is more ubiquitously expressed.[28]

The similarity between the two genes includes the presence of 3' isoforms with unique 5' exons, presumably driven by internal promoters located within 3' introns. The homologue of Dp116 is G-utrophin, which is a 5.5kb 3' transcript present in brain and sensory ganglia.[29] RT-PCR and 5'RACE analysis demonstrated the presence of paralogues of Dp71 and Dp140, called Up71 and Up140.[30] More recent studies have identified short isoforms in brain and testis using immunocytochemistry.[31] The translation of the short utrophin isoforms is relatively inefficient, especially Up71 and Up140. The function of these isoforms remains unknown.

Expression Pattern of Utrophin

Although utrophin and dystrophin show a high degree of sequence identity, the two proteins have distinct patterns of expression that may reflect their different functions.

Expression in Muscle

In normal adult muscle fibres, utrophin is localised to the cytoplasmic face of the myotendinous and neuromuscular junctions. In addition, it is also found in blood vessels (both arterial, venous and endomysial capillaries), intramuscular nerves, the capsule of muscle spindles and the sarcolemma of intrafusal fibres.[7,8,32-38] This contrasts with the distribution of dystrophin in muscle tissue, which is exclusively located within myofibres and localises to the whole of the sarcolemma, although is enriched at the NMJ and MTJ (see above).

In developing human muscle, utrophin levels are much higher than in mature muscle. The levels peak during the second trimester of pregnancy, at which point the protein is located at the sarcolemma. Levels fall thereafter commensurate with increasing levels of dystrophin, suggesting that the two proteins are coordinately regulated during embryogenesis. By 23 weeks, levels have fallen to their adult values.[39] In mouse muscle, the developmentally regulated decrease in utrophin expression is not complete until the end of the second post-natal week.[33]

In dystrophin-deficient muscle, the developmental profile of utrophin is similar to that of wild-type muscle.[39] However, once muscle necrosis has started, utrophin expression is upregulated (Fig. 3). In dystrophic muscle, utrophin is localised to the sarcolemma of

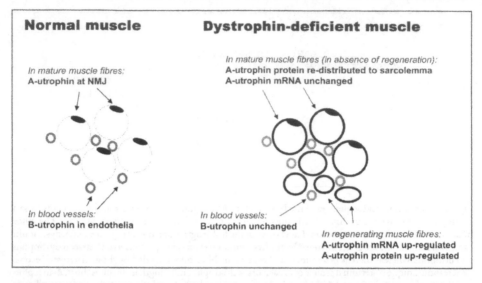

Figure 3. A schematic diagram is shown of the distribution of utrophin full-length isoforms in adult skeletal muscle. The figure depicts a transverse section across a muscle, with each of the muscle fibres and capillaries cut perpendicular to their long axis. In normal muscle, A-utrophin localises to the neuromuscular junction and B-utrophin localises to capillaries. In the absence of dystrophin, the localisation and expression level of B-utrophin is unchanged. However, A-utrophin is up-regulated in two different ways in dystrophin-null muscle. (i) First, in mature fibres, A-utrophin is re-distributed to the sarcolemma. This form of up-regulation occurs in the absence of muscle regeneration (for example in irradiated muscle and in the heart) and is not associated with any detectable increase in A-utrophin mRNA level. It is thought to be a post-transcriptional effect, perhaps related to enhanced A-utrophin stability in the absence of dystrophin, as a result of loss of competition for protein binding partners at the sarcolemma. (ii) Second, A-utrophin is transcriptionally up-regulated during muscle regeneration, so that an increase in A-utrophin mRNA and protein is detectable in the regenerating fibres that occur abundantly in dystrophic muscle. When A-utrophin mRNA and protein levels are measured in whole muscle, these two separate mechanisms result in the detection of greatly enhanced A-utrophin protein levels and slightly increased A-utrophin mRNA levels, the exact value of each presumably depending on the amount of muscle regeneration that is taking place.

regenerating fibres where it is associated with components of the DAPC.[8,25,38,40-44] This is consistent with observations suggesting that regenerating muscle fibres recapitulate the normal developmental patterns of gene expression. More recently, we have shown that it is A-utrophin levels which are increased in dystrophic muscle and that this may be due to increased stability of the protein rather than increased transcription.[45] The increased levels of utrophin were found to be independent of regeneration. Utrophin is also located at the sarcolemma in inflammatory myopathies.[38,40]

Expression in Nonmuscle Tissue

In contrast to dystrophic, utrophin has a wide distribution of expression in nonmuscle tissues. It is detected in most viscera, including spleen, kidney, liver, and stomach with highest levels in the lung. Utrophin is present at the cell membrane of myometrial and vascular smooth muscle, and in Schwann cells of the peripheral nervous system.[7,34,46,47] The association with vascular smooth muscle and endothelium may account in part for the wide tissue distribution.[8]

Within the heart, utrophin and dystrophin have distinct expression patterns.[48] Dystrophin is localised to the cytoplasmic face of the sarcolemma, the transverse tubules and the plasma membrane of the Purkinje system in mature cardiac myofibres. Utrophin is localised to the intercalated discs and cytoplasm of the Purkinje system. In dystrophin-deficient heart, however, utrophin is localised to the cardiomyocyte cell membrane.

In the brain, utrophin and dystrophin are expressed in different cell types. Utrophin expression is found in the choroid plexus, cerebral vasculature, ependyma, and the foot processes of astrocytes.[6] Dystrophin is mainly localised within neurons.[49]

The expression pattern of utrophin during development has been investigated by detection of the transcript using RNA in situ hybridization.[50] Expression in early development is concentrated around the neural tube and neural crest derivatives. Thereafter, utrophin mRNA is seen in numerous tissues; expression levels decline during gestation consistent with developmental regulation of the transcript. This study used a probe corresponding to the 3' UTR of utrophin, and thus is unable to distinguish full-length from truncated 3' transcripts.[29,30]

Utrophin Null Mutations

In contrast to the phenotype arising from dystrophin deficiency, targeted inactivation of utrophin produces an extremely mild phenotype.[51,52] Mice homozygous for a null allele displayed a normal clinical phenotype. However, abnormalities of the neuromuscular junction were demonstrable, in keeping with the localisation of utrophin in normal muscle. A reduction in post-synaptic folding was evident at the ultrastructural level, and the density of post-synaptic acetylcholine receptors was reduced. The functional deficits were, however, relatively minor. Although a reduction was detected in the nerve-evoked post-synaptic current, this was not sufficient to compromise neuromuscular transmission. Kneusel et al[53] have reported that utrophin may protect CNS neurons against pathological insults since the null mutant mice show increased vulnerability to kainite-induced seizures and there are studies which suggest that utrophin may be involved in actin-based motility (Holfziend and Bittner, personal communication).

No human diseases have been linked to primary defects in the utrophin gene, although some cases of myasthenia gravis, caused by autoimmune end-plate destruction[54] or inherited defects in acetylcholine receptors,[55] are associated with reduced post-synaptic utrophin. Although the neuromuscular transmission defects observed in murine utrophin null mutants were not sufficient to cause failure of post-synaptic activation, the safety factor for neuromuscular transmission in mice is very much greater than in humans. It remains possible that mutation of the human utrophin gene could cause a congenital myasthenic syndrome. There is one report of two families with inherited post-synaptic defects in neuromuscular transmission; utrophin was absent from the end plate regions of both, although it is not clear if this was a primary or secondary phenomenon.[56]

Utrophin and Dystrophin Double Knock Out Mice

The mild phenotypic changes seen in *mdx* mice and utrophin null mutant mice could conceivably arise from functional redundancy allowing each to compensate for absence of the other. In support of this, utrophin null-*mdx* mice have a very severe myopathic phenotype that is lethal within weeks of birth. The mice develop contractures, kyphoscoliosis and muscle weakness; the phenotype is more reminiscent of DMD than the *mdx* mouse.[57,58]

Utrophin Over-Expression in the *mdx* Mouse

The similarity in the sequence of dystrophin and utrophin led to the hypothesis that utrophin might be able to functionally replace dystrophin in dystrophin deficient muscle.[59] This question was addressed by expressing a utrophin transgene under the control of a constitutive muscle promoter on the *mdx* background. Compared with *mdx* mice, utrophin transgenic-*mdx* mice showed dramatic phenotypic improvement.[60] Histologically, there was little evidence of the fibrosis and necrosis normally seen in *mdx* diaphragm. Analysis showed a reduction in the proportion of myofibres in limb and diaphragm muscles with central nuclei, indicating a reduction in the amount of muscle regeneration compared with control *mdx* mice. In addition, the serum creatine kinase (a marker of sarcolemmal permeability and cell damage) was reduced to near control levels. Immunohistology showed that the truncated utrophin protein had localised to the sarcolemma, and that components of the DAPC had been restored to the sarcolemma.[60] In addition, several physiological parameters were improved. These included the mean normalised tetanic force, the force drop after eccentric contraction, measures of sarcolemmal disruption after eccentric contraction, and measures of Ca^{2+} homeostasis.[61]

In later studies, mice expressing a full-length utrophin transgene were generated.[62] The highest expressing lines were able to effect almost complete phenotypic rescue when bred onto the *mdx* background, both in morphological and physiological tests.[63-65] Quantitative analysis of the levels of protein expression indicated that morphological and functional recovery was achieved with levels of muscle protein expression that were 2 to 3 fold higher than wild-type muscle. This level of expression was about 50% of the normal wild type level found in the kidney and approximately 25% of the endogenous level in the lung.

Immuno-gold analyses of the localization of the utrophin transgene product in these mice showed that it was indistinguishable from the localization of dystrophin.[66] Furthermore, Ervasti and colleagues[22] demonstrated that utrophin binds laterally along filaments and can couple costameric actin with sarcolemma suggesting that utrophin can compensate for the lack of dystrophin.

Utrophin-Based Therapeutic Strategies

The observation that utrophin over-expression rescues the dystrophic phenotype in *mdx* muscle allows consideration of either delivering utrophin with viral vectors or determining pharmacological methods to increase the levels of utrophin in muscle. The transgenic mouse experiments showed that utrophin could compensate for dystrophin deficiency if given at birth. Wakefield et al[67] extended these studies using adenoviral delivery of utrophin and showed that delivery at birth could also prevent pathology. More detailed experiments using an inducible utrophin transgene supported these data and also showed that utrophin may well have some protective effect even at later stages of the disease.[68]

Utrophin Delivery to Muscle Using Viruses

DMD patients are not genetically deleted for the utrophin gene and might therefore be expected to be immunologically tolerant to the protein. Consequently, vector-mediated delivery of utrophin might be a preferable option to the use of a dystrophin transgene. Successful adenoviral delivery of utrophin to muscle has been demonstrated, with phenotypic improvement both in *mdx*[69] and *mdx*, utr[-/-] muscle.[67] In the latter study, introduction of the utrophin transgene after the onset of muscle necrosis and regeneration was able to correct the dystrophic phenotype. Furthermore, recent evidence suggests that expression of a utrophin transgene is more prolonged than expression of a dystrophin transgene following adenoviral delivery to the muscles of immune competent mice.[70] However, significant problems with viral delivery and long-term expression remain see Chapter 17.

Up-Regulation of the Endogenous Utrophin Gene

It may be possible to up-regulate endogenous utrophin production in skeletal muscle sufficient to effect the 2 to 3 fold increase in steady state protein levels necessary to prevent dystrophic pathology. One way of achieving this might be through transcriptional up-regulation of the endogenous utrophin gene, analogous to the transcriptional reactivation of γ-globin in β-globinopathies.[71] This approach has several inherent advantages. Use of a small chemical compound to interact with the gene promoter would obviate the need to develop a system for the stable expression of transgenes in skeletal muscle. Furthermore, chemical compounds of the type used to activate γ-globin expression are readily diffusible and therefore likely to be widely distributed throughout the skeletal muscle compartment. This would circumvent the considerable challenge imposed by the delivery of therapeutic transgenes to 50% of the total muscle mass.

Up-regulation by such pharmacological means would not necessarily be tightly controlled on a tissue-specific basis, or in terms of the amount of upregulation. Data from the transgenic experiments discussed above demonstrates that over-expression of utrophin in skeletal muscle is nontoxic. Furthermore, mice transgenic for the full-length utrophin coding sequence, driven by a ubiquitously expressed promoter (ubiquitin C) do not show any clinical or histopathological features of dysfunction in a number of tissues.[72]

Strategies to Identify Means of Effecting Utrophin Up-Regulation in Muscle

Up-regulation of utrophin in muscle might be effected by one of several means. First, screening of random small compound libraries against an easily quantifiable automated assay might identify molecules that increase utrophin levels by unknown mechanisms. An example might be to screen for transcriptional up-regulation of utrophin *cis*-acting regulatory sequences by using a reporter gene-cell culture system. These experiments are currently ongoing. This demands knowledge of the promoters and enhancers that are active in regulating transcription of the utrophin gene.

Another strategy might be to study the molecular physiology of utrophin expression in order to identify the various mechanisms that give rise to the complex expression pattern in healthy and diseased tissue. Detailed understanding of these processes might help identify routes by which pharmacological manipulation could effect therapeutic upregulation of utrophin. A rational approach for this strategy is to clone and characterise the *cis*-acting elements that determine transcriptional regulation of the utrophin gene. Identification of trans-acting factors and appropriate signalling pathways could follow, in addition to study of post-transcriptional phenomena.

The Transcriptional Regulation of Utrophin

The regulatory region in the 5' end of the utrophin gene is complex, as might be expected from the developmentally and topologically regulated expression profile of the two full-length isoforms. There are two independently regulated promoters that drive transcription of unique 5' exons that splice into a common mRNA at exon 3. We named these promoter A and promoter B, and the corresponding full-length transcripts and protein isoforms A-utrophin and B-utrophin[27] (Fig. 4). Following delineation of these *cis*-acting sequences, some trans-acting signals have been defined that interact with the regulatory region to govern the transcription of utrophin.

Promoter A

Promoter A is located within a CpG island at the most 5' end of the gene.[73] The unique region of the A-utrophin transcript consists of an untranslated first exon, 1A, and exon 2A, which contains the A-utrophin translational initiation codon. The long 5'UTR contains GC-rich regions. Exon 2A lies approximately 10kb 3' to exon 1A in the genomic map. Promoter A, contains no TATA or initiator elements. In common with a number of other TATA⁻ Inr⁻ GC-rich promoter regions, promoter A has multiple transcriptional start sites, as indicated by primer extension, RNase protection and 5'RACE analyses. A 1.3kb region spanning the transcriptional start sites was active in driving orientation-specific expression of a reporter gene in a range of cell lines.[73] Deletion analysis delineated a core 300bp fragment essential for basal promoter activity. This contained consensus Ap2 and Sp1 sites that were conserved between the human and mouse sequences.[73] Subsequent mutagenesis suggested that the conserved consensus Sp1 and Ap2 sites, which are clustered around the multiple transcription start sites, are functionally important in directing basal transcription from promoter A.[74] Mutation of these sites had a deleterious effect on the in vitro activity of promoter-reporter constructs in several cell lines. DNAse I footprinting and electrophoretic mobility shift assays confirmed that multiple consensus Sp1 and Ap2 sites were bound by the expected transcription factors.

The expression profile of A-utrophin is complex, and promoter A contains a number of important regulatory elements in addition to the core basal elements described above. Several

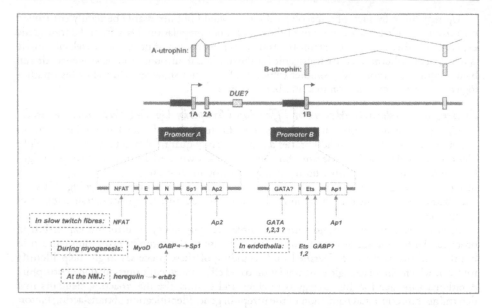

Figure 4. A diagrammatic summary of the transcriptional regulation of utrophin is shown. The upper part of the figure shows the utrophin 5' genomic map (not to scale), illustrating the locations of the promoters, putative enhancer and exons that give rise to the full-length A-utrophin and B-utrophin transcripts. In the lower part of the figure, promoter A and promoter B are expanded to illustrate details of transcription factor binding sites. Below each of these, the trans-acting factor that interacts with each of these cis-acting elements is shown, along with an indication of the specific physiological situations in which each pathway is thought to play an important part. The details of these pathways and the evidence to support each of them is discussed in the text.

features of promoter A have been examined in detail, in order to explain the expression pattern of A-utrophin in muscle:

A-Utrophin Is Upregulated during Myogenic Differentiation.

RNase protection analysis, using riboprobes able to distinguish A-utrophin and B-utrophin,[27] showed that myoblasts in vitro express only the A- isoform. A-utrophin mRNA is upregulated in regenerating muscle fibres following injury in vivo,[75] consistent with the observation that the A-utrophin transcript becomes more abundant during myogenic differentiation in vitro.[74] Differentiation in vitro was paralleled by increased activity of a reporter construct containing a 1.3kb fragment of utrophin promoter A. This promoter A fragment includes a consensus E-box that is conserved between the human and mouse sequences. Mutation of the E-box abolished the myogenic up-regulation of the reporter construct.[74] Separate cotransfections of myoD-, myogenin and mrf4-encoding plasmids with the promoter A—reporter plasmid in muscle cell lines caused transactivation of the promoter A construct, an effect that was abolished by E-box point mutations. Electrophoretic mobility shift assays confirmed that myoD and myogenin bound to the consensus E box in the context of the utrophin flanking sequence.[74] These data suggest that the myogenic upregulation of A-utrophin may involve binding of myogenic factors to the E-box upstream of the transcription start site in promoter A, with resulting transcriptional activation.

A-Utrophin Is Expressed at the Neuromuscular Junction

RNA in situ hybridization showed that utrophin mRNA is localised within myofibres to the region of the NMJ.[76-78] Utrophin expressed at the NMJ is A-utrophin,[28] and it seems very likely that the utrophin mRNA detected in these previous studies was A-utrophin. These studies raised the possibility that transcriptional regulation of promoter A in myonuclei in the region of the NMJ may be important. It was known from work on the expression of other muscle post-synaptic components that nerve-derived signals could enhance transcription of relevant genes, for example the acetylcholine receptor subunits, in the sub-junctional region of the myofibre. Ectopic synapses, formed in denervated muscle at sites remote from the original end plate[77] were shown to contain post-synaptic utrophin after two weeks, implying that nerve-derived signals induced utrophin synthesis. Using an in vivo plasmid expression assay, a 1.3kb fragment of utrophin promoter A drove expression of a reporter gene in a synapse-specific manner.[79] Mutation of a conserved N-box motif present within this sequence disrupted the synaptic induction of the reporter gene. The N-box is present within the promoters of other post-synaptic components, and is known to bond to GABP, an Ets-family transcription factor that is active in the post-synaptic region of myofibres. Cotransfection experiments showed that GABP could trans-activate the promoter A reporter construct, and further studies showed that the nerve-derived signals agrin and ARIA both enhanced reporter expression from a wild type, but not an N-box mutant, utrophin promoter A construct.[80] The role of the regulatory factor GABP at utrophin promoter A appears to be through a cooperative interaction with the basal factor Sp1, which has a key role in governing core promoter activity.[81,82]

A-Utrophin Is Expressed at Higher Levels in Slow Muscle Fibres

Utrophin expression varies with fibre type, in that slow twitch muscle fibres with high oxidative capacity express higher levels of utrophin.[83] The mechanism underlying this expression profile was recently investigated. Single fibre RT-PCR showed that the levels of A-utrophin mRNA correlated with markers of oxidative capacity, implying that a transcriptional mechanism may contribute to the distribution of protein. Physiological manoeuvres that favour the adoption of a slow twitch phenotype in muscle fibres resulted in up-regulation of A-utrophin. Other components of this adaptive response are dependent on calcineurin-NFAT signalling, and further examination of the role of calcineurin and NFAT in the regulation of promoter A was undertaken. Inhibition of calcineurin reduced A-utrophin expression. Conversely, A-utrophin expression was enhanced in the presence of active calcineurin. A NFAT consensus site was identified in promoter A and the recombinant factor shown to bind to the motif. Cotransfection studies suggested that NFAT transactivated a promoter A reporter construct, although specificity was not demonstrated as a mutagenesis study was not carried out. These data provide preliminary evidence that NFAT binds and activates utrophin promoter A, and that this mechanism contributes to the muscle fibre type-specific expression profile of A-utrophin.

Promoter B

RNase protection experiments carried out by Dr. C. Dennis (1996, unpublished) showed evidence of alternative splicing at the exon 2A—exon 3 boundary. We therefore screened a heart cDNA library for alternative 5' utrophin exons using 5'RACE.[27] Utrophin transcripts containing novel sequence 5' to the exon 3 splice boundary were cloned. Comparison with the genomic sequence showed that the novel sequence was a single exon, which we named exon 1B. Exon 1B contained a short 5'UTR and a translational start codon, predicting a unique 31 amino acid terminal to the utrophin protein. 5'RACE, primer extension and RT-PCR showed that there was single transcriptional start site adjacent to a consensus initiator sequence. A 1.5kb genomic fragment spanning the initiator drove orientation-specific expression of a reporter gene in several different cell lines. The level of expression of the reporter varied enormously between cell lines, in parallel with the endogenous exon 1B-containing transcript, implying that the 1.5kb fragment contained some important regulatory elements.

Subsequent immunohistochemical studies showed that B-utrophin expression is localized to vascular endothelium.[28] Further analysis of the promoter has thus been undertaken to elucidate the mechanisms responsible for this highly specific spatial expression pattern. Deletion analysis of the functional 1.5kb promoter fragment demarcated a minimal 300bp promoter fragment, containing conserved consensus Ap1, GATA and Ets sites.[27] Subsequent EMSA experiments showed that recombinant Ap1 binds the consensus site in the context of the utrophin promoter B flanking sequence.[64] Mutagenesis studies demonstrated the importance of the Ap1 and Ets sites in the basal activity of promoter B. A possible role for GATA family factors in facilitating the Ap1-dependent transactivation of the core element was suggested. This type of functional synergy between Ap1, Ets and Sp1 families has been reported previously in endothelial-specific promoters and it is possible that this mechanism contributes to the endothelial specificity of promoter B.

Downstream Utrophin Enhancer

A systematic search for cis-acting regulatory elements that may govern transcription from promoter A was undertaken, by cloning fragments of the utrophin 5' genomic region in a plasmid containing a promoter A-reporter construct and then screening for enhanced expression of the reporter.[84] This study located a 128-bp region lying 9kb 3' of exon 2A, which colocalised with a DNaseI hypersensitive site and conferred orientation-independent cis-activation of utrophin promoter A and the herpes simplex virus thymidine kinase promoter. The downstrean utrophin enhancer (DUE) shows moderate sequence conservation between human and mouse. A later study implied that this element may be involved in the transcriptional response of promoter A to muscle regeneration,[75] although we have been unable to show that the murine enhancer is active[45] and its role remains uncertain.

Conclusions

Compounds are currently being screened in a high through put assay for the up-regulation of utrophin.[65] In this assay, the utrophin promoter sequences are linked to a reporter gene (luciferase) and introduced into an *mdx* muscle cell line. Any drug able to increase luciferase levels is a good candidate for the up-regulation of the endogenous utrophin gene. L-arginine has been reported to increase levels of utrophin[85] but these studies have not yet been confirmed.

Other pathways increase the levels of utrophin but these probably act post transcriptionally. Adam 12 transgene mice for example increase the levels A-utrophin in muscle and alleviate the pathology but this is mediated through a post-transcriptional mechanism.[86] This may also be true in *mdx* mice over-expressing the cytotoxic T cell GalNAc transferase in skeletal muscle[87] which increases utrophin levels. It is has been shown that the sequences determining the stability of the utrophin mRNA reside in the 3'UTR[88] which could be targeted to increase levels of utrophin. Increasing levels of alpha 7 beta 1 intergrin in mice lacking both utrophin and dystrophin restores the viability of the mice suggesting other signaling pathways may be amenable to pharmacological intervention in DMD patients.[89]

Finally, all of the experiments on the use of utrophin for the therapy of DMD have been carried out in the mouse. It is therefore a concern that the data may not be replicated in animals with larger muscles. However, recent studies by Cerletti et al[90] suggests that delivery of utrophin to the dog model of the disease does show therapeutic benefit. This raises the hope that the mouse results are applicable to man. A pharmacological approach to DMD has the advantage of delivery of the utrophin to all muscles (including cardiac muscle) without the problems of the immune response inherent in the viral therapy routes for treatment. It remains to be seen whether a small drug can be found which will increase utrophin levels sufficiently for therapeutic benefit.

Acknowledgements

We are very grateful to Helen Blaber in the preparation of this manuscript. We would like to thank the Muscular Dystrophy Campaign (UK), the Muscular Dystrophy Association (USA) and the Association Francaise contre les Myopathies (France) for financial support.

References

1. Love DR, Hill DF, Dickson G et al. An autosomal transcript in skeletal muscle with homology to dystrophin. Nature 1989; 339(6219):55-58.
2. Buckle VJ, Guenet JL, Simon-Chazottes D et al. Localisation of a dystrophin-related autosomal gene to 6q24 in man, and to mouse chromosome 10 in the region of the dystrophia muscularis (dy) locus. Hum Genet 1990; 85(3):324-326.
3. Tinsley JM, Blake DJ, Roche A et al. Primary structure of dystrophin-related protein. Nature 1992; 360(6404):591-593.
4. Blake DJ, Weir A, Newey SE et al. Function and genetics of dystrophin and dystrophin-related proteins in muscle. Physiol Rev 2002; 82(2):291-329.
5. Matsumura K, Yamada H, Shimizu T et al. Differential expression of dystrophin, utrophin and dystrophin-associated proteins in peripheral nerve. FEBS Lett 1993; 334(3):281-285.
6. Khurana TS, Watkins SC, Kunkel LM. The subcellular distribution of chromosome 6-encoded dystrophin-related protein in the brain. J Cell Biol 1992; 119(2):357-366.
7. Love DR, Morris GE, Ellis JM et al. Tissue distribution of the dystrophin-related gene product and expression in the mdx and dy mouse. Proc Natl Acad Sci USA 1991; 88(8):3243-3247.
8. Nguyen TM, Ellis JM, Love DR et al. Localization of the DMDL gene-encoded dystrophin-related protein using a panel of nineteen monoclonal antibodies: Presence at neuromuscular junctions, in the sarcolemma of dystrophic skeletal muscle, in vascular and other smooth muscles, and in proliferating brain cell lines. J Cell Biol 1991; 115(6):1695-1700.
9. Nguyen TM, Le TT, Blake DJ et al. Utrophin, the autosomal homologue of dystrophin, is widely-expressed and membrane-associated in cultured cell lines. FEBS Lett 1992; 313(1):19-22.
10. Winder SJ, Hemmings L, Maciver SK et al. Utrophin actin binding domain: Analysis of actin binding and cellular targeting. J Cell Sci 1995; 108(Pt1):63-71.
11. Morris GE, Nguyen TM, Nguyen TN et al. Disruption of the utrophin-actin interaction by monoclonal antibodies and prediction of an actin-binding surface of utrophin. Biochem J 1999; 337(Pt1):119-123.
12. Winder SJ, Kendrick-Jones J. Calcium/calmodulin-dependent regulation of the NH2-terminal F-actin binding domain of utrophin. FEBS Lett 1995; 357(2):125-128.
13. Moores CA, Kendrick-Jones J. Biochemical characterisation of the actin-binding properties of utrophin. Cell Motil Cytoskeleton 2000; 46(2):116-128.
14. Moores CA, Keep NH, Kendrick Jones J. Structure of the utrophin actin-binding domain bound to F-actin reveals binding by an induced fit mechanism. J Mol Biol 2000; 297(2):465-480.
15. Keep NH, Winder SJ, Moores CA et al. Crystal structure of the actin-binding region of utrophin reveals a head-to-tail dimer. Structure 1999; 7:1539-1546.
16. Keep NH, Norwood FL, Moores CA et al. The 2.0 A structure of the second calponin homology domain from the actin-binding region of the dystrophin homologue, utrophin. J Mol Biol 1999; 285:1257-1264.
17. Sutherland-Smith AJ, Moores CA, Norwood FL et al. An atomic model for actin binding by the CH domains and spectrin-repeat modules of utrophin and dystrophin. J Mol Biol 2003; 23(329):15-33.
18. Galkin VE, Orlova A, VanLoock MS et al. Do the utrophin tandem calponin homology domains bind F-actin in a compact or extended conformation? J Mol Biol 2003; 331(5):967-972.
19. Pearce M, Blake DJ, Tinsley JM et al. The utrophin and dystrophin genes share similarities in genomic structure. Hum Mol Genet 1993; 2(11):1765-1772.
20. Amann KJ, Renley BA, Ervasti JM. A cluster of basic repeats in the dystrophin rod domain binds F-actin through an electrostatic interaction. J Biol Chem 1998; 273(43):28419-28423.
21. Amann KJ, Guo AW, Ervasti JM. Utrophin lacks the rod domain actin binding activity of dystrophin. J Biol Chem 1999; 274(50):35375-35380.
22. Rybakova IN, Patel JR, Davies KE et al. Utrophin binds laterally along actin filaments and can couple costameric actin with sarcolemma when overexpressed in dystrophin-deficient muscle. Mol Biol Cell May 2002; 13(5):1512-1521.
23. James M, Nguyen TM, Wise CJ et al. Utrophin-dystroglycan complex in membranes of adherent cultured cells. Cell Motil Cytoskeleton 1996; 33(3):163-174.
24. Peters MF, O'Brien KF, Sadoulet-Puccio HM et al. beta-dystrobrevin, a new member of the dystrophin family. Identification, cloning, and protein associations. J Biol Chem 1997; 272(50):31561-31569.
25. Matsumura K, Ervasti JM, Ohlendieck K et al. Association of dystrophin-related protein with dystrophin-associated proteins in mdx mouse muscle. Nature 1992; 360(6404):588-591.

26. Loh NY, Newey SE, Davies KE et al. Assembly of multiple dystrobrevin-containing complexes in the kidney. J Cell Sci 2000; 113(Pt 15):2715-2724.
27. Burton EA, Tinsley JM, Holzfeind PJ et al. A second promoter provides an alternative target for therapeutic up-regulation of utrophin in Duchenne muscular dystrophy. Proc Natl Acad Sci USA 1999; 96(24):14025-14030.
28. Weir AP, Burton EA, Harrod G et al. A- and B-utrophin have different expression patterns and are differentially up-regulated in mdx muscle. J Biol Chem 2002; 277(47):45285-45290.
29. Blake DJ, Schofield JN, Zuellig RA et al. G-utrophin, the autosomal homologue of dystrophin Dp116, is expressed in sensory ganglia and brain. Proc Natl Acad Sci USA 1995; 92(9):3697-3701.
30. Wilson J, Putt W, Jimenez C et al. Up71 and up140, two novel transcripts of utrophin that are homologues of short forms of dystrophin. Hum Mol Genet 1999; 8(7):1271-1278.
31. Jimenez-Mallebrera C, Torelli S, Brown SC et al. Profound skeletal muscle depletion of alpha-dystroglycan in Walker-Warburg syndrome. Eur J Paediatr Neurol 2003; 7(3):129-137.
32. Bewick GS, Young C, Slater CR. Spatial relationships of utrophin, dystrophin, beta-dystroglycan and beta-spectrin to acetylcholine receptor clusters during postnatal maturation of the rat neuromuscular junction. J Neurocytol 1996; 25(7):367-379.
33. Khurana TS, Byers TJ, Kunkel LM et al. Dystrophin detection in freeze-dried tissue. Lancet 1991; 338(8764):448.
34. Love DR, Byth BC, Tinsley JM et al. Dystrophin and dystrophin-related proteins: A review of protein and RNA studies. Neuromuscul Disord 1993; 3(1):5-21.
35. Ohlendieck K, Campbell KP. Dystrophin-associated proteins are greatly reduced in skeletal muscle from mdx mice. J Cell Biol 1991; 115(6):1685-1694.
36. Pons F, Nicholson LV, Robert A et al. Dystrophin and dystrophin-related protein (utrophin) distribution in normal and dystrophin-deficient skeletal muscles. Neuromuscul Disord 1993; 3(5-6):507-514.
37. Zhao J, Yoshioka K, Miyatake M et al. Dystrophin and a dystrophin-related protein in intrafusal muscle fibers, and neuromuscular and myotendinous junctions. Acta Neuropathol (Berl) 1992; 84(2):141-146.
38. Karpati G, Carpenter S, Morris GE et al. Localization and quantitation of the chromosome 6-encoded dystrophin-related protein in normal and pathological human muscle. J Neuropathol Exp Neurol 1993; 52(2):119-128.
39. Clerk A, Morris GE, Dubowitz V et al. Dystrophin-related protein, utrophin, in normal and dystrophic human fetal skeletal muscle. Histochem J 1993; 25(8):554-561.
40. Helliwell TR, Man NT, Morris GE et al. The dystrophin-related protein, utrophin, is expressed on the sarcolemma of regenerating human skeletal muscle fibres in dystrophies and inflammatory myopathies. Neuromuscul Disord 1992; 2(3):177-184.
41. Khurana TS, Watkins SC, Chafey P et al. Immunolocalization and developmental expression of dystrophin related protein in skeletal muscle. Neuromuscul Disord 1991; 1(3):185-194.
42. Lin S, Gaschen F, Burgunder JM. Utrophin is a regeneration-associated protein transiently present at the sarcolemma of regenerating skeletal muscle fibers in dystrophin-deficient hypertrophic feline muscular dystrophy. J Neuropathol Exp Neurol 1998; 57(8):780-790.
43. Mizuno Y, Nonaka I, Hirai S et al. Reciprocal expression of dystrophin and utrophin in muscles of Duchenne muscular dystrophy patients, female DMD-carriers and control subjects. J Neurol Sci 1993; 119(1):43-52.
44. Ohlendieck K, Ervasti JM, Matsumura K et al. Dystrophin-related protein is localized to neuromuscular junctions of adult skeletal muscle. Neuron 1991; 7(3):499-508.
45. Weir AP, Morgan JE, Davies KE. A-utrophin up-regulation in mdx skeletal muscle is independent of regeneration. Neuromuscul Disord 2003; In press.
46. Fabbrizio E, Latouche J, Rivier F et al. Reevaluation of the distributions of dystrophin and utrophin in sciatic nerve. Biochem J 1995; 312(Pt1):309-314.
47. Khurana TS, Hoffman EP, Kunkel LM. Identification of a chromosome 6-encoded dystrophin-related protein. J Biol Chem 1990; 265(28):16717-16720.
48. Pons F, Robert A, Fabbrizio E et al. Utrophin localization in normal and dystrophin-deficient heart. Circulation 1994; 90(1):369-374.
49. Lidov HG, Byers TJ, Watkins SC et al. Localization of dystrophin to postsynaptic regions of central nervous system cortical neurons. Nature 1990; 348(6303):725-728.
50. Schofield J, Houzelstein D, Davies K et al. Expression of the dystrophin-related protein (utrophin) gene during mouse embryogenesis. Dev Dyn 1993; 198(4):254-264.
51. Deconinck AE, Rafael JA, Skinner JA et al. Utrophin-dystrophin-deficient mice as a model for Duchenne muscular dystrophy. Cell 1997; 90(4):717-727.

52. Grady RM, Merlie JP, Sanes JR. Subtle neuromuscular defects in utrophin-deficient mice. J Cell Biol 1997; 136(4):871-882.
53. Knuesel I, Riban V, Zuellig RA et al. Increased vulnerability to kainate-induced seizures in utrophin-knockout mice. Eur J Neurosci 2002; 15(9):1474-1484.
54. Ito H, Yoshimura T, Satoh A et al. Immunohistochemical study of utrophin and dystrophin at the motor end-plate in myasthenia gravis. Acta Neuropathol (Berl) 1996; 92(1):14-18.
55. Slater CR, Young C, Wood SJ et al. Utrophin abundance is reduced at neuromuscular junctions of patients with both inherited and acquired acetylcholine receptor deficiencies. Brain 1997; 120(Pt9):1513-1531.
56. Sieb JP, Dorfler P, Tzartos S et al. Congenital myasthenic syndromes in two kinships with end-plate acetylcholine receptor and utrophin deficiency. Neurology 1998; 50(1):54-61.
57. Grady RM, Teng H, Nichol MC et al. Skeletal and cardiac myopathies in mice lacking utrophin and dystrophin: A model for Duchenne muscular dystrophy. Cell 1997; 90(4):729-738.
58. Deconinck AE, Potter AC, Tinsley JM et al. Postsynaptic abnormalities at the neuromuscular junctions of utrophin-deficient mice. J Cell Biol 1997; 136(4):883-894.
59. Tinsley JM, Davies KE. Utrophin: A potential replacement for dystrophin? Neuromuscul Disord 1993; 3(5-6):537-539.
60. Tinsley JM, Potter AC, Phelps SR et al. Amelioration of the dystrophic phenotype of mdx mice using a truncated utrophin transgene. Nature 1996; 384(6607):349-353.
61. Deconinck N, Tinsley J, De Backer F et al. Expression of truncated utrophin leads to major functional improvements in dystrophin-deficient muscles of mice. Nat Med 1997; 3(11):1216-1221.
62. Tinsley J, Deconinck N, Fisher R et al. Expression of full-length utrophin prevents muscular dystrophy in mdx mice. Nat Med 1998; 4(12):1441-1444.
63. Gillis JM. Multivariate evaluation of the functional recovery obtained by the overexpression of utrophin in skeletal muscles of the mdx mouse. Neuromuscul Disord 2002; 12(Suppl1):S90-94.
64. Perkins KJ, Davies KE. The role of utrophin in the potential therapy of Duchenne muscular dystrophy. Neuromuscul Disord 2002; 12(Suppl1):S78-89.
65. Khurana TS, Davies KE. Pharmacological strategies for muscular dystrophy. Nat Rev Drug Discov 2003; 2(5):379-390.
66. Culle MJ, Walsh JM, Tinsle JM et al. Immunogold confirmation that utrophin is localized to the normal position of dystrophin in dystrophin-negative transgenic mouse muscle. Histochem J 2001; 33(9-10):579-583.
67. Wakefield PM, Tinsley JM, Wood MJ et al. Prevention of the dystrophic phenotype in dystrophin/utrophin-deficient muscle following adenovirus-mediated transfer of a utrophin minigene. Gene Ther 2000; 7(3):201-204.
68. Squire S, Raymackers JM, Vandebrouck C et al. Prevention of pathology in mdx mice by expression of utrophin: Analysis using an inducible transgenic expression system. Hum Mol Genet 2002; 11(26):3333-3344.
69. Gilbert R, Nalbantoglu J, Petrof BJ et al. Adenovirus-mediated utrophin gene transfer mitigates the dystrophic phenotype of mdx mouse muscles. Hum Gene Ther 1999; 10(8):1299-1310.
70. Ebihara S, Guibinga GH, Gilbert R et al. Differential effects of dystrophin and utrophin gene transfer in immunocompetent muscular dystrophy (mdx) mice. Physiol Genomics 2000; 3(3):133-144.
71. Hoffbrand AV, Pettit JE. Essential Haematology. 2nd Ed. 1984.
72. Fisher R, Tinsley JM, Phelps SR et al. Nontoxic ubiquitous over-expression of utrophin in the mdx mouse. Neuromuscul Disord 2001; 11(8):713-721.
73. Dennis CL, Tinsley JM, Deconinck AE et al. Molecular and functional analysis of the utrophin promoter. Nucleic Acids Res 1996; 24(9):1646-1652.
74. Perkins KJ, Burton EA, Davies KE. The role of basal and myogenic factors in the transcriptional activation of utrophin promoter A: Implications for therapeutic up-regulation in Duchenne muscular dystrophy. Nucleic Acids Res 2001; 29(23):4843-4850.
75. Galvagni F, Cantini M, Oliviero S. The utrophin gene is transcriptionally up-regulated in regenerating muscle. J Biol Chem 2002; 277(21):19106-19113.
76. Vater R, Young C, Anderson LV et al. Utrophin mRNA expression in muscle is not restricted to the neuromuscular junction. Mol Cell Neurosci 1998; 10(5-6):229-242.
77. Gramolini AO, Dennis CL, Tinsley JM et al. Local transcriptional control of utrophin expression at the neuromuscular synapse. J Biol Chem 1997; 272(13):8117-8120.
78. Gramolini AO, Jasmin BJ. Expression of the utrophin gene during myogenic differentiation. Nucleic Acids Res 1999; 27(17):3603-3609.
79. Gramolini AO, Burton EA, Tinsley JM et al. Muscle and neural isoforms of agrin increase utrophin expression in cultured myotubes via a transcriptional regulatory mechanism. J Biol Chem 1998; 273(2):736-743.

80. Gramolini AO, Angus LM, Schaeffer L et al. Induction of utrophin gene expression by heregulin in skeletal muscle cells: Role of the N-box motif and GA binding protein. Proc Natl Acad Sci USA 1999; 96(6):3223-3227.
81. Galvagni F, Capo S, Oliviero S. Sp1 and Sp3 physically interact and cooperate with GABP for the activation of the utrophin promoter. J Mol Biol 2001; 306(5):985-996.
82. Gyrd-Hansen M, Krag TO, Rosmarin AG et al. Sp1 and the ets-related transcription factor complex GABP alpha/beta functionally cooperate to activate the utrophin promoter. J Neurol Sci 2002; 197(1-2):27-35.
83. Chakkalakal JV, Stocksley MA, Harrison MA et al. Expression of utrophin A mRNA correlates with the oxidative capacity of skeletal muscle fiber types and is regulated by calcineurin/NFAT signaling. Proc Natl Acad Sci USA 2003; 100(13):7791-7796.
84. Galvagni F, Oliviero S. Utrophin transcription is activated by an intronic enhancer. J Biol Chem 2000; 275(5):3168-3172.
85. Chaubourt E, Voisin V et al. Muscular nitric oxide synthase (muNOS) and utrophin. J Physiol Paris 2002; 96(1-2):43-52.
86. Moghadaszadeh B, Albrechtsen R, Guo LT et al. Compensation for dystrophin-deficiency: ADAM12 overexpression in skeletal muscle results in increased {alpha}7 integrin, utrophin and associated glycoproteins. Hum Mol Genet 2003; 12(19):2467-2479.
87. Nguyen HH, Jayasinha V, Xia B et al. Overexpression of the cytotoxic T cell GalNAc transferase in skeletal muscle inhibits muscular dystrophy in mdx mice. Proc Natl Acad Sci USA 2002; 99(8):5616-5621.
88. Gramolini AO, Belanger G, Thompson JM et al. Increased expression of utrophin in a slow vs. a fast muscle involves posttranscriptional events. Am J Physiol Cell Physiol 2001; 281(4):C1300-1309.
89. Burkin DJ, Wallace GQ, Nicol KJ et al. Enhanced expression of the alpha 7 beta 1 integrin reduces muscular dystrophy and restores viability in dystrophic mice. J Cell Biol 2001; 152(6):1207-1218.
90. Cerletti M, Negri T, Cozzi F et al. Dystrophic phenotype of canine X-linked muscular dystrophy is mitigated by adenovirus-mediated utrophin gene transfer. Gene Ther 2003; 10(9):750-757.

CHAPTER 4

Syntrophin:
A Molecular Adaptor Conferring a Signaling Role to the Dystrophin-Associated Protein Complex

Justin M. Percival, Marvin E. Adams and Stanley C. Froehner

Abstract

The dystrophin-associated protein complex (DAPC) plays a critical role in maintaining the structural integrity of the sarcolemma of skeletal muscle. In addition, several regulatory molecules, including neuronal nitric oxide synthase, are localized to the DAPC through the syntrophin family of adaptor proteins. The adaptor role of syntrophin is largely attributable to its PDZ (PSD-95 / Discs Large / Zona Occludens 1) domain, a modular domain used ubiquitously to create scaffolding complexes at sites of membrane specialization in metazoan organisms. Syntrophin plays important roles in the maturation of the neuromuscular synapse, nitric oxide-mediated vasomodulation and bidirectional water transport across the blood-brain barrier. By these and other mechanisms syntrophin may contribute to the pathology of muscular dystrophy.

Introduction

Adaptor proteins play a key role in the creation of signaling assemblies of polarized cells.[1] These proteins often contain the PDZ protein-protein interaction domain, abundantly occurring modules used to form multi-protein scaffolding complexes under specific domains of the plasma membrane. Signaling molecules are localized to these complexes via interactions with their PDZ domains and therefore create asymmetrically distributed signal transduction centers. The immobilization of signaling molecules and consequent spatial restriction of reaction products with downstream proteins can enhance the efficiency, specificity and sensitivity of a signaling cascade.[2]

PDZ proteins also play a key role in the organization of signaling proteins at synapses in the central nervous system. PSD-95 is localized under the dendritic membrane of the postsynaptic density and contains three PDZ domains which bind the NMDA (N-methyl-D-aspartate) glutamate receptor, nNOS and CRIPT proteins.[3-5] The localization of nNOS to the NMDA receptor complex is thought to facilitate efficient production of nitric oxide in response to Ca^{2+} influx through the open NMDAR channel. PSD-95 also plays a role in clustering of the AMPA (α-amino-3-hydroxy-5-methyl-4-isoxazole) receptor.[6] The AMPA receptor binds the seven PDZ domain protein GRIP (glutamate receptor-interacting protein) and the single PDZ PICK protein. This later interaction is implicated in long term depression in the cerebellum.[7,8]

PDZ proteins can also regulate the activity and trafficking of their interacting proteins. NHERF/EBP50 (Na^+/H^+ exchange regulatory factor/ezrin-radixin-moesin binding phosphoprotein of 50 kDa) contains two PDZ domains and regulates the activity of the Na^+/H^+ antiporter NHE3.[9,10] NHERF/EBP50 can also regulate the recycling and subcellular localization (and therefore the activity) of its binding partner the β-adrenergic receptor.[11] In epithelial cells of *C. elegans* a complex between PDZ proteins LIN-2, LIN-7 and LIN-10 is required for the basolateral targeting of the EGFR-like tyrosine kinase receptor LET23.[12]

High local concentrations of neurotransmitter receptors, ion channels and signaling proteins are found at the postsynaptic membrane of the neuromuscular synapse, where muscle fibers are innervated by motor neurons. These proteins are localized, anchored, stabilized or clustered by their interaction with cytoskeletal and scaffolding networks that lie immediately beneath the postsynaptic membrane. In skeletal muscle the synaptic cytoskeletal complex is formed by dystrophin (the protein encoded by the gene causing Duchenne Muscular Dystrophy) and a related protein, utrophin along with their associated proteins. The dystrophin complex is present underneath the muscle membrane (sarcolemma), but is highly concentrated postsynaptically at the neuromuscular junction. Dystrophin interacts with the syntrophin family of adaptor scaffold proteins. Syntrophins contain a PDZ domain and interact with a large number of kinases, ion channels and other signaling molecules. Thus syntrophins serve to create signal transduction complexes by localizing signaling molecules to the dystrophin-actin cytoskeleton at extrasynaptic and synaptic membranes. Dystrophin and syntrophin family members form similar complexes in other cell and tissue types including neurons, astrocytes and epithelia. In astrocytes of the brain, a dystrophin-syntrophin complex plays a key role in water transport across membranes of the blood-brain barrier. This review will focus on recent advances in understanding syntrophin's role in creating submembrane scaffolding complexes in different tissues.

Discovery and Cloning

Syntrophin was originally identified as a 58 kDa protein in acetylcholine receptor-rich postsynaptic membranes prepared from the electric organ of the cartilagenous fish *Torpedo*.[13] Syntrophin, as well as acetylcholine receptors, are found on the crests of postjunctional folds in *Torpedo* (see Fig. 2). It is similarly localized at neuromuscular junctions and the sarcolemma of skeletal muscle fibers of different vertebrate species including *Xenopus*, rat, mouse and human.[14-17] Ultrastructural analysis of syntrophin localization in rat myotubes showed that it was present on a network of filaments on the cytoplasmic side of the acetylcholine receptor clusters.[18] This suggested that syntrophin was a peripheral membrane protein.

Immunoscreening of *Torpedo* and mouse cDNA libraries led to the isolation of cDNAs for the electric organ protein and two distinct mouse syntrophins, syntrophin-1 (α-syntrophin) and syntrophin-2 (β2-syntrophin) encoded by separate genes.[19] *Torpedo* syntrophin protein is most homologous to mammalian α-syntrophin. A third distinct gene encoding human basic A1 syntrophin (β1) was subsequently reported.[20] The murine genes encoding α- and β2-syntrophin have also been characterized.[21] Human cDNAs encoding two more distantly related syntrophins, γ1 and γ2 were subsequently cloned.[22] Percentage identities between the different mouse syntrophin proteins are shown in Figure 1.

Syntrophin Is a Component of the Dystrophin-Associated Complex

Since its initial discovery in the electric organ syntrophin has had many names including 58K, 59DAP, A0, syntrophin-associated protein A1 and 59 kDa. These names invariably refer to syntrophin. Antibodies made against components of the dystrophin-associated complex from rabbit skeletal muscle revealed the existence of a 59 kDa dystrophin-associated peripheral membrane protein (59-DAP) that localized to the sarcolemma of skeletal muscle which was confirmed later as syntrophin.[23]

Syntrophin binds both dystrophin and the 87 kDa protein dystrobrevin purified from *Torpedo*.[17] Syntrophin localization at the sarcolemma of skeletal muscle is dystrophin-dependent since it is mislocalized in dystrophin-deficient *mdx* mice.[17] In *mdx* mice α-syntrophin levels are reduced by 50%.[4,17] Even when expressed at high levels α-syntrophin is unable to bind the sarcolemma in the absence of dystrophin, despite the presence of utrophin.[24] This may be due

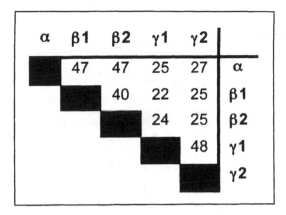

α	β1	β2	γ1	γ2	
	47	47	25	27	α
		40	22	25	β1
			24	25	β2
				48	γ1
					γ2

Figure 1. Percentage identity between syntrophin family proteins. Percentage identities represent homology between murine members of the syntrophin family of adaptor proteins.

in part to insufficient amounts of utrophin at the sarcolemma i.e., higher levels of utrophin may be required for syntrophin association with the sarcolemma.

α-Syntrophin also localizes to the neuromuscular junction (Fig. 2). However its localization at this site does not require dystrophin since neuromuscular synapses in *mdx* mice retain syntrophin. This raises the interesting question as to how syntrophin localizes to the neuromuscular junction. It is possible that syntrophin's binding partner is utrophin, a closely related paralog of dystrophin and/or dystrobrevin.

Syntrophin Distribution and Developmental Regulation

α-Syntrophin is present in large amounts in skeletal muscle and heart but is found in other tissues such as brain, liver, kidney and lung.[25,26] β1-syntrophin is enriched in liver where it localizes to sinusoids and the sinusoidal face of hepatocytes and found at low to moderate levels in muscle, intestine, brain, lung and kidney.[25,27] β2-syntrophin is found at low levels in skeletal muscle, heart, liver and brain; however it is present at high levels in intestine lung, kidney and testis.[25] γ1-Syntrophin protein expression is restricted to the brain; however γ2-syntrophin is found in both brain and skeletal muscle.[22]

In kidney, α-, β1- and β2-syntrophin are found in complexes with β-dystrobrevin, α-dystrobrevin 1, utrophin and a carboxyl terminus splice variant of Dp71 (an alternatively spliced form of dystrophin) called Dp71δC.[26] Similar complexes are seen in detergent-solubilized liver extracts; however α-syntrophin does not coimmuno-precipitate with β-dystrobrevin. Both α- and β2-syntrophin localize to the basal membrane of cortical renal tubules, Bowman's capsule and collecting ducts. β2-syntrophin also localizes to glomeruli.[26] β1-syntrophin is found in peritubular capillaries, blood vessels and smooth muscle.[26] The distribution of α- and β1-syntrophins is unchanged in kidneys of *mdx³ᶜᵛ* mice that lack all isoforms of dystrophin however; β2-syntrophin distribution is significantly disrupted suggesting that it's localization is dependent on a splice variant of Dp71.[26] In β-dystrobrevin-deficient mice, expression of all syntrophins in the kidney is either reduced or undetectable.[27]

Syntrophin proteins have unique and overlapping distributions in muscle.[25,28,29] α-syntrophin localizes to the sarcolemma and neuromuscular junction in all fiber types studied of both newborn and adult mice.[28,29] Rabbit α-syntrophin associates with T-tubules and the sarcolemma of cardiac muscle.[30] In mice, β1 is expressed at the sarcolemma and at the neuromuscular junction in most skeletal muscle fibers during the first six weeks of development after which its levels decrease. Subsequently, β1-syntrophin is restricted to the sarcolemma and presynaptic membranes of the junction of fast-twitch muscle fibers. These are the first fibers

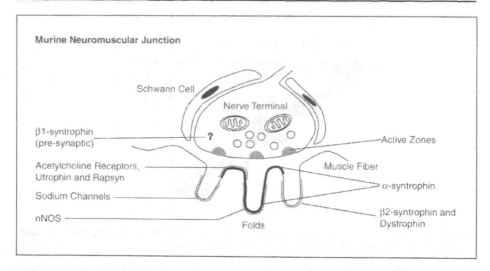

Figure 2. Murine neuromuscular junction. α-syntrophin is found on both the crests and troughs of junctional folds. β2-syntrophin, dystrophin, nNOS and sodium channels ($Na_v1.4$) are localized to the troughs of junctional folds only. Acetylcholine receptors and utrophin are found on the crests of folds only. β1-syntrophin is found presynaptically; however its exact localization is presently unclear.

to degenerate in Duchenne muscular dystrophy.[29] β2-syntrophin localizes exlusively to the neuromuscular junction in adult mice.[25,29] Although β2-syntrophin can bind utrophin this interaction is not necessary for localization of β2-syntrophin to the nerve-muscle junction.[32] In contrast to mice, β2-syntrophin is found at both the neuromuscular junction and the sarcolemma in human muscle (Fig. 2).[31]

In situ hybridization studies show γ1- and γ2-syntrophin transcripts to be highly expressed in neuronal cells of the central nervous system.[22] Neurons of the hippocampus, cerebral cortex, spinal cord and cerebellar cortex are positive for gamma syntrophin transcripts.[22] Transcript expression is paralleled by protein expression. γ2-Syntrophin is reported to associate with the sarcolemma in human muscle in a dystrophin-dependent manner.[22] However, other studies have been unable to corroborate this claim.[31]

Syntrophin Protein Domain Organization

Human and mouse syntrophins share a common highly conserved domain structure (Fig. 3).[21] They contain two domains characteristic of adaptor proteins, namely PH (Pleckstrin Homology) and PDZ domains suggesting that they function as molecular sockets into which signaling molecules 'plug in'. Indeed, the unique domain architecture of syntrophin has contributed to its characterization as a molecular scaffold/adaptor protein.

All syntrophins characterized to date consist of two pleckstrin homology domains (PH1 and PH2), one PDZ domain and a syntrophin unique (SU) domain that has no homology to other known domains in the databases (Fig. 3). The first pleckstrin homology domain (PH1) of about 100 amino acids is in the amino terminus of all syntrophins. Syntrophins have a second pleckstrin homology domain (PH2) that is immediately downstream of the PH1 domain. The SU domain lies at the carboxyl terminus of syntrophin proteins. γ-syntrophins have an additional feature not found in α-, β1- and β2-syntrophin: a p-loop containing an putative ATP/GTP nucleotide binding site motif.[22]

Pleckstrin homology domains (PH) form structurally conserved protein modules of approximately 100 amino acids and are present in many proteins involved in signaling and cytoskeletal organization.[33] They have a well described affinity for a class of glycerophospholipids called phosphoinositides and for βγ-subunits of heterotrimeric G-proteins.[34,35] The PH1

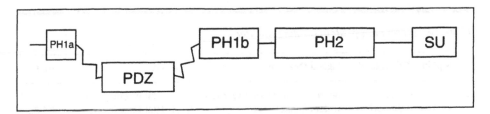

Figure 3. Syntrophin domain organization and family member homology. The general domain organization of syntrophin family members is shown. PH1a, PH1b and PH2 are pleckstrin homology domains. The PDZ domain is named after the PSD-95-Discs Large-Zona Occludens proteins in which it was first identified. The SU domain or syntrophin unique domain bears no homology to known protein domains. Note that the PDZ domain is inserted within the first PH domain.

domain of α1-syntrophin is reported to bind phosphatidylinositol 4,5-bisphosphate (PtdIns4,5P$_2$) with high affinity.[36] Sequences of the PH1b and SU domains of α-syntrophin bind both cardiac actin and skeletal actin isoforms with an affinity equal to or greater than dystrophin or utrophin in vitro.[37] A very interesting feature of syntrophin structure is that the PDZ domain is inserted within the PH1 domain itself, essentially splitting the PH1 domain into two halves (PH1a and PH1b). This unusual arrangement is also found in phospholipase C γ1 and γ2 in which SH2 and SH3 domains are inserted into the analogous position of its PH domain.[38] There is also a proline-rich region between the carboxyl terminus of the PDZ and PH1b domain. However, the split PH domain reconstitutes a functional structure which specifically and avidly binds PtdIns4,5P$_2$.[36] This is consistent with reports of split PH domains being able to form stable and functional structures when the two halves are near one another, suggesting that PH1a and PH1b of syntrophin are in close proximity in the folded protein.[39]

In biochemical experiments syntrophin protein binds dystrophin, utrophin, dystrobrevin and Dp71.[40-43] Both the PH2 (pleckstrin homology domain 2) and SU (syntrophin unique) domains are necessary for binding dystrophin and utrophin.[44,45] α-Syntrophin binds to sequence encoded by exons 73 and 74 in the cysteine rich carboxyl terminus of dystrophin.[41,42] Both γ1- and γ2-syntrophin interact with dystrophin.[22] Utrophin appears to have a low affinity for α-syntrophin in vitro and transgenically overexpressed syntrophin is unable to associate with the sarcolemma despite the presence of utrophin at the membrane in *mdx* mouse muscle.[24,25]

Ligands of Syntrophin

Most of the attention has focused on the PDZ domain of syntrophins since these domains play a central role in the assembly of multi-protein signaling complexes in metazoan organisms.[1] Interactors of the PDZ domains of syntrophin are presented in Table 1. There are several challenges inherent in studying the biology of syntrophin scaffold proteins and scaffold adaptor proteins in general. The first challenge is that measurements of syntrophin protein function are often indirect, because scaffold proteins do not possess any intrinsic activity of their own and are defined by their ability to affect the localization, interactions and activity of their target proteins. The ability of the best-characterized syntrophin (α-syntrophin) to act as a scaffold has come from studies of localization and activity of its ligands nNOS (in muscle) and the water channel aquaporin 4 (in brain) in genetically-manipulated α-syntrophin mutant mice (see below). However, α-syntrophin may have an intrinsic structural activity since α-syntrophin-deficient mice have aberrant neuromuscular junctions, although this phenotype may also be potentially due to the loss of a currently unidentified protein partner.

A second challenge inherent in studying scaffold proteins is that the signaling pathways they organize can be very complex. For example, the PDZ domain of α-syntrophin is known to associate with at least ten distinct proteins and its mode of binding to ligands such as nNOS leave the PDZ domain of nNOS free for additional protein-protein interactions. Furthermore some of these interactions may not be of significance in vivo. Indeed, studies of α-syntrophin-null

Table 1A. Syntrophin protein binding partners

Syntrophin	PDZ Ligand	Ligand Description	Tissue	Refs.
α	nNOS	neuronal nitric oxide synthase	skeletal muscle, brain	4,24,32
	Aquaporin 4	water channel	skeletal muscle, brain	24,82
	SAPK3	stress-activated protein kinase-3	skeletal muscle	46
	Grb2	growth factor bound 2 adaptor	skeletal muscle	83
	Na$_v$1.4	voltage-gated sodium channel	skeletal muscle, brain	84,85
	Na$_v$1.5	voltage-gated sodium channel	skeletal muscle	84
β1	nNOS	neuronal nitric oxide synthase	skeletal muscle	84
	Na$_v$1.4	voltage-gated sodium channel	skeletal muscle	84
	Na$_v$1.5	voltage-gated sodium channel	skeletal muscle	84
β2	nNOS	neuronal nitric oxide synthase	skeletal muscle	84
	ICA512	receptor tyrosine phosphatase-like protein	pancreas	50
	ABCA1	ATP-binding cassette transporter 1	liver, leukemia cell line	86
	MAST205	microtubule-associated serine/ threonine kinase	skeletal muscle	87
	SAST	syntrophin-associated serine/ threonine kinase	brain	87
	Na$_v$1.4	voltage-gated sodium channel	skeletal muscle	84
	Na$_v$1.5	voltage-gated sodium channel	skeletal & cardiac muscle	84
γ1	DGKz	diacylglycerol kinase-zeta	brain	88
γ2	Na$_v$1.5	voltage-gated sodium channel	intestinal smooth muscle	89

Proteins that Interact with syntrophin PDZ domains are shown in Table 1a. Additional syntrophin protein binding partners that interact with other domains are listed in Table 1b.

mice have exposed the potential promiscuity of PDZ interactions in vitro. The PDZ domains of α-, β1- and β2-syntrophin are all able to bind nNOS in vitro; however in the absence of α-syntrophin, despite the upregulation of both β1- and β2-syntrophin, nNOS does not associate with the sarcolemma or neuromuscular junction. Thus α-syntrophin is the only synthrophin isoform of relevance for nNOS localization in vivo.

Syntrophin has a class I PDZ domain that binds to the consensus sequence R/K E/D S/T X Φ (where X is any amino acid and Φ denotes a hydrophobic amino acid, V, I or L) at the carboxyl terminal tail of its ligands. Such interactions can be seen between the carboxyl terminal amino acids of skeletal muscle voltage-gated sodium channels (Na$_v$1.4) and the PDZ domains of α-, β1- and β2-syntrophin or between the KETXL sequence in the tail of the stress-activated protein kinase SAPK3 and the PDZ domain of α-syntrophin.[46]

Table 1B. Non-PDZ domain syntrophin interactors

Syntrophin	Interactor	Ligand Description	References
α	PtIns 4,5 P$_2$	phosphatidylinositol 4,5-bisphosphate	36
	Ca^{2+}/calmodulin	Ca^{2+}-mediator or Ca^{2+}-dependent modulator	37, 81
	muscle actin	cardiac and skeletal muscle actin	37

However, syntrophin's PDZ domain can bind other ligand types including internal SXV sequences that are topologically constrained.[47] The best characterized ligand of syntrophin, and of syntrophins in general, is the enzyme neuronal nitric oxide synthase (nNOS). nNOS produces the messenger molecule nitric oxide in a calcium-dependent manner and regulates blood flow to exercising muscle.[4,48] Studies of α-syntrophin and nNOS revealed a novel mode of PDZ-ligand interaction. The PDZ domain of α-syntrophin binds the H L E T T F sequence within the β hairpin loop immediately downstream from the PDZ domain of nNOS.[49] This is the first structural determination of the binding mode of nonterminal peptides to PDZ domains in vitro. This interaction has been confirmed in vivo.[24]

Another example of atypical binding of a ligand to the PDZ domain of syntrophin comes from studies of ICA512 (islet cell autoantigen 512 kDa) found in neuroendocrine cells.[50,51] ICA512 is a receptor tyrosine phosphatase-like protein that binds the PDZ domain of β2 syntrophin.[50] The last 29 residues of the carboxyl terminus of ICA512 and 37 amino acids upstream of the protein tyrosine phosphatase domain are both required for strong binding to β2-syntrophin; however the residues at the end of the carboxyl terminus of ICA512 do not conform to the expected E S/T X Φ consensus binding sequence. Furthermore, ICA512 preferentially binds the phosphorylated form of syntrophin, suggesting a role for post-translational regulation of the interaction between syntrophin and ICA512.

Another important potential ligand of syntrophin is the water channel aquaporin 4 (Aqp4). The carboxyl terminus of Aqp4 contains the consensus SXV sequence and its localization to the dystrophin-glycoprotein complex in skeletal muscle requires the PDZ domain of α-syntrophin.[24] The SSV tail of Aqp4 does appear to mediate the stability of the protein.[52] However, despite major effort, a direct interaction between syntrophin and aquaporin has not been observed. This suggests that additional factors and/or proteins are involved in the interaction.[52,53]

Animal Models and Syntrophin Function

Two independent and distinct targeting strategies were used to produce α-syntrophin-null mice. The first strategy involves an in-frame insertion of a neomycin cassette into exon 2 which contains the PDZ domain.[54] The second approach involves the targeted deletion of exon 1 and surrounding flanking sequence.[32] However, subsequent detection of α-syntrophin and utrophin at the neuromuscular junction in mice produced by the first strategy suggests that a truncated PDZ-less form of α-syntrophin has been produced and therefore this mutant mouse is not a true null.[53,54]

Studies of these mice reveal that α-syntrophin plays an important role in normal neuromuscular synapse maturation.[32,53] Mutant mice deficient in α-syntrophin show structurally compromised neuromuscular junctions. Acetylcholine receptor protein is reduced to 35 % of wild type levels. Receptor distribution is patchy with branch borders that have a frayed appearance with needle-like processes extending outwards from postsynaptic boundaries.[32] Furthermore, receptors are found in discontinuous clusters that lie outside nerve-muscle contacts. The postsynaptic membranes of α-syntrophin-deficient mice have shallow gutters, and fewer folds than wild type junctions. In addition, the number of junctional folds opening into the synaptic cleft is reduced.[32] Due to the absence of myopathy in α-syntrophin-deficient muscle these abnormal junctions are not a secondary consequence of myopathic changes in muscle.

The junctions of mutant mice expressing a truncated PDZ-less form of α-syntrophin, like the knockout, are still aberrantly organized even in the presence of utrophin.[32,53] Interestingly, despite these substantial abnormalities at the junction there is no detectable impact on mouse mobility as measured by voluntary running wheel assays. Studies of muscle contractility properties reveal no difference between wild type and mice expressing a mutant α syntrophin that lacked its PDZ domain.[53,54] This suggests that there may be functional compensation occurring at the synapse. Indeed, Adams and coworkers report that the levels of acetylcholine esterase (an enzyme localized in the synaptic cleft that hydrolyzes acetylcholine) are reduced by half in the α-syntrophin-deficient mice.[32] Acetylcholine may persist in the synapse (presumably due to a decrease in AchE activity) and therefore activate more receptors. The organization of the neuromuscular junction in α-syntrophin-deficient mice is similar to that seen in α-dystrobrevin null mice.[55] Similarly, utrophin-deficient mice have reduced levels of acetylcholine receptors, although not as dramatically decreased as the α-syntrophin-null mice and a 50 % reduction in postsynaptic membrane folding.[56] Taken together these data suggest that α-syntrophin is essential for the maturation of the neuromuscular synapse.

Syntrophin and nNOS

Activation of the sympathetic nervous system during exercise causes vasoconstriction in visceral organs and inactive muscles resulting in a redirection of blood flow to active muscles. Nitric oxide produced at the sarcolemma by nNOS in skeletal muscle is involved in the regulation of blood flow in exercising muscle by modulation of α-adrenergic receptor-mediated vasoconstriction.[57,58] Modulation of α-adrenergic-mediated blood vessel constriction is disrupted in skeletal muscle of DMD patients and *mdx* mice.[57,58] Studies in α-syntrophin mutant mice demonstrate that the PDZ domain of α-syntrophin is required for nNOS to associate with the sarcolemma.[24,54] Studies of the ability of NO to override α-adrenergic-mediated vasoconstriction in α-syntrophin mutant mice show that the sarcolemmal localization of nNOS is required for correct vasomodulation.[59]

The localization of nNOS is important for its normal physiological activity in other tissues; however nNOS is not mislocalized in the brains of α-syntrophin null mice consistent with observations that the dystrophin complex is not required for the correct localization of nNOS in the brain.[60] Mice expressing low levels of catalytically active nNOS lacking its amino terminal PDZ domain in neurons show behavioral abnormalities.[4,61,62] α-Syntrophin-deficient mice and mice expressing a mutant α-syntrophin lacking its PDZ domain exhibit aberrant modulation of norepinephrine-mediated vasoconstriction in contracted hind limb muscle.[59] The activity of soluble nNOS was largely unaffected in all the syntrophin mutant mice. Thus the association of nNOS via α-syntrophin with the dystrophin complex is essential for proper regulation of blood flow to exercising muscle. These results provide perhaps the strongest evidence to date of a signaling role for the dystrophin complex.

Investigations of cardiotoxin-induced muscle regeneration in mice expressing a mutant α-syntrophin missing its PDZ domain show an increased number of hypertrophied muscle fibers early in the regeneration process.[53] Although there is no difference in muscle contractile properties between wild type and mutant mice under steady state conditions, there is a two-fold decrease in maximal tetanic force and peak twitch force in the regenerating muscles of syntrophin mutant mice compared with regenerating wild type mice. Furthermore, muscles of mutant mice are significantly weaker than controls in wire net holding strength tests early during regeneration.[53] Muscle hypertrophy and compromised contractile activity are characteristic of dystrophic muscle, for example muscle of dystrophin-deficient or α-sarcoglycan deficient mice.[63,64] These data suggest that the absence of a PDZ interactor of α-syntrophin results in abnormal regeneration of skeletal muscle fibers.

The absence of α-syntrophin results in changes in the levels of other dystrophin-associated proteins including an increase in β1-syntrophin levels and a decrease in α-dystrobrevin 2 levels at the sarcolemma. However, the most surprising change in the α-syntrophin null mice is the loss of utrophin from the crests of junctional folds at the synapse.[32] It is presently unclear how the absence of syntrophin affects the localization of utrophin to the junction. Since utrophin upregulation is a promising therapeutic approach for Duchenne Muscular Dystrophy, it is important to understand the role of α-syntrophin in utrophin membrane association.

The worm *Caenorhabditis elegans* has a syntrophin isoform with 35% identity to human β1 syntrophin.[65] RNAi-mediated inactivation of the syntrophin gene F30A10.8 results in worms with exaggerated head bending during forward movement, slight hyperactivity and a tendency to hypercontract.[65] This phenotype is also characteristic of worms with loss of function mutations in their dystrophin-like gene.[66] There was no observable worsening of the phenotype of mutant dys-1 worms when the syntrophin gene was also simultaneously inactivated.[65]

Syntrophin and Aquaporin 4

Recently, α-syntrophin has been implicated in the normal subcellular localization of aquaporin 4 in both skeletal muscle and brain, highlighting further uncharacterized roles of the dystrophin complex in other tissues. Aquaporin 4 (Aqp4) is a high capacity water-selective channel found in cellular membranes. It has six transmembrane helices and forms tetramers which are organized into large macromolecular orthogonal arrays in membranes that are depleted in *mdx* mice.[67] Aqp4 is found in a diverse variety of tissues and is expressed at high levels in the brain, especially in the end feet of astrocytes and also ependymal cells that make up the lining of brain ventricles and pia.[68] Aqp4 is present at much lower levels (*ca.* 20-fold less than in brain) at the extrasynaptic sarcolemma of fast twitch fibers of skeletal muscle.[69-71]

In adult *mdx*, Duchenne and sarcoglycan-deficient muscle, Aqp4 is lost from the sarcolemma and its expression is substantially reduced.[69,71-73] This is consistent with studies of skeletal muscle of α-syntrophin-deficient mice which showed that Aqp4 is dependent on α-syntrophin for sarcolemmal localization despite the normal distribution of dystrophin.[32] In the absence of α-syntrophin Aqp4 protein levels are reduced dramatically in skeletal muscle.[52] The PDZ domain of α-syntrophin is required for the targeting of Aqp4 to the sarcolemma of skeletal muscle since Aqp4 does not associate with the sarcolemma of syntrophin mutant mice expressing a PDZ-less mutant α-syntrophin.[24]

Aqp4 is concentrated in astrocyte endfeet membranes facing the basal lamina of endothelial cells of blood vessels forming part of the blood-brain barrier. In rat cerebellum, dystrophin (Dp71), Aqp4, syntrophin and β-dystroglycan could be coimmunoprecipitated as a complex after chemical crosslinking.[52] In astrocytes of the cerebellum and cerebral cortex of α-syntrophin-deficient mice Aqp4 is improperly localized and accumulated in membranes facing the neuropil, a fibrous network of nonmyelinated nerve fibres.[52] Astrocyte dystrophin (Dp71) is similarly required for the localization of Aqp4 to astrocyte endfeet membranes at the blood-brain and blood-cerebrospinal fluid-brain interface.[74] Association of Aqp4 with membranes of other cell types including ependymal cells is unaffected demonstrating a specific depletion of the perivascular astrocytic pool of Aqp4 in syntrophin-deficient mice. The perivascular and subpial astrocyte endfeet are swollen suggesting that mislocalization of the aquaporin channel reduced water outflow generated by normal metabolism.

Interestingly, α-syntrophin deficient mice show attenuated brain swelling or edema in response to cerebral ischemia induced by blocking the middle cerebral artery.[60] In *mdx* β*geo* mice (which lack all alternatively spliced forms of dystrophin) water influx into the brain is delayed after induction of edema by the intraperitoneal injection of water and 8-deamino-arginine vasopressin.[74] During ischemia in wild type mice, Aqp4 is lost from the endfoot membrane, while syntrophin remains. Therefore the pool of α-syntrophin associated with Aqp4 in astrocytes plays an important role in the bidirectional movement of water through the blood-brain barrier. The delay in swelling seen in the α-syntrophin null mouse led to the proposal that specific disruption of the syntrophin-Aqp4 interaction may be of therapeutic benefit in treating brain edema in the clinic.[60]

The association of Aqp4 with membranes is not as straightforward as initially thought. Aqp4 can associate with membranes in the absence of a dystrophin complex. In prenecrotic *mdx* muscle of one to two week old mice Aqp4 localizes to the sarcolemma in the absence of dystrophin.[71] Furthermore, Aqp4 does not show sarcolemmal targeting in *mdx* mice expressing truncated dystrophin transgenes that reconstitute the dystrophin-associated protein complex.[71] Moreover, in direct contrast to studies in genetically modified mice, levels of AQP4 do not closely correlate with those of α-syntrophin in Becker muscular dystrophy patients.[75] Together, these data suggest that Aqp4 localization in muscle is not solely dependent on the dystrophin complex and its association with the sarcolemma may be achieved by additional mechanisms.

Syntrophin and Intracellular Trafficking

Recent reports describe the presence of syntrophin on carrier vesicles of neuroendocrine insulin-secreting cells and *Torpedo* electric organ. The PDZ domain of β2-syntrophin interacts with ICA512, a receptor tyrosine phosphatase-like protein that associates with secretory granules of neuroendocrine cells.[50] β2-syntrophin and ICA512 colocalize on intracellular secretory granules in rat insulinoma cells.[51] ICA512 preferentially binds phosphorylated β2- syntrophin and this interaction protects ICA512 from calpain-mediated degradation. Stimulation of insulin secretion promotes dephosphorylation of β2 syntrophin and degradation of mature ICA512; thus releasing secretory granules from immobilization on the actin cytoskeleton through β2 syntrophin and utrophin.[51]

In studies of the targeting of dystrophin complex members to the postsynaptic membrane in the *Torpedo* electric organ, Marchand and coworkers[76] show by biochemical isolation and immunoelectron microscopy the presence of β-dystroglycan, acetylcholine receptors and rapsyn on the same vesicles thought to represent a pool of exocytotic Golgi-derived carriers bound for the postsynaptic membrane.[76] Interestingly, syntrophin associated with these vesicles only transiently and did so in the absence of both dystrophin and dystrobrevin.[76] Together these data suggest that syntrophin has a role in vesicle trafficking independent of the dystrophin-associated protein complex at the plasma membrane.

Syntrophins and Muscular Dystrophies

Mice lacking α-syntrophin do not show any of the characteristics of dystrophic muscle such as fiber degeneration, centrally located nuclei, fibrosis and fiber size variability.[32,54] Thus, at least in mice, α-syntrophin does not appear to play a causal role in muscular dystrophy. It has been proposed that aberrantly localized nNOS may result in additional free radical damage thereby exacerbating the dystrophic phenotype of *mdx* mice.[77] However, there is no dystrophy in muscle of nNOS-deficient mice expressing a catalytically inactive form of nNOS and the dystrophic characteristics of *mdx* mice lacking functional nNOS do not differ significantly from *mdx* mice.[78,79] These data suggest that nNOS does not contribute to pathogenesis in dystrophin mutant mice. Consistent with these findings, mice expressing mutant syntrophin without its PDZ domain do not have dystrophic muscle.[24,54] However, the role of nNOS in muscular dystrophies is far from resolved since transgenic overexpression of nNOS in *mdx* muscle dramatically improves the dystrophic phenotype and increases muscle membrane integrity.[80] This effect was attributed to the anti-inflammatory properties of nitric oxide. Direct manipulation of nNOS and thus the dystrophin signaling complex provides support for the hypothesis that deregulated signal transduction processes may play a key role in the pathology of some muscular dystrophies (see Chapters 5, 16 and 19).

Characterization of a cohort of patients with myopathies of unknown etiology with normal spectrin, dystrophin, sarcoglycans, β-dystroglycan and laminin-α2 led to the identification of a group that were negative for β2-syntrophin. This group also showed abnormal and decreased α-dystrobrevin labeling and had a severe clinical phenotype.[31] One patient lacking both β2-syntrophin and α dystrobrevin displayed a severe congenital myopathy and died soon after birth.[31] Sequencing analysis of coding regions did not reveal any disease-causing mutations in either of the genes. Nevertheless, abnormalities in β2-syntrophin and dystrobrevin may contribute to the pathology of specific classes of human myopathies.

Aquaporin 4 sarcolemmal localization is often severely impaired or lost in the absence of dystrophin in Duchenne patients and *mdx* mice suggesting it may contribute to the dystrophic phenotype.[75] However aquaporin 4-deficient mice have normal muscle histopathology and do not have dystrophy and have normal contractility and water permeability.[70] This is consistent with the phenotype of muscle of α-syntrophin-deficient mice which have no Aqp4. Thus it seems unlikely that Aqp4 contributes to the dystrophic phenotype in *mdx* mice.

Concluding Remarks

Syntrophin adaptor proteins play a key role in tethering a variety of molecules to the dystrophin complex in muscle and nonmuscle tissues. Syntrophins act as scaffold molecules to facilitate the assembly of multi-protein complexes underneath specialized plasma membrane domains including the neuromuscular junction and astrocyte endfeet. These complexes play key roles in vasomodulation, water transport and neuromuscular junction maturation. Syntrophins' ability to localize signaling molecules to the dystrophin complex provide an important additional role for the complex in normal muscle physiology. The absence of syntrophin results in dysregulated signal transduction pathways and suggests that syntrophin may contribute to the dystrophic phenotype in muscle. Further characterization of the long list of binding partners of syntrophin will undoubtedly expand the repertoire of pathways that require a syntrophin scaffold.

References

1. Harris BZ, Lim WA. Mechanism and role of PDZ domains in signaling complex assembly. J Cell Sci 2001; 114(Pt 18):3219-31.
2. Zhang M, Wang W. Organization of signaling complexes by PDZ-domain scaffold proteins. Acc Chem Res 2003; 36(7):530-8.
3. Kornau HC, Schenker LT, Kennedy MB et al. Domain interaction between NMDA receptor subunits and the postsynaptic density protein PSD-95. Science 1995; 269(5231):1737-40.
4. Brenman JE, Chao DS, Gee SH et al. Interaction of nitric oxide synthase with the postsynaptic density protein PSD-95 and alpha1-syntrophin mediated by PDZ domains. Cell 1996; 84(5):757-67.
5. Niethammer M, Valtschanoff JG, Kapoor TM et al. CRIPT, a novel postsynaptic protein that binds to the third PDZ domain of PSD-95/SAP90. Neuron 1998; 20(4):693-707.
6. El-Husseini AE, Schnell E, Chetkovich DM et al. PSD-95 involvement in maturation of excitatory synapses. Science 2000; 290(5495):1364-8.
7. Dong H, O'Brien RJ, Fung ET et al. GRIP: A synaptic PDZ domain-containing protein that interacts with AMPA receptors. Nature 1997; 386(6622):279-84.
8. Xia J, Chung HJ, Wihler C et al. Cerebellar long-term depression requires PKC-regulated interactions between GluR2/3 and PDZ domain-containing proteins. Neuron 2000; 28(2):499-510.
9. Lamprecht G, Weinman EJ, Yun CH. The role of NHERF and E3KARP in the cAMP-mediated inhibition of NHE3. J Biol Chem 1998; 273(45):29972-8.
10. Weinman EJ, Steplock D, Donowitz M et al. NHERF associations with sodium-hydrogen exchanger isoform 3 (NHE3) and ezrin are essential for cAMP-mediated phosphorylation and inhibition of NHE3. Biochemistry 2000; 39(20):6123-9.
11. Hall RA, Premont RT, Chow CW et al. The beta2-adrenergic receptor interacts with the Na+/H+-exchanger regulatory factor to control Na+/H+ exchange. Nature 1998; 392(6676):626-30.
12. Kaech SM, Whitfield CW, Kim SK. The LIN-2/LIN-7/LIN-10 complex mediates basolateral membrane localization of the C. Elegans EGF receptor LET-23 in vulval epithelial cells. Cell 1998; 94(6):761-71.
13. Froehner SC. Peripheral proteins of postsynaptic membranes from Torpedo electric organ identified with monoclonal antibodies. J Cell Biol 1984; 99(1 Pt 1):88-96.
14. Froehner SC, Murnane AA, Tobler M et al. A postsynaptic Mr 58,000 (58K) protein concentrated at acetylcholine receptor-rich sites in Torpedo electroplaques and skeletal muscle. J Cell Biol 1987; 104(6):1633-46.
15. Carr C, Fischbach GD, Cohen JB. A novel 87,000-Mr protein associated with acetylcholine receptors in Torpedo electric organ and vertebrate skeletal muscle. J Cell Biol 1989; 109(4 Pt 1):1753-64.
16. Chen Q, Sealock R, Peng HB. A protein homologous to the Torpedo postsynaptic 58K protein is present at the myotendinous junction. J Cell Biol 1990; 110(6):2061-71.
17. Butler MH, Douville K, Murnane AA et al. Association of the Mr 58,000 postsynaptic protein of electric tissue with Torpedo dystrophin and the Mr 87,000 postsynaptic protein. J Biol Chem 1992; 267(9):6213-8.
18. Bloch RJ, Resneck WG, O'Neill A et al. Cytoplasmic components of acetylcholine receptor clusters of cultured rat myotubes: The 58-kD protein. J Cell Biol 1991; 115(2):435-46.
19. Adams ME, Butler MH, Dwyer TM et al. Two forms of mouse syntrophin, a 58 kd dystrophin-associated protein, differ in primary structure and tissue distribution. Neuron 1993; 11(3):531-40.

20. Ahn AH, Yoshida M, Anderson MS et al. Cloning of human basic A1, a distinct 59-kDa dystrophin-associated protein encoded on chromosome 8q23-24. Proc Natl Acad Sci USA 1994; 91(10):4446-50.
21. Adams ME, Dwyer TM, Dowler LL et al. Mouse alpha 1- and beta 2-syntrophin gene structure, chromosome localization, and homology with a discs large domain. J Biol Chem 1995; 270(43):25859-65.
22. Piluso G, Mirabella M, Ricci E et al. Gamma1- and gamma2-syntrophins, two novel dystrophin-binding proteins localized in neuronal cells. J Biol Chem 2000; 275(21):15851-60.
23. Ervasti JM, Campbell KP. Membrane organization of the dystrophin-glycoprotein complex. Cell 1991; 66(6):1121-31.
24. Adams ME, Mueller HA, Froehner SC. In vivo requirement of the alpha-syntrophin PDZ domain for the sarcolemmal localization of nNOS and aquaporin-4. J Cell Biol 2001; 155(1):113-22.
25. Peters MF, Adams ME, Froehner SC. Differential association of syntrophin pairs with the dystrophin complex. J Cell Biol 1997; 138(1):81-93.
26. Loh NY, Newey SE, Davies KE et al. Assembly of multiple dystrobrevin-containing complexes in the kidney. J Cell Sci 2000; 113(Pt 15):2715-24.
27. Loh NY, Nebenius-Oosthuizen D, Blake DJ et al. Role of beta-dystrobrevin in nonmuscle dystrophin-associated protein complex-like complexes in kidney and liver. Mol Cell Biol 2001; 21(21):7442-8.
28. Peters MF, Kramarcy NR, Sealock R et al. Beta 2-Syntrophin: Localization at the neuromuscular junction in skeletal muscle. Neuroreport 1994; 5(13):1577-80.
29. Kramarcy NR, Sealock R. Syntrophin isoforms at the neuromuscular junction: Developmental time course and differential localization. Mol Cell Neurosci 2000; 15(3):262-74.
30. Klietsch R, Ervasti JM, Arnold W et al. Dystrophin-glycoprotein complex and laminin colocalize to the sarcolemma and transverse tubules of cardiac muscle. Circ Res 1993; 72(2):349-60.
31. Jones KJ, Compton AG, Yang N et al. Deficiency of the syntrophins and alpha-dystrobrevin in patients with inherited myopathy. Neuromuscul Disord 2003; 13(6):456-67.
32. Adams ME, Kramarcy N, Krall SP et al. Absence of alpha-syntrophin leads to structurally aberrant neuromuscular synapses deficient in utrophin. J Cell Biol 2000; 150(6):1385-98.
33. Blomberg N, Baraldi E, Nilges M et al. The PH superfold: A structural scaffold for multiple functions. Trends Biochem Sci 1999; 24(11):441-5.
34. Harlan JE, Hajduk PJ, Yoon HS et al. Pleckstrin homology domains bind to phosphatidylinositol-4,5-bisphosphate. Nature 1994; 371(6493):168-70.
35. Touhara K, Inglese J, Pitcher JA et al. Binding of G protein beta gamma-subunits to pleckstrin homology domains. J Biol Chem 1994; 269(14):10217-20.
36. Chockalingam PS, Gee SH, Jarrett HW. Pleckstrin homology domain 1 of mouse alpha 1-syntrophin binds phosphatidylinositol 4,5-bisphosphate. Biochemistry 1999; 38(17):5596-602.
37. Iwata Y, Pan Y, Yoshida T et al. Alpha1-syntrophin has distinct binding sites for actin and calmodulin. FEBS Lett 1998; 423(2):173-7.
38. Rebecchi MJ, Pentyala SN. Structure, function, and control of phosphoinositide-specific phospholipase C. Physiol Rev 2000; 80(4):1291-335.
39. Sugimoto K, Mori Y, Makino K et al. Functional reassembly of a split PH domain. J Am Chem Soc 2003; 125(17):5000-4.
40. Kramarcy NR, Vidal A, Froehner SC et al. Association of utrophin and multiple dystrophin short forms with the mammalian M(r) 58,000 dystrophin-associated protein (syntrophin). J Biol Chem 1994; 269(4):2870-6.
41. Suzuki A, Yoshida M, Ozawa E. Mammalian alpha 1- and beta 1-syntrophin bind to the alternative splice-prone region of the dystrophin COOH terminus. J Cell Biol 1995; 128(3):373-81.
42. Yang B, Jung D, Rafael JA et al. Identification of alpha-syntrophin binding to syntrophin triplet, dystrophin, and utrophin. J Biol Chem 1995; 270(10):4975-8.
43. Ahn AH, Freener CA, Gussoni E et al. The three human syntrophin genes are expressed in diverse tissues, have distinct chromosomal locations, and each bind to dystrophin and its relatives. J Biol Chem 1996; 271(5):2724-30.
44. Ahn AH, Kunkel LM. Syntrophin binds to an alternatively spliced exon of dystrophin. J Cell Biol 1995; 128(3):363-71.
45. Kachinsky AM, Froehner SC, Milgram SL. A PDZ-containing scaffold related to the dystrophin complex at the basolateral membrane of epithelial cells. J Cell Biol 1999; 145(2):391-402.
46. Hasegawa M, Cuenda A, Spillantini MG et al. Stress-activated protein kinase-3 interacts with the PDZ domain of alpha1-syntrophin. A mechanism for specific substrate recognition. J Biol Chem 1999; 274(18):12626-31.

47. Gee SH, Sekely SA, Lombardo C et al. Cyclic peptides as noncarboxyl-terminal ligands of syntrophin PDZ domains. J Biol Chem 1998; 273(34):21980-7.
48. Stamler JS, Meissner G. Physiology of nitric oxide in skeletal muscle. Physiol Rev 2001; 81(1):209-237.
49. Hillier BJ, Christopherson KS, Prehoda KE et al. Unexpected modes of PDZ domain scaffolding revealed by structure of nNOS-syntrophin complex. Science 1999; 284(5415):812-5.
50. Ort T, Maksimova E, Dirkx R et al. The receptor tyrosine phosphatase-like protein ICA512 binds the PDZ domains of beta2-syntrophin and nNOS in pancreatic beta-cells. Eur J Cell Biol 2000; 79(9):621-30.
51. Ort T, Voronov S, Guo J et al. Dephosphorylation of beta2-syntrophin and Ca2+/mu-calpain-mediated cleavage of ICA512 upon stimulation of insulin secretion. EMBO J 2001; 20(15):4013-23.
52. Neely JD, Amiry-Moghaddam M, Ottersen OP et al. Syntrophin-dependent expression and localization of Aquaporin-4 water channel protein. Proc Natl Acad Sci USA 2001; 98(24):14108-13.
53. Hosaka Y, Yokota T, Miyagoe-Suzuki Y et al. Alpha1-syntrophin-deficient skeletal muscle exhibits hypertrophy and aberrant formation of neuromuscular junctions during regeneration. J Cell Biol 2002; 158(6):1097-107.
54. Kameya S, Miyagoe Y, Nonaka I et al. Alpha1-syntrophin gene disruption results in the absence of neuronal-type nitric-oxide synthase at the sarcolemma but does not induce muscle degeneration. J Biol Chem 1999; 274(4):2193-200.
55. Grady RM, Grange RW, Lau KS et al. Role for alpha-dystrobrevin in the pathogenesis of dystrophin-dependent muscular dystrophies. Nat Cell Biol 1999; 1(4):215-20.
56. Grady RM, Merlie JP, Sanes JR. Subtle neuromuscular defects in utrophin-deficient mice. J Cell Biol 1997; 136(4):871-82.
57. Thomas GD, Sander M, Lau KS et al. Impaired metabolic modulation of alpha-adrenergic vasoconstriction in dystrophin-deficient skeletal muscle. Proc Natl Acad Sci USA 1998; 95(25):15090-5.
58. Sander M, Chavoshan B, Harris SA et al. Functional muscle ischemia in neuronal nitric oxide synthase-deficient skeletal muscle of children with Duchenne muscular dystrophy. Proc Natl Acad Sci USA 2000; 97(25):13818-23.
59. Thomas GD, Shaul PW, Yuhanna IS et al. Vasomodulation by skeletal muscle-derived nitric oxide requires alpha-syntrophin-mediated sarcolemmal localization of neuronal nitric oxide synthase. Circ Res 2003; 92(5):554-60.
60. Amiry-Moghaddam M, Otsuka T, Hurn PD et al. An alpha-syntrophin-dependent pool of AQP4 in astroglial end-feet confers bidirectional water flow between blood and brain. Proc Natl Acad Sci USA 2003; 100(4):2106-11.
61. Huang PL, Dawson TM, Bredt DS et al. Targeted disruption of the neuronal nitric oxide synthase gene. Cell 1993; 75(7):1273-86.
62. Nelson RJ, Demas GE, Huang PL et al. Behavioural abnormalities in male mice lacking neuronal nitric oxide synthase. Nature 1995; 378(6555):383-6.
63. Coulton GR, Curtin NA, Morgan JE et al. The mdx mouse skeletal muscle myopathy: II. Contractile properties. Neuropathol Appl Neurobiol 1988; 14(4):299-314.
64. Duclos F, Straub V, Moore SA et al. Progressive muscular dystrophy in alpha-sarcoglycan-deficient mice. J Cell Biol 1998; 142(6):1461-71.
65. Grisoni K, Martin E, Gieseler K et al. Genetic evidence for a dystrophin-glycoprotein complex (DGC) in Caenorhabditis elegans. Gene 2002; 294(1-2):77-86.
66. Bessou C, Giugia JB, Franks CJ et al. Mutations in the Caenorhabditis elegans dystrophin-like gene dys-1 lead to hyperactivity and suggest a link with cholinergic transmission. Neurogenetics 1998; 2(1):61-72.
67. Wakayama Y, Jimi T, Misugi N et al. Dystrophin immunostaining and freeze-fracture studies of muscles of patients with early stage amyotrophic lateral sclerosis and Duchenne muscular dystrophy. J Neurol Sci 1989; 91(1-2):191-205.
68. Borgnia M, Nielsen S, Engel A et al. Cellular and molecular biology of the aquaporin water channels. Annu Rev Biochem 1999; 68:425-58.
69. Frigeri A, Nicchia GP, Verbavatz JM et al. Expression of aquaporin-4 in fast-twitch fibers of mammalian skeletal muscle. J Clin Invest 1998; 102(4):695-703.
70. Yang B, Verbavatz JM, Song Y et al. Skeletal muscle function and water permeability in aquaporin-4 deficient mice. Am J Physiol Cell Physiol 2000; 278(6):C1108-15.
71. Crosbie RH, Dovico SA, Flanagan JD et al. Characterization of aquaporin-4 in muscle and muscular dystrophy. FASEB J 2002; 16(9):943-9.
72. Frigeri A, Nicchia GP, Nico B et al. Aquaporin-4 deficiency in skeletal muscle and brain of dystrophic mdx mice. FASEB J 2001; 15(1):90-98.
73. Liu JW, Wakayama Y, Inoue M et al. Immunocytochemical studies of aquaporin 4 in the skeletal muscle of mdx mouse. J Neurol Sci 1999; 164(1):24-8.

74. Vajda Z, Pedersen M, Fuchtbauer EM et al. Delayed onset of brain edema and mislocalization of aquaporin-4 in dystrophin-null transgenic mice. Proc Natl Acad Sci USA 2002; 99(20):13131-6.
75. Frigeri A, Nicchia GP, Repetto S. Altered aquaporin-4 expression in human muscular dystrophies: A common feature FASEB J 2002; 16(9):1120-2.
76. Marchand S, Stetzkowski-Marden F, Cartaud J. Differential targeting of components of the dystrophin complex to the postsynaptic membrane. Eur J Neurosci 2001; 13(2):221-9.
77. Brenman JE, Chao DS, Xia H et al. Nitric oxide synthase complexed with dystrophin and absent from skeletal muscle sarcolemma in Duchenne muscular dystrophy. Cell 1995; 82(5):743-52.
78. Chao DS, Silvagno F, Bredt DS. Muscular dystrophy in mdx mice despite lack of neuronal nitric oxide synthase. J Neurochem 1998; 71(2):784-9.
79. Crosbie RH, Straub V, Yun HY et al. mdx muscle pathology is independent of nNOS perturbation. Hum Mol Genet 1998; 7(5):823-9.
80. Wehling M, Spencer MJ, Tidball JG. A nitric oxide synthase transgene ameliorates muscular dystrophy in mdx mice. J Cell Biol 2001; 155(1):123-31.
81. Newbell BJ, Anderson JT, Jarrett HW. Ca2+-calmodulin binding to mouse alpha1 syntrophin: Syntrophin is also a Ca2+-binding protein. Biochemistry 1997; 36(6):1295-305.
82. Yokota TY, Miyagoe Y, Hosaka K et al. Aquaporin-4 is absent at the sarcolemma and perivascular astrocyte endfeet in alpha1-syntrophin knockout mice. Proc Jpn Acad 2000; 76:22–2759.
83. Oak SA, Russo K, Petrucci TC et al. Mouse alpha1-syntrophin binding to Grb2: Further evidence of a role for syntrophin in cell signaling. Biochemistry 2001; 40(37):11270-8.
84. Gee SH, Madhavan R, Levinson SR et al. Interaction of muscle and brain sodium channels with multiple members of the syntrophin family of dystrophin-associated proteins. J Neurosci 1998; 18(1):128-37.
85. Schultz J, Hoffmuller U, Krause G et al. Specific interactions between the syntrophin PDZ domain and voltage-gated sodium channels. Nat Struct Biol 1998; 5(1):19-24.
86. Buechler C, Boettcher A, Bared SM et al. The carboxyterminus of the ATP-binding cassette transporter A1 interacts with a beta2-syntrophin/utrophin complex. Biochem Biophys Res Commun 2002; 293(2):759-65.
87. Lumeng C, Phelps S, Crawford GE et al. Interactions between beta 2-syntrophin and a family of microtubule-associated serine/threonine kinases. Nat Neurosci 1999; 2(7):611-7.
88. Hogan A, Shepherd L, Chabot J et al. Interaction of gamma 1-syntrophin with diacylglycerol kinase-zeta. Regulation of nuclear localization by PDZ interactions. J Biol Chem 2001; 276(28):26526-33.
89. Ou Y, Strege P, Miller SM et al. Syntrophin gamma 2 regulates SCN5A gating by a PDZ domain-mediated interaction. J Biol Chem 2003; 278(3):1915-23.

CHAPTER 5

Molecular and Functional Diversity of Dystrobrevin-Containing Complexes

Derek J. Blake and Roy V. Sillitoe

Introduction

I t is now well established that mutations in the gene encoding the large sarcolemmal-associated protein dystrophin cause Duchenne and Becker muscular dystrophies (DMD and BMD). The core dystrophin glycoprotein complex (DGC) as described by the Campbell and Ozawa groups,[1,2] is assembled around dystrophin, the lack of which has a dramatic effect upon the skeletal muscle of patients lacking the protein and also affects the heart and brain.[3] In muscle, the core DGC is composed of at least 12 proteins (dystrophin, α- and β-dystroglycan, the sarcoglycans, sarcospan, the syntrophins and the α-dystrobrevins) that form a molecular bridge between the actin-based cytoskeleton and laminin in the extracellular matrix (ECM). Patients lacking dystrophin have severely reduced levels of the DGC components at the sarcolemma whereas the levels of total cellular protein often remain unaltered.[3] The effects of these alterations on the normal physiological localization of the DGC may directly contribute to the complex pathology typically observed in dystrophin-deficient muscle.[3]

The DGC in muscle is organized into at least three sub-complexes; the sarcoglycan:sarcospan sub-complex, the dystroglycan sub-complex and the cytoplasmic sub-complex.[4] The latter complex contains the dystrobrevins, dystrophin and the syntrophins, all of which are peripheral membrane proteins located on the cytoplasmic face of the sarcolemma. Dystrobrevin was originally identified as an 87kDa phosphoprotein in the electric organ (a model of the mammalian neuromuscular junction (NMJ)) of *Torpedo californica* that binds to the *Torpedo* orthologues of syntrophin and dystrophin.[5,6] In a separate study, the Ozawa group showed that a 94kDa component in the purified mammalian DGC (assigned as A0) was in fact a homologue of the *Torpedo* 87kDa protein.[7] Whilst this study provided only a partial characterization of dystrobrevin and its association with the DGC, a more complete description of the dystrobrevin family of proteins in muscle was presented in two independent papers from the Davies and Kunkel laboratories.[8,9] The subsequent identification of a second mammalian dystrobrevin, β-dystrobrevin, completes the family of dystrophin-related proteins which is composed of dystrophin, utrophin (also known as dystrophin-related protein (DRP) or DMD-like (DMDL), dystrophin-related protein-2 (DRP2), α- and β-dystrobrevin.[10,11]

The Dystrobrevin Family of Proteins

The dystrobrevins are both dystrophin-related and -associated proteins that have sequence similarity to the C-terminal region of dystrophin. Dystrophin and the dystrobrevins contain two to four EF-hand motifs that are predicted to bind Ca^{2+}, a ZZ-domain[12] that in dystrophin is required for binding to β-dystroglycan[13] and two adjacent sequences that are predicted to form coiled-coils separated by a proline-rich linker.[14] This configuration of motifs and domains is conserved in all members of the dystrophin-related protein family (see ref. 3 for a more detailed description).

In mammals, two different genes encode all of the known dystrobrevin isoforms. The α-dystrobrevin gene (*dtna* in the mouse) is located on human chromosome 18q12.1-12.2 (mouse chromosome 18) and encodes several different protein isoforms that represent C-terminal truncations of the full size protein, α-dystrobrevin-1.[8,9,15] In addition to the sequence motifs that are the hallmark of the dystrophin-related protein family, α-dystrobrevin-1 has an extended C-terminus that contains several tyrosine residues that can be phosphorylated by *src* and related kinases.[6,16] Skeletal and cardiac muscle are the only tissues that contain all three major α-dystrobrevin isoforms; α-dystrobrevins-1, -2 and -3.[8,9,15] The transcription of the major α-dystrobrevin isoforms is driven by at least three independent promoters that are active in different tissues. The muscle promoter is active in both skeletal and cardiac muscle (and unusually, to a lesser extent in the brain) and is able to drive the expression of all three α-dystrobrevin isoforms.[17] This promoter contains several sites for muscle regulatory factors such as myogenin and MEF-1 and is activated during myoblast differentiation.[17] By contrast, only α-dystrobrevins-1 and -2 are expressed in the brain from all three promoters.[8,17,18] The complexity in promoter organization is mirrored by the array of different 5'-untranslated region (UTR) exons that precede the common first coding exon.[17,19] At least seven different 5' UTR exons, designated A-G, have been described.[17] Some transcripts from the brain contain up to three different 5' UTR exons that are spliced together before the common first coding exon.[17] This regulatory complexity may reflect differential transcription from the numerous cell types and lineages expressing the different α-dystrobrevin isoforms. This may be particularly important in the brain where α-dystrobrevin isoforms are found in neurons, glia and vascular endothelial cells. Each site of expression could be determined by specific promoter and 5' UTR usage, which would not be apparent from the analysis of transcripts identified in whole brain preparations.[20]

Further complexity is added to the α-dystrobrevin transcriptome by extensive alternative splicing within the coding region. Three alternatively spliced sites, variable regions (vr) 1, 2 and 3 have been reported.[9,19-21] Alternative splicing at vr3 occurs only in skeletal and cardiac muscle.[15,17,21] The insertion of two addition exons encoding 57 amino acids only occurs in α-dystrobevins-1 and -2 transcripts. The function of this alternatively spliced site is to provide an additional binding site for the syntrophin family proteins thereby modulating the stoichiometry of the DGC in skeletal and cardiac muscle.[21] The function of the alterative spliced exons at vr1 and vr2 is unknown.

The gene encoding the second member of the dystrobrevin family, β-dystrobrevin (*dtnb* in the mouse), is found on human chromosome 2p22-23 (mouse chromosome 12).[10,11,22] β-dystrobrevin is expressed in several nonmuscle tissues including liver, kidney, and brain but never appears to be found in the same cells type as the α-dystrobrevins suggesting that these proteins play complementary, cell-specific roles.[10] Only one major isoform is encoded by the β-dystrobrevin gene although the C-terminus of the protein and regions homologous to α-dystrobrevin's vr2 and vr3 are alternatively spliced.[10,22] Moreover, there appears to be only two alternative 5' UTR exons that are used differentially.[22] Both α- and β-dystrobrevin are present in DGC-like complexes found in different tissues and cell types.[20,23] The role of each protein and its association with the different complexes is considered below.

Dystrophin and several other components of the DGC including dystrobrevin are found in model organisms such as the nematode *Caenorhabditis elegans* and the fruit fly *Drosophila melanogaster*. *C. elegans* dystrobrevin (*dyb-1*) mutants phenotypically resemble dystrophin (*dys-1*) mutants in that they are enigmatically hyperactive and hypercontractile.[24] Muscle degeneration occurs when either the dystrobrevin or the dystrophin mutants are crossed with a mild *MyoD/hlh-1* allele (a gene required for muscle regeneration).[24,25] Both mutants are sensitive to acetylcholine and acetylcholinesterase inhibitors such as aldicarb indicative of a defect in cholinergic neurotransmission.[25] Over-expression of dystrobrevin in *dys-1* mutants delays the onset of muscle degeneration[26] suggesting that both proteins participate in a similar pathway.[26] This hypothesis is supported by the finding that dystrobrevin must bind directly to dystrophin in order to rescue the *dyb-1* mutant phenotype.[27] Interestingly, *C. elegans* syntrophin mutants (*stn-1*) are phenotypically similar to the *dyb-1* and *dys-1* mutants in that they are hypercontracted and hypersensitive to aldicarb.[28] By contrast to the situation in mammals, *C. elegans* syntrophin cannot bind directly to dystrophin but is anchored to the DGC by dystrobrevin, which in turn binds directly to dystrophin.[28] Unsurprisingly, *C. elegans* dystrophin lacks the consensus syntrophin binding-site defined in mammals by Newey and colleagues.[21] *C. elegans* dystrobrevin also lacks the conserved C-terminal tyrosines that are found in mammalian α-dystrobrevin-1 and is most similar to mouse β-dystrobrevin. Each of the *C. elegans* DGC-associated proteins are expressed in neurons as well as in the musculature a feature that is conserved in mammals. These findings make it tempting to speculate that the ancestral role of dystrobrevin may be as an adaptor protein linking syntrophin and its associated proteins to the DGC.

The Functions of α-Dystrobrevin in Muscle

In muscle, each isoform of α-dystrobrevin is found at the sarcolemma, NMJ and myotendinous junction (MTJ).[15,18,29] In the case of α-dystrobrevin-3 this localization has been inferred since there are no specific antibodies for this isoform.[18] Like other members of the DGC, the α-dystrobrevins are severely reduced in DMD muscle and in the dystrophin-deficient *mdx* mouse model of muscular dystrophy that is caused by a point mutation in the gene encoding murine dystrophin.[30,31] Interestingly however, the α-dystrobrevins, in common with several other components of the DGC, persist at the NMJ indicating that the mechanism(s) for anchoring these complexes at the postsynaptic sarcolemma differ from those at the extra-junctional sarcolemma.[15,31]

α-dystrobrevin-1 and -2 bind directly to dystrophin, its isoforms (Dp71, Dp140 etc) and utrophin through reciprocal coiled-coil-dependent interactions involving homologous regions in the C-termini of both proteins.[14,32] Although α-dystrobrevin-3 lacks the C-terminal coiled-coil region and syntrophin-binding sites it still appears to be a DGC-associated protein and is severely reduced in extracts of total cellular protein from the *mdx* mouse.[18] These data suggest that the α-dystrobrevins may be anchored to several components of the DGC or to as yet unidentified proteins. Yoshida and colleagues have shown that the α-dystrobrevins interact with the sarcoglycan:sarcospan complex and that α-dystrobrevin-3 is specifically depleted from the sarcolemma in mice lacking β-sarcoglycan.[33] This model fits nicely with the experimental findings that dystrophin and other components of the cytoplasmic DGC are retained at the sarcolemma of β-sarcoglycan-deficient mice probably through a ternary complex of dystrophin, the syntrophins and α-dystrobrevins -1 and -2. Since α-dystrobrevin-3 lacks the binding sites that mediate these associations it is selectively depleted when β-sarcoglycan is lost.

α-dystrobrevins -1 and -2 (and β-dystrobrevin, see below) are syntrophin-binding proteins. The syntrophin family of PDZ-domain-containing proteins is composed of five members, α1-syntrophin, β1 and β2-syntrophin and γ1 and γ2 syntrophin (see ref. 3 for review and key references). The syntrophins bind to a variety of different molecules in muscle and brain, including nNOS and some voltage-gated sodium channels, and can be considered as adaptors linking signaling proteins or transmembrane receptors to the DGC.[3,34,35] There are two syntrophin-binding sites in the muscle-specific splice variants (vr3) of α-dystrobrevins -1 and -2 and a further two sites on dystrophin each formed by two homologous adjacent α-helices.[21]

Several studies have attempted to gain insights into the function of α-dystrobrevin using genetic and biochemical techniques aimed at identifying novel dystrobrevin-binding proteins that are not members of the core DGC. Using the yeast-two hybrid system, α-dystrobrevin has been shown to bind to a novel intermediate filament (IF)-like protein named syncoilin[36] and to desmuslin the orthologue of the chick IF protein synemin.[37,38] Syncoilin is most similar to the family of type IV IF proteins such as the neurofilament triplet and α-internexin but is expressed predominantly in cardiac and skeletal muscle.[36] In normal skeletal muscle, syncoilin is localized to the NMJ and costameres whereas desmuslin is found at the Z-line.[37] Since syncoilin and desmuslin both bind to α-dystrobrevin, it is reasonable to hypothesize that these interactions form a physical link between the DGC and the IF network in muscle. This idea is supported by the α paper published by Poon and colleagues who showed that in addition to α-dystrobrevin, syncoilin binds to desmin at costameres and the NMJ.[39] However, by contrast to desmin, syncoilin cannot form homomeric filaments in vitro or heteromeric filaments with desmin when coexpressed in the same cells.[39] Thus, syncoilin may regulate IF assembly in cardiac and skeletal muscle.

A major clue to the function of α-dystrobrevin in muscle arose from the production of α-dystrobrevin-deficient mice (*dtna-/-*) created by homologous recombination.[40] These mice which lack all of the major α-dystrobrevin isoforms, have revealed a dual role for α-dystrobrevin in the pathogenesis of muscular dystrophy and in AChR-cluster stabilization at the NMJ (see below).[40,41] α-dystrobrevin-deficient mice develop a form of mild muscular dystrophy without apparently perturbing the assembly of the DGC at the sarcolemma; a feature evident in mice lacking dystrophin or any one of the sarcoglycans (see ref. 42 for example). These mice develop a milder form of muscular dystrophy compared to the *mdx* mouse and have relatively low levels of sarcolemmal damage that is a cardinal feature of the pathology in the *mdx* mouse.[40] The only immunocytochemical abnormality in these mice is the reduction in the levels of neuronal nitric oxide synthase (nNOS) at the sarcolemma that is thought to impair nNOS-dependent signally by altering the levels of cGMP in muscle[40] and a reduction in the levels of the syntrophins and nNOS at the NMJ.[41] This important study provided the first evidence that the DGC may be involved in dynamic signaling processes and is not merely a membrane scaffolding complex.[43] Whilst several groups have now postulated roles for the DGC in intracellular signal transduction[44,45] the role of α-dystrobrevin in this process has remained elusive.[31]

The specific roles of α-dystrobrevin-1 and -2 in muscle have been dissected by Grady et al, who used transgenes expressing different dystrobrevin isoforms to rescue the myopathic phenotype of *adbn-/-* mice.[46] This study found that both α-dystrobrevins-1 and -2 could rescue the myopathic changes in dystrobrevin-deficient muscles including the diaphragm and quadriceps.[46] *Adbn-/-* mice also have abnormal MTJs that appear to have fewer and shallower invaginations when viewed under the electron microscope.[46] Transgenic expression of either α-dystrobrevins-1 or -2 corrected this defect although α-dystrobrevin-1 had greater efficacy in this assay.[46]

Although *adbn-/-* mice have a mild myopathy, mutations in the human α-dystrobrevin gene have not been conclusively linked to any disease. A point mutation in *DTNA* has been reported in a family with left ventricular noncompaction (LVNC) and congenital heart disease (CHD).[47] This mutation changes the amino acid proline at position 121 to leucine. Whilst cardiomyopathic features have been described in the *dtna-/-* mouse,[40] it is unlikely that α-dystrobrevin mutations will contribute greatly to the aetiology of LVNC and CHD. Kenton and colleagues were unable to find any dystrobrevins mutations in a cohort of 48 patients with LVNC.[48] Similarly, a study on dystrobrevin immunostaining in 172 patients with idiopathic myopathies failed to find anomalies in sarcolemmal α-dystrobrevin immuoreactivity that could suggest its involvement in muscle disease.[49]

α-Dystrobrevin and NMJ Formation

Several components of the DGC are involved in the formation and stabilization of NMJs. For example, α-dystroglycan is a receptor for agrin and laminin-1 in the synaptic basal lamina and, together with the receptor tyrosine kinase MuSK (muscle-specific kinase) can recruit AChRs into clusters during synapse formation in muscle.[50] As mentioned previously, α-dystrobrevin is one of the major phosphoproteins in the *Torpedo* electric organ and is phosphorylated in mammalian muscle.[6,15,16] The individual α-dystrobrevin isoforms are located at different parts of the NMJ. α-dystrobrevin-2 is concentrated at the depths of the junctional folds and preferentially copurifies with dystrophin whereas α-dystrobrevin-1 is found at the crests of the junctional folds (alongside the AChRs) and preferentially copurifies with utrophin.[29]

α-dystrobrevin-deficient mice have abnormal NMJs characterized by reduction in the extent and numbers of junctional folds and rearrangement of AChRs and their associated proteins.[41] *Adbn-/-* myotubes respond to agrin but produce reduced numbers of AChR clusters compared to normal myotubes that are unstable in the absence of agrin.[41] Furthermore, α-dystrobrevin-deficient myotubes do not respond to the plant lectin VVA-B4 that clusters AChRs in a similar manner to agrin.[41] These data show that α-dystrobrevin is required for the postnatal maintenance of junctional stability.

Further evidence supporting a role for α-dystrobrevin at the NMJ comes from the analysis of agrin-induced clustering of post-synaptic proteins in the absence of AChRs. In C2 myotube variants that lack AChRs or where the receptor has been down regulated with antibodies, agrin is still able to cluster MuSK, α-dystrobrevin-1 and utrophin whereas dystroglycan and syntrophin are not clustered.[51] These data suggest that α-dystrobrevin-1 may be part of a scaffolding or signaling complex involved in the assembly of the postsynaptic membrane during NMJ formation.

Transgenic expression of α-dystrobrevin-1 (and to a lesser extent α-dystrobrevin-2) is able to rescue the synaptic defects in *adbn-/-* muscle.[46] Importantly, a transgene expressing α-dystrobrevin-1 carrying mutations in the three sites for tyrosine phosphorylation in the C-terminal tail of this isoform is unable to completely rescue the synaptic defects in *adbn-/-* mice.[46] These studies demonstrate the importance of α-dystrobrevin-1 as a tyrosine kinase substrate and also explain why α-dystrobrevin-2 is unable to completely rescue the synaptic stabilization defects in *adbn-/-* mice.

Although MuSK is a tyrosine kinase that is critically required for synapse formation, it is unable to phosphorylate α-dystrobrevin-1 in agrin-treated myotubes.[15] The tyrosine kinases *src* and *fyn* are required for maintenance of AChR clusters after laminin-withdrawal[52] and stabilize agrin-induced AChR clusters.[53,54] *Src* and *fyn* may stabilize AChR clusters through their adaptor activities mediating protein interactions using their SH2 or SH3 domains[54] or by phosphorylating other proteins such as α-dystrobrevin-1. Agrin-induced AChR clusters are unstable in the absence of α-dystrobrevin[41] raising the possibility that *src* and *fyn* stabilize these clusters by linking the receptor-based post-synaptic cytoskeleton to components of the DGC, specifically α-dystrobrevin-1.

Finally, several components of the NMJ, including the α-dystrobrevins and utrophin are preferentially transcribed from sub-synaptic nuclei in muscle.[31] Newey et al, showed that the transcripts encoding α-dystrobrevin-1 and -2 accumulated at the NMJ whereas the α-dystrobrevin-3 mRNA was found throughout the muscle fibre.[18] Given that all α-dystrobrevin isoforms originate from the same promoter in muscle, these data are consistent with a post-transcriptional mechanism for modulating synaptic gene expression probably through RNA transport to post-synaptic sites.[18]

The Dystrobrevin Protein Family in the Brain

Mild, nonprogressive cognitive impairment is evident in about one third of patients with DMD suggesting that dystrophin and the DGC have an important function in the brain.[55,56]

Whilst the sarcoglycan:sarcospan complex is predominantly found in skeletal, cardiac and smooth muscle, the widespread distribution of both α- and β-dystrobrevin in neurons and glia suggest a variety of functions in the brain.[20,55] Several DGC-like complexes exist in neurons and glia that may be involved in the neuropathology of DMD and related disorders.[20] α-dystrobrevin-1 and several components of the DGC, including α-dystroglycan, Dp71, the syntrophins and laminin are associated with the cerebral microvasculature and are enriched at the glial:vascular interface particularly in perivascular astrocytes and vascular endothelial cells.[20,57,58] The DGC-like complexes in perivascular astrocytes could potentially form a membrane-spanning link between the ECM and the cytoskeleton of glial processes similar to that found in muscle. This complex could contribute to the maintenance of the blood:brain barrier.

A complex of β-dystrobrevin, dystrophin and syntrophin is enriched in the postsynaptic density (PSD) fraction of forebrain.[20,59] β-dystrobrevin is also found in the dendrites, somata and axons of various neuronal cell types but not in glia.[20] Furthermore, in the chick retina β-dystrobrevin, dystrophin and Dp260 are found exclusively in the outer plexiform layer; the site of synaptic contacts between the photoreceptors and bipolar neurons and horizontal cells.[60] The synaptic location of β-dystrobrevin and its association with dystrophin and syntrophin defines a neuronal DGC-like complex that is almost certainly involved in the neuropathology of DMD. Another important component of this is likely to be nNOS. Reciprocal PDZ domains mediate the interaction between α-syntrophin and nNOS.[34,61-63] nNOS binds to the PSD proteins PSD-93 and PSD-95, which are both involved in clustering N-methyl D-aspartate (NMDA) receptors.[64] This interaction places nNOS in proximity to the NMDA receptor at the postsynaptic membrane of excitatory synapses.[65,66] Thus, nNOS and syntrophin are shared components of two macromolecular complexes located at postsynaptic sites in neurons. Whilst the NMDA-receptor complex and the dystrophin:β-dystrobrevin-based neuronal DGC may be distinct entities it is possible that through their shared components each complex could interact. These interactions, coupled with the role of nitric oxide in the Ca^{2+}-dependent processes of excitotoxicity, could render dystrophin-deficient neurons more susceptible to metabolic or physiological stress[67] due to reduced regulation of ionic fluxes. This hypothesis is supported by the findings that dystrophin-deficient neurons have an increased susceptibility to hypoxia-induced loss of synaptic transmission[68] and have increased intracellular Ca^{2+}.[69]

Whilst muscle is a contractile cell type, neurons are not subject to the same physical forces, consequently, there is little or no evidence for neuronal loss or damage in DMD. Although dystrophin and the DGC are involved in the maintenance of sarcolemmal integrity during exercise, it is unlikely that this is the function of the DGC in neurons. Moreover, the DGC in neurons has a different composition to the muscle DGC lacking the sarcoglycan sub-complex. This raises the following questions; what is the role of the DGC in the brain and how is this function compromised in DMD patients with cognitive impairment? We hypothesized that the dystrobrevins may play a role in this process for the following reasons: First, *dtna-/-* mice have muscle disease without membrane damage and DGC perturbation. Second, DGC-like complexes in brain and muscle differ in the dystrobrevin isoforms that they contain. Third, the dystrobrevins are thought to be involved in intracellular signaling that could be important in transducing signals from the synapse to the neuronal cytoskeleton.

To search for genes associated with cognitive impairment in DMD and further define the role of the dystrobrevins in the brain, a yeast-two hybrid screen was performed using the entire β-dystrobrevin protein as a bait.[70] This screen identified dysbindin, a novel, widely expressed coil-coiled-containing protein that interacts with both α- and β-dystrobrevin.[70] In common with syncoilin, dysbindin is up-regulated in *mdx* muscle[70] but is also up-regulated in subsets of neuronal synapses in the cerebella of *mdx* mice.[71] Dysbindin is now known to be involved in the biogenesis of lysosome-related organelles and is mutated in Hermansky-Pudlak syndrome type 7 (HPS7) and the *sandy* mouse, an animal model of the disease in humans.[72] HPS patients have a hypo-pigmentation and a bleeding disorder caused by defects in the formation of secretory, lysosome-related organelles such as melanocytes and platelet dense granules. Several protein complexes contribute to the formation of lysosome-related organelles. Dysbindin is part of BLOC-1 (biogenesis of lysosome-related organelles complex-1) that contains at least six other proteins including pallidin and the SNAP25-associated protein, snapin.[73] Although very little is know about dysbindin in the brain, it is possible that BLOC-1 could play a role in protein trafficking, delivering cargos such as the dystrobrevins to different parts of the neuron.

Variation in the gene encoding dysbindin in humans (DTNBP1, dystrobrevin-binding protein-1) has been associated with susceptibility to schizophrenia (see ref. 74 for recent review). Several groups have found highly significant linkage between single nucleotide polymorphisms in DTNBP1 and schizophrenia in different populations.[74] Moreover, reduction in the levels of the dysbindin transcript and protein have been found in post-mortem tissue from the prefrontal cortex and hippocampal formation of patients with schizophrenia.[75,76] Perhaps the most relevant feature of DTNBP1 association with schizophrenia to DMD is the preliminary observation that dysbindin variation may be associated with cognitive dysfunction that is common in patients with schizophrenia (and DMD). Williams et al, have found that at-risk haplotypes for schizophrenia correlated with lower educational achievement and low IQ.[77]

In addition to dysbindin, β-dystrobrevin has been shown to interact with KIF5A, a neuronal member of the kinesin superfamily of proteins (KIFs) that consists of the heavy chains of conventional kinesin.[78] The authors suggest a novel function for β-dystrobrevin as a motor protein receptor that might play a major role in the transport of components of the DGC complex to specific sites in the neuron.[78] In a separate study, Albrecht and Froehner found an interesting interaction between α-dystrobrevin and a novel protein called DAMAGE.[79] DAMAGE is found in neurons in the cerebellum and hippocampus and binds to a region encompassing the ZZ-domain and extended N-terminus of α-dystrobrevin.[79] DAMAGE is a MAGE (melanoma antigen-encoding gene) protein with similarity to another protein called NRAGE which has previously been shown to mediate neuronal apoptosis.[80] NRAGE directly interacts with the cytoplasmic tail of the p75 neurotrophin receptor causing caspase activation and cell death through a JNK-dependent mitochondrial apoptotic pathway.[81] Any one of these new dystrobrevin-binding partners could be part of the neuronal DGC and would therefore by implicated in the neuropathology of DMD.

A Role for the Dystrobrevins in Other Tissues

The second member of the dystrobrevin family, β-dystrobrevin is expressed in most nonmuscle tissues where it is present in a bewildering variety of DGC-like complexes. The kidney contains a multitude of different DGC-like complexes that are located in the different cell types contained within the organ.[82] For example, the basal membrane of the renal tubule contains a complex of β-dystrobrevin, utrophin and β2-syntrophin whereas β-dystrobrevin is replaced by α-dystrobrevin-1 in the glomerulus.[82] These complexes differ from the DGC-like complexes that exist in the collecting ducts which contain Dp71δC (an alternatively spliced form of Dp71 with a different C-terminus) and α1-syntrophin as well as β2-syntrophin and β-dystrobrevin.[82] Similar complexes are found in the liver where β-dystrobrevin is localized to the sinusoids and/or at the sinusoidal face of hepatocytes.[83] Whilst these interactions demonstrate the importance of DGC-like complexes to specialized epithelial cells, it is not know why so many different complexes are produced by the different cell types. One possible clue to the function of these complexes has emerged from the finding that β2-syntrophin, dystrobrevin and utrophin have a polarized distribution in MDCK (Madin-Darby canine kidney) cells.[23] The DGC-like complex in these cells is restricted to the basolateral membrane allowing the PDZ-domain of β2-syntrophin to recruit additional proteins into polarized membranes.[23] Thus, the dystrobrevin-containing DGC-like complexes in different epithelial cells may form the sites for the assembly sub-membranous specializations (cf. α-dystrobrevin-1 at the NMJ).

Mice lacking β-dystrobrevin (*dtnb-/-*) are phenotypically normal but do show alterations in the assembly of the different DGC-like complexes in kidney and liver.[83] No abnormality was detected in the ultrastructure of the kidney and liver or in the renal function of these mice, suggesting a primarily anchoring or scaffolding role for β-dystrobrevin in these tissues.[83] The levels of Dp71δC, the syntrophins and utrophin are all severely reduced in cortical renal tubules of *dtnb-/-* mice.[83] However, α-dystrobrevin-1 levels are elevated in *dtnb-/-* kidney suggesting a possible compensatory mechanism.[83] It is not known whether *dtnb-/-* mice have subtle CNS defects nor has the phenotype of the *dtna-/-:dtnb-/-* double mutant been described. This mutant could potentially provide important clues to the function of the two proteins. It is important to remember that mice lacking utrophin have mild NMJ abnormalities whereas the double mutant lacking both dystrophin (*mdx*) and utrophin has a severe skeletal and cardiac myopathies causing the mouse to die prematurely.[84,85]

Concluding Remarks

The dystrobrevins are conserved, integral components of the muscle DGC and DGC-like protein complexes found in the majority of cell types that have been postulated to be involved in signalling. It is now becoming increasingly accepted that the complex pathology of DMD is in part due to alterations in intracellular signaling which could be mediated by the dystrobrevins and their associated proteins. Whilst there is a paucity of evidence demonstrating a conclusive role for the dystrobrevins in cell signalling there is a distinct possibility that this protein family mediates these events. In addition to the efforts aimed at the replacement of dystrophin and up-regulation of so-called "booster genes"[86] in DMD, targeting the interaction and regulation of dystrobrevin with cellular signalling pathways could potentially yield novel therapeutic strategies for the treatment of DMD and BMD and may shed light upon the role of the DGC in the brain.[31]

Acknowledgements

DJB is supported by grants from the Wellcome Trust and is a Wellcome Trust Senior Fellow. RVS was a recipient of a Fellowship from the Alberta Heritage Foundation for Medical Research.

References

1. Campbell KP, Kahl SD. Association of dystrophin and an integral membrane glycoprotein. Nature 1989; 338(6212):259-262.
2. Yoshida M, Ozawa E. Glycoprotein complex anchoring dystrophin to sarcolemma. J Biochem (Tokyo) 1990; 108(5):748-752.
3. Blake DJ, Weir A, Newey SE et al. Function and genetics of dystrophin and dystrophin-related proteins in muscle. Physiol Rev 2002; 82(2):291-329.
4. Yoshida M, Suzuki A, Yamamoto H et al. Dissociation of the complex of dystrophin and its associated proteins into several unique groups by n-octyl beta-D-glucoside. Eur J Biochem 1994; 222(3):1055-1061.
5. Carr C, Fischbach GD, Cohen JB. A novel 87,000-Mr protein associated with acetylcholine receptors in Torpedo electric organ and vertebrate skeletal muscle. J Cell Biol 1989; 109(4 Pt 1):1753-1764.
6. Wagner KR, Cohen JB, Huganir RL. The 87K postsynaptic membrane protein from Torpedo is a protein- tyrosine kinase substrate homologous to dystrophin. Neuron 1993; 10(3):511-522.
7. Yoshida M, Yamamoto H, Noguchi S et al. Dystrophin-associated protein A0 is a homologue of the Torpedo 87K protein. FEBS Lett 1995; 367(3):311-314.
8. Blake DJ, Nawrotzki R, Peters MF et al. Isoform diversity of dystrobrevin, the murine 87-kDa postsynaptic protein. J Biol Chem 1996; 271(13):7802-7810.
9. Sadoulet-Puccio HM, Khurana TS, Cohen JB et al. Cloning and characterization of the human homologue of a dystrophin related phosphoprotein found at the Torpedo electric organ post-synaptic membrane. Hum Mol Genet 1996; 5(4):489-496.
10. Blake DJ, Nawrotzki R, Loh NY et al. beta-dystrobrevin, a member of the dystrophin-related protein family. Proc Natl Acad Sci USA 1998; 95(1):241-246.
11. Peters MF, O'Brien KF, Sadoulet-Puccio HM et al. beta-dystrobrevin, a new member of the dystrophin family. Identification, cloning, and protein associations. J Biol Chem 1997; 272(50):31561-31569.
12. Ponting CP, Blake DJ, Davies KE et al. ZZ and TAZ: New putative zinc fingers in dystrophin and other proteins. Trends Biochem Sci 1996; 21(1):11-13.
13. Ishikawa-Sakurai M, Yoshida M, Imamura M et al. ZZ domain is essentially required for the physiological binding of dystrophin and utrophin to beta-dystroglycan. Hum Mol Genet 2004; 13(7):693-702.
14. Blake DJ, Tinsley JM, Davies KE et al. Coiled-coil regions in the carboxy-terminal domains of dystrophin and related proteins: potentials for protein-protein interactions. Trends Biochem Sci 1995; 20(4):133-135.
15. Nawrotzki R, Loh NY, Ruegg MA et al. Characterisation of alpha-dystrobrevin in muscle. J Cell Sci 1998; 111(Pt 17):2595-2605.
16. Balasubramanian S, Fung ET, Huganir RL. Characterization of the tyrosine phosphorylation and distribution of dystrobrevin isoforms. FEBS Lett 1998; 432(3):133-140.

17. Holzfeind PJ, Ambrose HJ, Newey SE et al. Tissue-selective expression of alpha-dystrobrevin is determined by multiple promoters. J Biol Chem 1999; 274(10):6250-6258.
18. Newey SE, Gramolini AO, Wu J et al. A novel mechanism for modulating synaptic gene expression: Differential localization of alpha-dystrobrevin transcripts in skeletal muscle. Mol Cell Neurosci 2001; 17(1):127-140.
19. Ambrose HJ, Blake DJ, Nawrotzki RA et al. Genomic organization of the mouse dystrobrevin gene: Comparative analysis with the dystrophin gene. Genomics 1997; 39(3):359-369.
20. Blake DJ, Hawkes R, Benson MA et al. Different dystrophin-like complexes are expressed in neurons and glia. J Cell Biol 1999; 147(3):645-658.
21. Newey SE, Benson MA, Ponting CP et al. Alternative splicing of dystrobrevin regulates the stoichiometry of syntrophin binding to the dystrophin protein complex. Curr Biol 2000; 10(20):1295-1298.
22. Loh NY, Ambrose HJ, Guay-Woodford LM et al. Genomic organization and refined mapping of the mouse beta-dystrobrevin gene. Mamm Genome 1998; 9(11):857-862.
23. Kachinsky AM, Froehner SC, Milgram SL. A PDZ-containing scaffold related to the dystrophin complex at the basolateral membrane of epithelial cells. J Cell Biol 1999; 145(2):391-402.
24. Gieseler K, Bessou C, Segalat L. Dystrobrevin- and dystrophin-like mutants display similar phenotypes in the nematode Caenorhabditis elegans. Neurogenetics 1999; 2(2):87-90.
25. Gieseler K, Mariol MC, Bessou C et al. Molecular, genetic and physiological characterisation of dystrobrevin-like (dyb-1) mutants of Caenorhabditis elegans. J Mol Biol 2001; 307(1):107-117.
26. Gieseler K, Grisoni K, Mariol MC et al. Overexpression of dystrobrevin delays locomotion defects and muscle degeneration in a dystrophin-deficient Caenorhabditis elegans. Neuromuscul Disord 2002; 12(4):371-377.
27. Grisoni K, Gieseler K, Segalat L. Dystrobrevin requires a dystrophin-binding domain to function in Caenorhabditis elegans. Eur J Biochem 2002; 269(6):1607-1612.
28. Grisoni K, Gieseler K, Mariol MC et al. The stn-1 syntrophin gene of C.elegans is functionally related to dystrophin and dystrobrevin. J Mol Biol 2003; 332(5):1037-1046.
29. Peters MF, Sadoulet-Puccio HM, Grady MR et al. Differential membrane localization and intermolecular associations of alpha-dystrobrevin isoforms in skeletal muscle. J Cell Biol 1998; 142(5):1269-1278.
30. Metzinger L, Blake DJ, Squier MV et al. Dystrobrevin deficiency at the sarcolemma of patients with muscular dystrophy. Hum Mol Genet 1997; 6(7):1185-1191.
31. Blake DJ. Dystrobrevin dynamics in muscle-cell signalling: a possible target for therapeutic intervention in Duchenne muscular dystrophy? Neuromuscul Disord 2002; 12(Suppl 1):S110-117.
32. Sadoulet-Puccio HM, Rajala M, Kunkel LM. Dystrobrevin and dystrophin: An interaction through coiled-coil motifs. Proc Natl Acad Sci USA 1997; 94(23):12413-12418.
33. Yoshida M, Hama H, Ishikawa-Sakurai M et al. Biochemical evidence for association of dystrobrevin with the sarcoglycan-sarcospan complex as a basis for understanding sarcoglycanopathy. Hum Mol Genet 2000; 9(7):1033-1040.
34. Brenman JE, Chao DS, Xia H et al. Nitric oxide synthase complexed with dystrophin and absent from skeletal muscle sarcolemma in Duchenne muscular dystrophy. Cell 1995; 82(5):743-752.
35. Gee SH, Madhavan R, Levinson SR et al. Interaction of muscle and brain sodium channels with multiple members of the syntrophin family of dystrophin-associated proteins. J Neurosci 1998; 18(1):128-137.
36. Newey SE, Howman EV, Ponting CP et al. Syncoilin, a novel member of the intermediate filament superfamily that interacts with alpha-dystrobrevin in skeletal muscle. J Biol Chem 2001; 276(9):6645-6655.
37. Mizuno Y, Thompson TG, Guyon JR et al. Desmuslin, an intermediate filament protein that interacts with alpha -dystrobrevin and desmin. Proc Natl Acad Sci USA 2001; 98(11):6156-6161.
38. Blake DJ, Martin-Rendon E. Intermediate filaments and the function of the dystrophin-protein complex. Trends Cardiovasc Med 2002; 12(5):224-228.
39. Poon E, Howman EV, Newey SE et al. Association of syncoilin and desmin: Linking intermediate filament proteins to the dystrophin-associated protein complex. J Biol Chem 2002; 277(5):3433-3439.
40. Grady RM, Grange RW, Lau KS et al. Role for alpha-dystrobrevin in the pathogenesis of dystrophin-dependent muscular dystrophies. Nat Cell Biol 1999; 1(4):215-220.
41. Grady RM, Zhou H, Cunningham JM et al. Maturation and maintenance of the neuromuscular synapse: Genetic evidence for roles of the dystrophin—glycoprotein complex. Neuron 2000; 25(2):279-293.
42. Hack AA, Ly CT, Jiang F et al. Gamma-sarcoglycan deficiency leads to muscle membrane defects and apoptosis independent of dystrophin. J Cell Biol 1998; 142(5):1279-1287.

43. Bredt DS. Knocking signalling out of the dystrophin complex. Nat Cell Biol 1999; 1(4):E89-91.
44. Langenbach KJ, Rando TA. Inhibition of dystroglycan binding to laminin disrupts the PI3K/AKT pathway and survival signaling in muscle cells. Muscle Nerve 2002; 26(5):644-653.
45. Spence HJ, Dhillon AS, James M et al. Dystroglycan, a scaffold for the ERK-MAP kinase cascade. EMBO Rep 2004; 5(5):484-489.
46. Grady RM, Akaaboune M, Cohen AL et al. Tyrosine-phosphorylated and nonphosphorylated isoforms of alpha-dystrobrevin: Roles in skeletal muscle and its neuromuscular and myotendinous junctions. J Cell Biol 2003; 160(5):741-752.
47. Ichida F, Tsubata S, Bowles KR et al. Novel gene mutations in patients with left ventricular noncompaction or Barth syndrome. Circulation 2001; 103(9):1256-1263.
48. Kenton AB, Sanchez X, Coveler KJ et al. Isolated left ventricular noncompaction is rarely caused by mutations in G4.5, alpha-dystrobrevin and FK Binding Protein-12. Mol Genet Metab 2004; 82(2):162-166.
49. Ishikawa H, Nonaka I, Nishino I. Negative result in search for human alpha-dystrobrevin deficiency. Muscle Nerve 2003; 28(3):387-388.
50. Sanes JR, Lichtman JW. Induction, assembly, maturation and maintenance of a postsynaptic apparatus. Nat Rev Neurosci 2001; 2(11):791-805.
51. Marangi PA, Forsayeth JR, Mittaud P et al. Acetylcholine receptors are required for agrin-induced clustering of postsynaptic proteins. Embo J 2001; 20(24):7060-7073.
52. Marangi PA, Wieland ST, Fuhrer C. Laminin-1 redistributes postsynaptic proteins and requires rapsyn, tyrosine phosphorylation, and Src and Fyn to stably cluster acetylcholine receptors. J Cell Biol 2002; 157(5):883-895.
53. Ferns M, Deiner M, Hall Z. Agrin-induced acetylcholine receptor clustering in mammalian muscle requires tyrosine phosphorylation. J Cell Biol 1996; 132(5):937-944.
54. Smith CL, Mittaud P, Prescott ED et al. Src, Fyn, and Yes are not required for neuromuscular synapse formation but are necessary for stabilization of agrin-induced clusters of acetylcholine receptors. J Neurosci 2001; 21(9):3151-3160.
55. Blake DJ, Kroger S. The neurobiology of duchenne muscular dystrophy: learning lessons from muscle? Trends Neurosci 2000; 23(3):92-99.
56. Mehler MF. Brain dystrophin, neurogenetics and mental retardation. Brain Res Brain Res Rev 2000; 32(1):277-307.
57. Ueda H, Baba T, Terada N et al. Immunolocalization of dystrobrevin in the astrocytic endfeet and endothelial cells in the rat cerebellum. Neurosci Lett 2000; 283(2):121-124.
58. Moukhles H, Carbonetto S. Dystroglycan contributes to the formation of multiple dystrophin-like complexes in brain. J Neurochem 2001; 78(4):824-834.
59. Kim TW, Wu K, Xu JL et al. Detection of dystrophin in the postsynaptic density of rat brain and deficiency in a mouse model of Duchenne muscular dystrophy. Proc Natl Acad Sci USA 1992; 89(23):11642-11644.
60. Blank M, Blake DJ, Kroger S. Molecular diversity of the dystrophin-like protein complex in the developing and adult avian retina. Neuroscience 2002; 111(2):259-273.
61. Brenman JE, Chao DS, Gee SH et al. Interaction of nitric oxide synthase with the postsynaptic density protein PSD-95 and alpha1-syntrophin mediated by PDZ domains. Cell 1996; 84(5):757-767.
62. Hashida-Okumura A, Okumura N, Iwamatsu A et al. Interaction of neuronal nitric-oxide synthase with alpha1-syntrophin in rat brain. J Biol Chem 1999; 274(17):11736-11741.
63. Hillier BJ, Christopherson KS, Prehoda KE et al. Unexpected modes of PDZ domain scaffolding revealed by structure of nNOS-syntrophin complex. Science 1999; 284(5415):812-815.
64. Brenman JE, Christopherson KS, Craven SE et al. Cloning and characterization of postsynaptic density 93, a nitric oxide synthase interacting protein. J Neurosci 1996; 16(23):7407-7415.
65. Kornau HC, Schenker LT, Kennedy MB et al. Domain interaction between NMDA receptor subunits and the postsynaptic density protein PSD-95. Science 1995; 269(5231):1737-1740.
66. Niethammer M, Kim E, Sheng M. Interaction between the C terminus of NMDA receptor subunits and multiple members of the PSD-95 family of membrane-associated guanylate kinases. J Neurosci 1996; 16(7):2157-2163.
67. Christopherson KS, Bredt DS. Nitric oxide in excitable tissues: Physiological roles and disease. J Clin Invest 1997; 100(10):2424-2429.
68. Mehler MF, Haas KZ, Kessler JA et al. Enhanced sensitivity of hippocampal pyramidal neurons from mdx mice to hypoxia-induced loss of synaptic transmission. Proc Natl Acad Sci USA 1992; 89(6):2461-2465.
69. Hopf FW, Steinhardt RA. Regulation of intracellular free calcium in normal and dystrophic mouse cerebellar neurons. Brain Res 1992; 578(1-2):49-54.

70. Benson MA, Newey SE, Martin-Rendon E et al. Dysbindin, a novel coiled-coil-containing protein that interacts with the dystrobrevins in muscle and brain. J Biol Chem 2001; 276(26):24232-24241.
71. Sillitoe RV, Benson MA, Blake DJ et al. Abnormal dysbindin expression in cerebellar mossy fiber synapses in the mdx mouse model of Duchenne muscular dystrophy. J Neurosci 2003; 23(16):6576-6585.
72. Li W, Zhang Q, Oiso N et al. Hermansky-Pudlak syndrome type 7 (HPS-7) results from mutant dysbindin, a member of the biogenesis of lysosome-related organelles complex 1 (BLOC-1). Nat Genet 2003; 35(1):84-89.
73. Starcevic M, Dell'Angelica EC. Identification of snapin and three novel proteins (BLOS1, BLOS2, and BLOS3/reduced pigmentation) as subunits of biogenesis of lysosome-related organelles complex-1 (BLOC-1). J Biol Chem 2004; 279(27):28393-28401.
74. Benson MA, Sillitoe RV, Blake DJ. Schizophreina genetics: Dysbindin under the microscope. Trends Neurosci 2004; in press.
75. Weickert CS, Straub RE, McClintock BW et al. Human dysbindin (DTNBP1) gene expression in normal brain and in schizophrenic prefrontal cortex and midbrain. Arch Gen Psychiatry 2004; 61(6):544-555.
76. Talbot K, Eidem WL, Tinsley CL et al. Dysbindin-1 is reduced in intrinsic, glutamatergic terminals of the hippocampal formation in schizophrenia. J Clin Invest 2004; 113(9):1353-1363.
77. Williams NM, Preece A, Morris DW et al. Identification in 2 independent samples of a novel schizophrenia risk haplotype of the dystrobrevin binding protein gene (DTNBP1). Arch Gen Psychiatry 2004; 61(4):336-344.
78. Macioce P, Gambara G, Bernassola M et al. Beta-dystrobrevin interacts directly with kinesin heavy chain in brain. J Cell Sci 2003; 116(Pt 23):4847-4856.
79. Albrecht DE, Froehner SC. DAMAGE, a novel alpha-dystrobrevin-associated MAGE protein in dystrophin complexes. J Biol Chem 2004; 279(8):7014-7023.
80. Salehi AH, Roux PP, Kubu CJ et al. NRAGE, a novel MAGE protein, interacts with the p75 neurotrophin receptor and facilitates nerve growth factor-dependent apoptosis. Neuron 2000; 27(2):279-288.
81. Salehi AH, Xanthoudakis S, Barker PA. NRAGE, a p75 neurotrophin receptor-interacting protein, induces caspase activation and cell death through a JNK-dependent mitochondrial pathway. J Biol Chem 2002; 277(50):48043-48050.
82. Loh NY, Newey SE, Davies KE et al. Assembly of multiple dystrobrevin-containing complexes in the kidney. J Cell Sci 2000; 113(Pt 15):2715-2724.
83. Loh NY, Nebenius-Oosthuizen D, Blake DJ et al. Role of beta-dystrobrevin in nonmuscle dystrophin-associated protein complex-like complexes in kidney and liver. Mol Cell Biol 2001; 21(21):7442-7448.
84. Deconinck AE, Rafael JA, Skinner JA et al. Utrophin-dystrophin-deficient mice as a model for Duchenne muscular dystrophy. Cell 1997; 90(4):717-727.
85. Grady RM, Teng H, Nichol MC et al. Skeletal and cardiac myopathies in mice lacking utrophin and dystrophin: a model for Duchenne muscular dystrophy. Cell 1997; 90(4):729-738.
86. Engvall E, Wewer UM. The new frontier in muscular dystrophy research: Booster genes. Faseb J 2003; 17(12):1579-1584.

CHAPTER 6

Commonalities and Differences in Muscular Dystrophies:
Mechanisms and Molecules Involved in Merosin-Deficient Congenital Muscular Dystrophy

Markus A. Ruegg

Abstract

Congenital muscular dystrophies are autosomal recessive diseases characterized by generalized hypotonia, delayed motor milestones and involvement of the brain. A large subgroup of this rather heterogeneous disease is due to mutations in one of the chains of the extracellular matrix molecules laminin-2 and laminin-4. The detailed knowledge of the binding partners of these laminins together with studies in mouse models have provided insights into the mechanisms that underlie this disease. Interestingly, similar mechanisms and molecules are likely at work not only in this but also in several, phenotypically distinct muscular dystrophies. These insights have led to the development of new strategies for the treatment of this and several other types of muscular dystrophies.

Introduction

Muscular dystrophies are characterized by a generalized and progressive loss of muscle mass. They are caused by mutations in many different genes but show a striking similarity in phenotypes. The wasting of muscle has a big impact on the quality of life because it often leads to the inability to walk. Moreover, severe forms lead to the death of the patient due to respiratory failure. All of these genetic diseases are rare and the similarity in phenotypes makes it difficult to diagnose the disease without molecular markers.

In the past 15 years we have seen tremendous progress in mapping the mutations to different genes. This development was initiated 1986 by determining the *dystrophin* gene as the primary cause for Duchenne muscular dystrophy (DMD).[1] A most recent highlight of gene mapping is the discovery that mutations in glycosylation enzymes are the cause for severe forms of congenital muscular dystrophy such as Muscle-eye-brain disease (MEB) or Walker-Warburg syndrome (WWS);[2] see also Table 1. Interestingly, most of the mutations causing muscular dystrophy appear to affect the function of molecules that belong to the dystrophin-glycoprotein complex (DGC).[3] For example, loss of dystrophin prevents the localization of the transmembrane components of the DGC to the sarcolemma and one of the main targets of the glycosylation defects in a subgroup of CMD is α-dystroglycan. Consequently, α-dystroglycan is not able to bind to its extracellular ligands such as laminin, agrin, perlecan and the neurexins.[4] Thus, alterations in the integrity of this protein complex result in the weakening and subsequent degeneration of muscle fibers.

Table 1. Classification of congenital muscular dystrophies

Muscular Dystrophies	Mode of Inheritance	Gene Location	Symbol (Gene Product)
Fukuyama CMD	AR	9q31-q33	**FCMD** (Fukutin)
Muscle-eye-brain disease	AR	1p3	**MEB** (POMGnT)
Walker-Warburg syndrome	AR	9q-34	**WWS** (POMT1)
MDC 1A	AR	6q2	**LAMA2** (Laminin-α2 chain)
MDC 1B	AR	1q42	**unknown**
MDC 1C	AR	19q13.3	**FKRP** (Fukutin-related protein 1)
MDC 1D	AR	22q-12	**LARGE**
CMD with rigid spine	AR	1p35-36	**SEPN1** (Selenoprotein N)
CMD with integrin deficiency	AR	12q13	**ITGA7** (Integrin α7)
Ullrich scleroatonic MD	AR	21q2	**UCMD** (α1-, α-2, α-3 chain
or		21q2	of collagen VI)
Bethlem-Myopathy		2q3	

All diseases of this class are autosomal recessive (AR). The majority of the cases is caused by mutations in the gene encoding the laminin-α2 chain (MDCIA). A recently emerging subgroup of congenital muscular dystrophies, which includes FMCD, MEB, WWS, FKRP, and LARGE, affects enzymes that glycosylate α-dystroglycan (see also Chapter 7).

Despite this tremendous progress in elucidating the mechanisms of disease, muscular dystrophies are still orphan diseases and currently no curative treatments are available. However, the fact that many different muscular dystrophies affect molecules that are associated in only a few complexes and that they converge in similar signaling pathways could lead to new concepts for the treatment of these diseases. In this review, I will discuss both the mechanistic and the therapeutic aspects of muscular dystrophies using the merosin-deficient congenital muscular dystrophy (MDC1A) as example.

Congenital Muscular Dystrophies

Congenital muscular dystrophy (CMD) is a heterogeneous group of muscle disorders that is often difficult to diagnose. The feature that makes it distinct from other muscular dystrophies, in particular to limb-girdle muscular dystrophies, is the early onset of symptoms at birth or within the first 6 months of life. CMDs are all autosomal recessive diseases. In addition to the involvement of muscle, some CMDs also show cerebral malformations and are often associated with mental retardation. The recent progress in deciphering the molecular origin of this disease implies dysregulation of α-dystroglycan as one of the key feature in CMD. Today, CMD is subgrouped into ten distinct diseases based on their genetic origin (Table 1). The prevalence of CMD is low (approx. 1-2.5 x 10^{-5}; refs. 5, 6). Merosin-deficient CMD, which was recently renamed MDC1A, is one of main subgroups (Table 1). Its prevalence varies greatly between countries. For example, approximately half of all the CMD patients in Europe suffer from MDC1A, while only ~6% of CMD patients belong to this subgroup in Japan. In Japan, the vast majority suffers from Fukuyama-type of CMD that originated from an ancestral founder mutation in fukutin, an enzyme that is likely to be involved in the glycosylation of α-dystroglycan[7] (see Chapter 7).

MDC1A is caused by mutations in the gene encoding the laminin α2 chain (previously called merosin) and shows a rather homogenous clinical picture with severe neonatal hypotonia associated with joint contracture and inability to stand or walk. This severe muscular

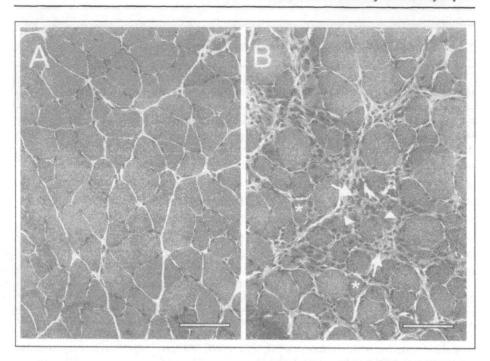

Figure 1. Haematoxylin and eosin staining of muscle cross-sections from wild-type (A) and a mouse model for MDC1A (B). Muscle fibers in (A) show the characteristic polygonal shape and the peripheral localization of their myonuclei. Muscle fibers from mice models of MDC1A (B) are often round-shaped, small (asterix) and their myonuclei are occasionally localized in the center of the fiber (arrow head). In addition, a many mononucleated cells (arrows) fill the spaces between muscle fibers. These infiltrating, non-muscle cells are often positive for markers indicative of inflammation. Bar = 100 μM.

dystrophy is accompanied by a peripheral neuropathy originating from demyelination in the peripheral[8,9] and central nervous system.[10] Creatine kinase levels in the serum of these patients are high and magnetic resonance imaging often shows abnormalities of white matter in the brain. However, no mental retardation is observed in most patients. On the cellular and structural level, muscle of MDC1A patients is characterized by marked variation in muscle fiber size, extensive fibrosis and proliferation of adipose tissue (see Fig. 1 as an example). Moreover, the basement membrane that surrounds the muscle fibers is disrupted.[11]

Laminins

The laminins are heterotrimeric molecules consisting of α, β and γ chains (Fig. 2). Today, five α (α1 - α5), three β (β1 -β3) and three γ (γ1 - γ3) chains, giving rise to at least fifteen different laminins (laminin-1 to laminin-15) have been identified in mammals (for review see refs.12, 13). Laminins are phylogenetically highly conserved and orthologues have been identified in *Drosophila* and *C. elegans*.[14] Laminins are essential components of basement membranes whose expression is tightly regulated during embryogenesis. They are expressed in many tissues and inactivation of individual chains often causes severe defects. For example, knockout of the γ1 chain, which is a component of 10 different laminin isoforms, causes embryonic lethality at the pre-implantation stage.[15] Inactivation of the laminin-α5 chain, which is the mammalian orthologue of one of the ancestral α chains found in *Drosophila* and *C. elegans*, causes embryonic death around embryonic day fourteen. These mice exert extensive ectopia of the brain, defects in vasculature and in the development of kidneys.[16] Like in human MDC1A patients, inactivation of laminin-α2 chain synthesis causes severe muscular dystrophy (see below).

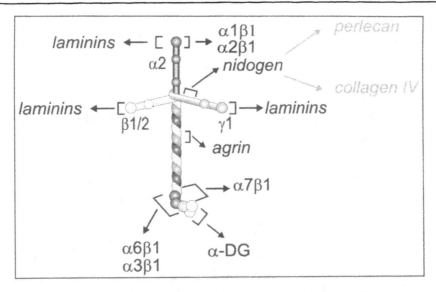

Figure 2. Structure and binding sites of the laminin-2/4 isoforms. Laminins are cruciform molecules that form from α, β and γ chains by a specific coiled-coil interaction. Muscle fibers express mainly laminin-2 (α2, β1, γ1) and laminin-4 (α2, β2, γ1). The interactions of laminin-2/4 with extracellular matrix components (names in italics) include the binding to other laminins, binding to nidogen via the γ1 chain, which in turn can also bind to collagen IV and perlecan, and binding to agrin via its coiled-coiled region. The major cellular receptors in muscle are the integrins (α1β1, α2β1, α6β1, α3β1, α7β1) and α-dystroglycan (αDG).

The central role of laminins in development can be explained by their dual function in (i) organizing a structured basement membrane and (ii) linking basement membranes to apposing cells via cell surface receptors (Fig. 2). The organizing role of laminins in the formation of basement membranes is thanks to their capability to self-polymerize with the amino-terminal regions of individual laminin chains (Fig. 2) and their binding to nidogen, which in turn forms a network with perlecan and type IV collagens.[17] In this way, a large network of proteins is created that serves as a scaffold and a reservoir for growth factors. In its strict sense, this model predicts that binding of nidogen to laminin is absolutely required for basement membranes to form. However, mice that carry a specific mutation of the nidogen-binding site in the laminin γ1 chain still form basement membranes at early stages.[18] It is only later that some basement membranes, such as the pial basement membrane, are disrupted resulting in ectopias at random sites.[19] The two major cell surface receptors of laminins are dystroglycan[20,21] and the integrins.[22] Mutations in laminin chains leave these cell receptors without a ligand and, therefore, affect the signaling and the transduction of mechanical forces. In the next section, I will discuss the significance of these receptors by considering the phenotype of mice that carry a targeted deletion.

Dystroglycan and Integrins

Alpha-dystroglycan is a peripheral membrane protein associated with the transmembrane component β-dystroglycan (Fig. 3A). Both proteins are derived from the common precursor protein dystroglycan and they are generated by post-translational cleavage. The constitutive inactivation of dystroglycan in mice causes embryonic lethality because of the disruption of Reichert's membrane, which is the first extra-embryonic basement membrane formed in rodents.[23] Embryonic death is overcome in mice in which dystroglycan synthesis is only disrupted in fully differentiated muscle fibers. Consistent with the notion that dystroglycan is important for the stability of muscle, dystroglycan-deficient muscle fibers degenerate.[24] In the

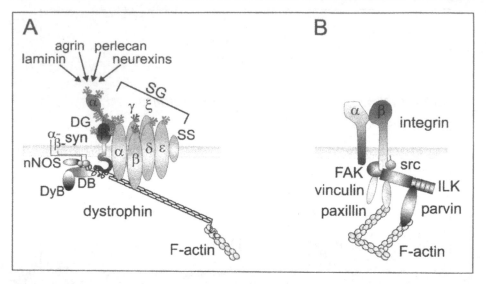

Figure 3. Receptor complexes active in mediating laminin function in muscle, CNS and PNS. A) The dystrophin-glycoprotein complex (DGC) includes the peripheral membrane protein α-dystroglycan (α) and the transmembrane protein β-dystroglycan (β), which originate from posttranslational cleavage of its precursor dystroglycan (DG). Binding of α-dystroglycan to laminin, agrin, perlecan and neurexins requires the carbohydrate moiety. In the membrane, dystroglycan associates with the sarcoglycans (SG), which are composed of α, β, δ, γ, ε, ζ subunits and sarcospan (SS). Intracellularly, dystrophin binds to β-dystroglycan and this association also involves putative signaling components and adaptors, including α1 and β1 synthrophin (α-, β-syn), neuronal nitric synthase (nNOS), dystrobrevin (DB) and dysbindin (DyB). Binding of dystrophin to the f-actin links the entire complex to the cytoskeleton. B) Integrin that bind to laminin are composed of α- and β-subunits. Some of the intracellular signaling molecules and adaptors that have been implicated in laminin-mediated functions are src-like kinases (src), focal adhesion kinase (FAK) and integrin-linked kinase (ILK). They are known to be involved in the linkage to the F-actin cytoskeleton via parvin, paxillin or vinculin.

membrane, dystroglycan forms a large complex with the sarcoglycans (α-, β-, γ, δ-, ε, and ζ-sarcoglycan) and with sarcospan (Fig. 3A). Mutations in the sarcoglycans result in muscular dystrophies of the limb-girdle type[25] (see Chapter 10). The only exception are mutations in ε-sarcoglycan, a close homologue of α-sarcoglycan that is expressed in muscle at low levels. Such mutations cause the myoclonus-dystonia syndrome, an autosomal-dominant disease characterized by brief, often alcohol-responsive myoclonic jerks.[26] Intracellularly, β-dystroglycan binds to dystrophin, which in turn binds to the syntrophins (α1, β1), α-dystrobrevin and neuronal nitric oxide synthase (nNOS). Mutations in dystrophin and in some of the other intracellular components cause muscular dystrophies. Thus, the evidence is strong that dystroglycan and its associated proteins are functionally involved in maintaining muscle tissue. Suggestive of additional roles of the dystrophin-glycoprotein complex outside of muscle is the recent finding that mutations in the α- and β-dystrobrevin-binding protein dysbindin cause Hermansky-Budlack syndrome type 7.[27] This disease is characterized by albinism, prolonged bleeding and pulmonary dysfunction.

Laminin-2 and laminin-4 can also bind to α1β1, α2β1, α3β1, α6β1, α6β4 and α7β1 integrins[12] (Fig. 3B). Muscle cells and their precursors express most of these integrins at some stage of development.[28] The expression of integrins during myogenesis is under tight control. The most prominent integrin expressed in adult muscle is the α7β1 form. Consistent with a function of integrins in mediating laminin-2 and laminin-4 function, α7 integrin knockout mice also display a mild muscular dystrophy that is most prominent at the myotendinous junction (MTJ).[29] Moreover, muscle-specific inactivation of $\beta1$ integrins has a major impact on the fusion of myoblasts, the organization of sarcomeres and the maintenance of MTJs.[30] This phenotype is not due to the lack of deposition of laminin-2 and laminin-4 in muscle basement membrane, as the laminin-$\alpha2$ chain is still deposited around $\beta1$ integrin-deficient muscle fibers.

Animal Models of MDC1A

Almost fifty years ago, the dystrophia muscularis (*dy/dy*) mouse was identified in the Jackson Laboratories.[31] This and another spontaneous mutant, called *dy2J/dy2J*,[32] are both hypomorphs for the laminin $\alpha2$ chain.[33] Another natural, although extinct mutant is the *dyPAS/dyPAS* mouse, in which the laminin-$\alpha2$ gene was inactivated by a retrotransposal insertion.[34] In addition, two mice models, called *dyw/dyw*[35] and *dy3K/dy3K*,[36] are based on homologous recombination. Moreover, laminin-$\alpha2$-deficient dogs and cats have been described.[37,38]

There is a constant debate whether mice are good models for human diseases. For example, the *mdx* mouse, which is the model for DMD, is much less affected than human patients.[39] Such difference to the MDC1A patients is not seen in the mice deficient of the laminin $\alpha2$ chain. Like in humans, the phenotype is much more severe than in *mdx* mice. For example, mice die early (i.e., 3-16 weeks) after birth, they grow at a much slower rate than wild-type littermates and all mice develop scoliosis. Three weeks after birth, muscle strength is significantly lower compared to wild-type mice, but, unlike *mdx* mice, no decrease of fatigue resistance is observed.[40] The histology of affected muscles is very similar to that of human patients and is characterized by great variation in fiber size, extensive fibrosis, infiltration of adipose tissue (Fig. 1), and high levels of creatine kinase in the blood. In addition, the hindlegs of laminin-$\alpha2$-deficient mice are paralyzed after a few weeks and abnormal myelination can be observed in the central nervous system. Thus, these mice are a good model for discovering the potential molecular mechanisms underlying the disease.

Staining for other laminin chains in the mouse models for MDC1A has revealed that expression of the $\gamma1$ chain is only slightly or not at all altered.[41,42] Thus, ablation of laminin-$\alpha2$ chain expression must result in a compensatory expression of another laminin-α chain. Indeed, the laminin-$\alpha4$ chain is strongly upregulated in mature dystrophic mice.[41,43] The laminin-$\alpha4$ chain is, however, not able to compensate for the loss of laminin-$\alpha2$. Besides being truncated at the amino-terminal end (see ref. 12 for review), the laminin-$\alpha4$ chain does not bind to α-dystroglycan with high affinity.[44] Muscle fiber membranes of MDC1A patients and mice models thereof contain also significantly lower levels of $\alpha7\beta1$ integrins[45-47] and α-dystroglycan.[42] Another laminin α chain that has been suggested to be upregulated in MDC1A is the laminin-$\alpha5$ chain (e.g., see refs. 48, 43), but only a very slight increase in laminin-$\alpha5$ was observed in four week-old mice.[42] The laminin-$\alpha5$ chain contains amino-terminal regions important for self-polymerization.[12] However, its binding affinity to α-dystroglycan and to integrins is much lower than that of the laminin-$\alpha2$ chain.[49,50] Studies in *dy2J/dy2J* mice indicate that a failure to form the laminin network is alone sufficient to trigger the disease. These mice synthesize a truncated form of laminin-$\alpha2$ that lacks the amino-terminal domain VI[51] but still exert a dystrophic phenotype in skeletal muscle and peripheral nerve.[52] Interestingly, their phenotype is considerably less severe than in mice that lack the laminin-$\alpha2$ chain entirely. In summary, both the formation of basal lamina and receptors-mediated responses are important in the maintenance of muscle. In MDC1A, impairment of both functions contributes to the severity of the disease.

Muscle Fiber Damage and Regeneration

There are several reasons that may account for the pronounced muscle wasting observed in MDC1A and other muscular dystrophies. First, lack of ECM-cytoskeletal linkage and/or intracellular signaling could result in the inability of muscle fibers to stand the mechanical forces imposed on them during contraction and lead to their degeneration. This mechanism is likely to contribute largely to the disease phenotype in the DMD and its *mdx* mouse model as evidenced by the fact that muscle fibers take up immunoglobulins and extracellularly applied dyes, such as Evans blue.[53,54] Interestingly, *dy/dy* and *dy2J/dy2J* mice show only little accumulation of Evans blue in their skeletal muscle suggesting that membrane leakage is not the main cause of the disease.[54] Secondly, the mutations in a particular gene may hinder successful regeneration of skeletal muscle after injury. New muscle fibers are formed from quiescent, mononucleated satellite cells localized between the basement membrane and the sarcolemma of muscle fibers.[55,56] Indicative of successful regeneration is the presence of centrally located myonuclei in newly formed muscle fibers. While the high number of centrally located myonuclei is a distinctive feature of *mdx* mice[57] and mice with a targeted disruption of dystroglycan in adult skeletal muscle fibers,[24] laminin α2-deficient mice display only few centrally located nuclei. Rather, the number of apoptotic fibers and the extent of fibrotic tissue is increased.[36,58] Together with the fact that laminin-α2 synthesis is upregulated during muscle regeneration in wild-type mice,[59] these data implicate laminin-α2 in the survival and the regeneration of muscle. This mechanistic difference to DMD may add to the severity of MDC1A.

Binding and Signaling through Laminin-2/4 Receptors

The DGC and the integrins could transduce the function of laminin-2 and laminin-4 to the inside of muscle fibers either by providing a structural link or by activating an intracellular signaling cascade. While the evidence is strong that structural changes in the DGC cause muscular dystrophy, only little direct evidence is available for a signaling role of the DGC. The co-localization with several putative signaling molecules, such as nNOS, α-dystrobrevin or Grb2, has long been used to argue for a signaling function of the DGC (see Fig. 3A). Only recent experiments have substantiated this hypothesis. For example, α-dystrobrevin-deficient mice show a mild muscular dystrophy although the structure of the DGC is not affected.[60] Because nNOS is not anymore localized at the sarcolemmal membrane and cGMP levels are considerably lower, this phenotype may reflect a defect in DGC-mediated signaling. However, neither α-synthrophin, which lacks nNOS at the sarcolemma[61,62] nor nNOS knockout mice[63,64] exert dystrophic symptoms. However, transgenic overexpression of nNOS in *mdx* mice ameliorates the dystrophy.[65] In cultured muscle cells, the DGC also affects the organization of the cytoskeleton. For example, binding of laminin to the DGC recruits the Rho family GTPase Rac1 and this in turn phosphorylates c-Jun via a particular c-Jun NH2-terminal kinase (JNK).[66] Although these results suggest a signaling function of the DGC, it is not known if dysfunction of any of these signaling pathways underlies the pathology in muscular dystrophies.

In contrast to the DGC, the signaling function of integrins is well established.[67] The binding of several adaptor and signaling molecules appears to mediate this function. Today, approximately 20 molecules are known that bind directly to the intracellular portion of integrins. Among those, focal adhesion kinase (FAK), integrin-linked-kinase (ILK) and Src-like-kinases (Fig. 3B) are particularly intriguing examples that mediate integrin signaling. Integrins are also different from most other signaling receptors because they can signal to the outside (inside-out signaling) that affects the affinity by which they bind to their ligands (see refs. 22,68).

Several lines of evidence indicate that integrins and dystroglycan share some structural function. This 'shared' function has become particularly obvious during the analysis of mice where either integrins or dystroglycan were knocked out in the same tissue. For example, brain-specific knockouts of β1 integrin and dystroglycan are both characterized by the "smoothening" of the brain surface, retraction of glial endfeet and a particularly striking disruption or weakening of pial basement membranes that results in localized ectopias.[69,70] Interestingly, similar phenotypes have been described for mice devoid of the extracellular matrix components laminin-α5 and perlecan, and upon deletion of the nidogen-binding site in the laminin-γ1 chain.[16,19,71] All these phenotypes are reminiscent of the congenital muscular dystrophies including FCMD, MEB and WWS (see Table 1 and Chapter 7). Thus, integrins and dystroglycan are responsible for the clustering of their extracellular ligands and the subsequent organizing of basal lamina. How integrins and dystroglycan exert this organizing role for basement membranes is not clearly understood (see ref. 13). The finding that brain-specific inactivation of FAK (see Fig.

3B) also results in aberrant basement membranes and localized ectopias,[72] strongly indicates that integrin signaling also affects the organization of basal lamina.

Although MDC1A is mainly a myopathic disease, neuropathic symptoms remain after muscle-selective expression of the laminin-α2 chain in *dyw/dyw* mice.[73] The peripheral nervous system of *dy/dy* and *dy3K/dy3K* mice shows an impairment of axonal sorting in the proximal nerve,[74] myelination abnormalities in the paranodes of distal portions of the nerves and a patchy appearance of basal lamina in spinal roots.[75,76] These disturbances result in a decrease of the conduction velocity in the peripheral nerve.[76] The main laminin-2 receptors expressed by Schwann cells during early differentiation are two integrins (α6β1 and α6β4) and dystroglycan.[77] The major integrin receptor for laminin-2 in muscle (α7β1) does not play a role in peripheral nerve because it is expressed only late in development and the peripheral nerve in α7-knockout mice is normal.[78] However, Schwann cell-specific inactivation of β1 integrins and dystroglycan reproduce the phenotype of laminin α2-deficient mice. Mice with β1 integrin-deficient Schwann cells are impaired in the radial sorting of axons and in peripheral myelination.[79] Schwann cells deficient of dystroglycan show no deficit in axonal sorting but nerve conduction is grossly abnormal.[80] These results allow the conclusion that α6β1 integrin is the laminin-2 receptor important for early steps in myelination and that dystroglycan is rather involved in the maintenance of myelination and the organization of local specializations important for action potential propagation.

Most of the ataxic phenotype of laminin-α2 chain deficient mice has been attributed to the problems in the peripheral nervous system described above. Although MRI scans of MDC1A patients often show differences in white matter, only the analysis of CNS myelin in *dy/dy* mice has now shed some light on the mechanisms that might be involved.[10] According to these data, the alterations in white matter are not ubiquitous and are based on a specific impairment of the cell spreading of oligodendrocytes. These deficits can be reproduced by dominant-negative constructs of integrin-linked kinase (ILK) and involve the PI3K/Akt- but not the MAPK pathway.[10] Thus, myelination in the CNS and PNS requires interactions with integrins, in particular with α6β1 integrin. In contrast, dystroglycan seems to be important at later stages of development. Moreover, the binding of laminin-2 to integrins triggers an intracellular signaling via FAK and ILK and this signaling is important for the formation of a structured basement membrane.

Are There Any Possibilities of Treatment?

Today, there is no effective treatment available for any muscular dystrophies. The method of choice for many years was somatic gene therapy. However, this technique has suffered major setbacks in 1999 and 2002 because one patient died during a gene therapy trial and two children, whose blood stem cells were treated with retroviral vectors, developed leukemia-like symptoms. Moreover, gene therapy aimed at re-inserting the laminin-α2 chain in MDC1A patients would encounter several additional difficulties. First, the multiple organs that are affected in the disease will require infection of skeletal muscle, PNS and CNS. Second, the early onset of disease requires treatment of small children. Third, the laminin-α2 chain is more than 300 kDa in size and its proper incorporation into the laminin heterotrimer requires co-expression of laminin-β and γ chains. Fourth, the cDNA encoding laminin-α2 is more than 9 kb in size, which prevents the use of improved viral vectors such as the adeno-associated virus (AAV). Fifth, de novo expression of laminin α2 is likely to trigger immune responses. Thus, it is difficult to envisage that gene therapy inserting the laminin-α2 chain will be the method of choice for the treatment of MDC1A.

Replacement Therapies as a New Strategy

A new, promising approach to the treatment of diseases in general and muscular dystrophies in particular is to replace disease-causing genes with homologous proteins or proteins that share functional properties. An exciting example of such strategy is the proof-of-concept in mice that utrophin, which is an autosomally-encoded homologue of dystrophin, is able to ameliorate DMD-like symptoms.[81,82] This work has created new hope that upregulation of an endogenous gene may suffice to ameliorate the symptoms in human patients. So far, however, no chemical entities have been identified that increase utrophin at the sarcolemma (reviewed in ref. 83; see also Chapter 3).

Because of the high degree of conservation between dystrophin and utrophin, it is not surprising that utrophin is able to substitute for dystrophin. Because all the transmembrane components of the DGC are strongly reduced or absent from the sarcolemmal membrane of DMD patients and *mdx* mice, it has also been argued that the disease may be caused by a disconnection between α-dystroglycan and laminin-2. If this would indeed contribute to the disease, increased expression of another laminin-2 receptor may help ameliorate disease (see Fig. 4). Indeed, transgenic expression of α7 integrin in *mdx* mice reduced muscular dystrophy, although the extent of amelioration was not 100% (Fig. 4B;[84] see also review by ref. 85).

Similar treatment strategies have also been evaluated for MDC1A. The finding that transgenic expression of the human laminin-α2 chain in *dyw/dyw* mice improved longevity, locomotory behavior and the histological status of muscle significantly indicates that prevention of disease in muscle alone is sufficient to substantially improve the overall health.[35] Based on these encouraging results, my laboratory has gone a step further and made a deletion construct of a protein, termed agrin, to test whether such a designed protein could also ameliorate the disease when overexpressed in skeletal muscle (Fig. 4C). In muscle, the main source of agrin are motor neurons that innervate muscle fibers. The amount of agrin found in non-synaptic regions of the muscle is low, although this can vary from muscle to muscle.[86] In *dyw/dyw* mice, agrin levels are altered but the changes are not consistent. For example, in the slow-twitch soleus muscle, where agrin is expressed at considerably high levels in wild-type mice, *dyw/dyw* mice express less of the protein. In a fast-twitch muscle, such as gastrocnemius, which normally expresses low levels of agrin, the amount of agrin detected in *dyw/dyw* mice increases.[86] Thus, agrin is not consistently up- or downregulated by the loss of laminin-2 and its level in non-synaptic regions of skeletal muscle is low, suggesting that it is not directly involved in disease progression in MDC1A. Instead, agrin is well-known as a key organizer of the nerve-muscle synapse.[87,88] Agrin mRNA undergoes tissue-specific, alternative mRNA splicing near its 3' end. The splice version synthesized in motor neurons can induce postsynaptic differentiation while the isoforms synthesized in many non-neuronal cells and muscle lack this function. While the function of agrin to induce postsynapstic structures at the nerve-muscle synapse involves the muscle-specific receptor tyrosine kinase MuSK, agrin isoforms that are unable to induce postsynaptic structures, bind to α-dystroglycan.[89] The binding to α-dystroglycan is of nanomolar affinity and can compete with the binding of laminin-1 and laminin-2.[90,91] In addition, the amino-terminal region of agrin, termed NtA, binds with nanomolar affinity to the coiled-coil region of laminins containing the γ1 chain.[92-94] The afore-mentioned binding properties of agrin to α-dystroglycan and to laminin suggested to us that a minigene encoding only these domains could be a means to restore the linkage between muscle basement membrane and the sarcolemma in MDC1A. In our model, the laminin-8 and laminin-9, which are overexpressed in MDC1A, would serve as the binding partner for the NtA domain and the carboxy-terminal, laminin G-like modules would bind to α-dystroglycan in the sarcolemmal membrane (Fig. 4C). When we tested this prediction in *dyw/dyw* mice, we found that the mice over-expressing the agrin mini-gene in skeletal muscle showed an improved locomotory activity and survived much longer. In addition, muscle histology and basement membrane structure was substantially improved.[42] Because the transgene in these mice was not expressed in the CNS and the PNS, neurological and neuropathic symptoms were still present.

This work has therefore proven that the use of genes that are structurally not closely related to the mutated, disease-causing gene can ameliorate the disease. This can be considered a major advancement in a potential treatment of MDC1A because gene therapy with this mini-agrin would offer several advantages over re-insertion of the laminin-α2 chain. First, the cDNA encoding the agrin mini-gene is small enough to allow the use of AAV. Second, the notoriously low efficacy in the infection of muscle, which is a major drawback when gene therapy is used to

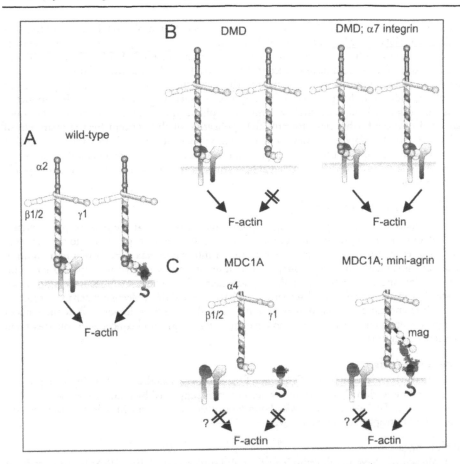

Figure 4. Molecular mechanisms involved in the linkage of basement membrane and cytoskeleton. A) In wild type mice, laminin-2/4 are linked via integrins (left) and dystroglycan (right). B) Dystrophin deficiency in DMD and mouse models thereof is accompanied by the loss of dystroglycan from the sarcolemma and the loss of its linkage to F-actin (left). Overexpression of α7 integrin increases the available binding sites for laminin-2/4 and thus strengthens the linkage to F-actin (right). C) In laminin α2-deficient mice, the level of α-dystroglycan at the sarcolemma is diminished although β-dystroglycan is still expressed. As a compensation for the missing α2 chain, the α4 chain is overexpressed giving rise to laminin-8/9 (left). These laminin isoforms lack domains important for the binding to α-dystroglycan and for the formation of a laminin network. Consequently, the linkage to F-actin is abrogated. Transgenic expression of a minigene for agrin (mag) in muscle, restabilizes α-dystroglycan at the sarcolemma, thus linking laminin-8/9 to F-actin (right).

bring back intramuscular genes (e.g., dystrophin), is not a problem because the protein is secreted from infected muscle fibers and could therefore also reach neighboring, non-infected muscle fibers. Third, thanks to MDC1A patients expressing agrin, immunological rejection of the mini-agrin protein would be minimal. In addition and most importantly, the mini-agrin protein could directly be applied to MDC1A patients. Thus, the potential of a mini-agrin-based therapy is high but all these predictions must be carefully validated in animal models of MDC1A before any of these concepts can be applied to patients.

According to the hypothesis presented in Figure 4C, the binding of the mini-agrin to laminin-8/9 and to α-dystroglycan results in the restoration of the linkage between basement membrane and sarcolemma and is thus the basis of the observed amelioration. If this were true, modules from other extracellular matrix components could be used as linker between laminin-8/9 and α-dystroglycan. For example fusion of modules from nidogen, a high affinity ligand for laminin[95] and perlecan, which binds to α-dystroglycan,[91] could create another chimera with the potential to ameliorate disease in MDC1A. Mutations in perlecan are the cause of the rare, autosomal recessive Schwartz-Jampel syndrome.[96] This disease manifests itself in a pronounced skeletal dysplasia. Moreover, the patients show signs of a permanent myotonia indicating that perlecan is also required to warrant muscle function. In support of this, perlecan knockout mice also have a muscular phenotype.[97] Future work in my laboratory is aimed to make further steps towards the application of this concept for the treatment of MDC1A. Another important question that can now be addressed is whether the pure binding of α-dystroglycan and its physical linkage to muscle basement membranes is sufficient for the observed effect or whether specific signaling pathways become activated by the binding of a particular ligand.

Conclusions and Outlook

The many functional studies using knockout mice suggest that the two receptor systems, the integrin and the dystroglycan complex, link muscle basement membrane to the cytoskeleton. Moreover, the finding that overexpression of integrin in *mdx* mice or of a mini-agrin in *dyw* mice ameliorates the disease, suggests that integrin and dystroglycan receptors share similar function and can, at least partially, replace each other. The last 15 years have been earmarked by many fundamental discoveries that have revealed the molecular mechanisms underlying different muscular dystrophies. It is my hope that the next decades will witness major progress in applying this knowledge to develop new strategies and concepts for the treatment of these devastating diseases.

Acknowledgments

I thank Mrs. P. Barzaghi and S. Meinen for pictures and all the members in my laboratory for fruitful discussion. The work in my laboratory is supported by grants from the Swiss National Science Foundation, the Swiss Foundation for Research on Muscle Diseases and the Muscular Dystrophy Association USA.

References

1. Monaco AP, Neve RL, Colletti-Feener C et al. Isolation of candidate cDNAs for portions of the Duchenne muscular dystrophy gene. Nature 1986; 323(6089):646-650.
2. Martin-Rendon E, Blake DJ. Protein glycosylation in disease: new insights into the congenital muscular dystrophies. Trends Pharmacol Sci 2003; 24(4):178-183.
3. Durbeej M, Campbell KP. Muscular dystrophies involving the dystrophin-glycoprotein complex: an overview of current mouse models. Curr Opin Genet Dev 2002; 12(3):349-361.
4. Michele DE, Barresi R, Kanagawa M et al. Post-translational disruption of dystroglycan ligand interactions in congenital muscular dystrophies. Nature 2002; 418(6896):417-421.
5. Darin N, Tulinius M. Neuromuscular disorders in childhood: a descriptive epidemiological study from western Sweden. Neuromuscul Disord 2000; 10(1):1-9.
6. Mostacciuolo ML, Miorin M, Martinello F et al. Genetic epidemiology of congenital muscular dystrophy in a sample from north-east Italy. Hum Genet 1996; 97(3):277-279.
7. Kobayashi O, Hayashi Y, Arahata K et al. Congenital muscular dystrophy: Clinical and pathologic study of 50 patients with the classical (Occidental) merosin-positive form. Neurology 1996; 46(3):815-818.
8. Di Muzio A, De Angelis MV, Di Fulvio P et al. Dysmyelinating sensory-motor neuropathy in merosin-deficient congenital muscular dystrophy. Muscle Nerve 2003; 27(4):500-506.
9. Matsumura K, Yamada H, Saito F et al. Peripheral nerve involvement in merosin-deficient congenital muscular dystrophy and dy mouse. Neuromusc Disord 1997; 7:7-12.
10. Chun SJ, Rasband MN, Sidman RL. Integrin-linked kinase is required for laminin-2-induced oligodendrocyte cell spreading and CNS myelination. J Cell Biol 2003; 163(2):397-408.
11. Osari S, Kobayashi O, Yamashita Y et al. Basement membrane abnormality in merosin-negative congenital muscular dystrophy. Acta Neuropathol 1996; 91(4):332-336.
12. Colognato H, Yurchenco PD. Form and function: the laminin family of heterotrimers. Dev Dyn 2000; 218(2):213-234.

13. Li S, Edgar D, Fassler R et al. fThe role of laminin in embryonic cell polarization and tissue organization. Dev Cell 2003; 4(5):613-624.
14. Hutter H, Vogel BE, Plenefisch JD et al. Conservation and novelty in the evolution of cell adhesion and extracellular matrix genes. Science 2000; 287(5455):989-994.
15. Smyth N, Vatansever HS, Murray P et al. Absence of basement membranes after targeting the LAMC1 gene results in embryonic lethality due to failure of endoderm differentiation. J Cell Biol 1999; 144(1):151-160.
16. Miner JH, Cunningham J, Sanes JR. Roles for laminin in embryogenesis: exencephaly, syndactyly, and placentopathy in mice lacking the laminin alpha5 chain. J Cell Biol 1998; 143(6):1713-1723.
17. Timpl R, Brown JC. Supramolecular assembly of basement membranes. Bio Essays 1996; 18(2):123-131.
18. Willem M, Miosge N, Halfter W et al. Specific ablation of the nidogen-binding site in the laminin gamma1 chain interferes with kidney and lung development. Development 2002; 129(11):2711-2722.
19. Halfter W, Dong S, Yip YP et al. A critical function of the pial basement membrane in cortical histogenesis. J Neurosci 2002; 22(14):6029-6040.
20. Henry MD, Campbell KP. Dystroglycan inside and out. Curr Opin Cell Biol 1999; 11(5):602-607.
21. Meier T, Ruegg MA. The role of dystroglycan and its ligands in physiology and disease. News Physiol Sci 2000; 15:255-259.
22. Danen EH, Sonnenberg A. Integrins in regulation of tissue development and function. J Pathol 2003; 200(4):471-480.
23. Williamson RA, Henry MD, Daniels KJ et al. Dystroglycan is essential for early embryonic development: disruption of Reichert's membrane in Dag1-null mice. Hum Mol Genet 1997; 6(6):831-841.
24. Cohn RD, Henry MD, Michele DE et al. Disruption of DAG1 in differentiated skeletal muscle reveals a role for dystroglycan in muscle regeneration. Cell 2002; 110(5):639-648.
25. Hack AA, Groh ME, McNally EM. Sarcoglycans in muscular dystrophy. Microsc Res Tech 2000; 48(3-4):167-180.
26. Zimprich A, Grabowski M, Asmus F et al. Mutations in the gene encoding epsilon-sarcoglycan cause myoclonus-dystonia syndrome. Nature Genet 2001; 29(1):66-69.
27. Li W, Zhang Q, Oiso N et al. Hermansky-Pudlak syndrome type 7 (HPS-7) results from mutant dysbindin, a member of the biogenesis of lysosome-related organelles complex 1 (BLOC-1). Nature Genet 2003; 35(1):84-89.
28. Gullberg D, Velling T, Lohikangas L et al. Integrins during muscle development and in muscular dystrophies. Front Biosci 1998; 3.D1039-1050.
29. Mayer U, Saher G, Fassler R et al. Absence of integrin α7 causes a novel form of muscular dystrophy. Nature Genet 1997; 17:318-323.
30. Schwander M, Leu M, Stumm M et al. Beta1 integrins regulate myoblast fusion and sarcomere assembly. Dev Cell 2003; 4(5):673-685.
31. Michelson AM, Russell ES, Harman PJ. Dystrophia muscularis: a hereditary primary myopathy in the house mouse. Proc Natl Acad Sci USA 1955; 41:1079-1084.
32. Meier H, Southard JL. Muscular dystrophy in the mouse caused by an allele at the dy-locus. Life Sci 1970; 9:137-144.
33. Guo LT, Zhang XU, Kuang W et al. Laminin alpha2 deficiency and muscular dystrophy; genotype-phenotype correlation in mutant mice. Neuromuscul Disord 2003; 13(3):207-215.
34. Besse S, Allamand V, Vilquin JT et al. Spontaneous muscular dystrophy caused by a retrotransposal insertion in the mouse laminin alpha2 chain gene. Neuromuscul Disord 2003; 13(3):216-222.
35. Kuang W, Xu H, Vachon PH et al. Merosin-deficient congenital muscular dystrophy. Partial genetic correction in two mouse models. J Clin Invest 1998; 102(4):844-852.
36. Miyagoe Y, Hanaoka K, Nonaka I et al. Laminin alpha2 chain-null mutant mice by targeted disruption of the Lama2 gene: a new model of merosin (laminin 2)-deficient congenital muscular dystrophy. FEBS Lett 1997; 415(1):33-39.
37. O'Brien DP, Johnson GC, Liu LA et al. Laminin alpha 2 (merosin)-deficient muscular dystrophy and demyelinating neuropathy in two cats. J Neurol Sci 2001; 189(1-2):37-43.
38. Shelton GD, Liu LA, Guo LT et al. Muscular dystrophy in female dogs. J Vet Intern Med 2001; 15(3):240-244.
39. Allamand V, Campbell KP. Animal models for muscular dystrophy: valuable tools for the development of therapies. Hum Mol Genet 2000; 9(16):2459-2467.
40. Connolly AM, Keeling RM, Mehta S et al. IThree mouse models of muscular dystrophy: the natural history of strength and fatigue in dystrophin-, dystrophin/utrophin-, and laminin alpha2-deficient mice. Neuromuscul Disord 2001; 11(8):703-712.
41. Patton BL, Miner JH, Chiu AY et al. Distribution and function of laminins in the neuromuscular system of developing, adult, and mutant mice. J Cell Biol 1997; 139(6):1507-1521.

42. Moll J, Barzaghi P, Lin S et al. An agrin minigene rescues dystrophic symptoms in a mouse model for congenital muscular dystrophy. Nature 2001; 413(6853):302-307.

43. Ringelmann B, Roder C, Hallmann R et al. Expression of laminin alpha1, alpha2, alpha4, and alpha5 chains, fibronectin, and tenascin-C in skeletal muscle of dystrophic 129ReJ dy/dy mice. Exp Cell Res 1999; 246(1):165-182.

44. Talts JF, Sasaki T, Miosge N et al. Structural and functional analysis of the recombinant G domain of the laminin alpha 4 chain and its proteolytic processing in tissues. J Biol Chem 2000; 275(45):35192-35199.

45. Vachon PH, Xu H, Liu L et al. Integrins (α7β1) in muscle function and survival. Disrupted expression in merosin-deficient congenital muscular dystrophy. J Clin Invest 1997; 100(7):1870-1881.

46. Hodges BL, Hayashi YK, Nonaka I et al. Altered expression of the alpha7beta1 integrin in human and murine muscular dystrophies. J Cell Sci 1997; 110(Pt 22):2873-2881.

47. Cohn RD, Mayer U, Saher G et al. Secondary reduction of alpha7B integrin in laminin alpha2 deficient congenital muscular dystrophy supports an additional transmembrane link in skeletal muscle. J Neurol Sci 1999; 163(2):140-152.

48. Tome FMS, Guicheney P, Fardeau M. Congenital muscular dystrophies. In: Emery AEH, ed. Neuromuscular Disorders: Clinical and Molecular Genetics West Sussex: John Wiley & Sons; 1998:21-57.

49. Ferletta M, Kikkawa Y, Yu H et al. Opposing roles of integrin alpha6Abeta1 and dystroglycan in laminin-mediated extracellular signal-regulated kinase activation. Mol Biol Cell 2003; 14(5):2088-2103.

50. Yu H, Talts JF. Beta1 integrin and alpha-dystroglycan binding sites are localized to different laminin-G-domain-like (LG) modules within the laminin alpha5 chain G domain. Biochem J 2003; 371(Pt 2):289-299.

51. Colognato H, Yurchenco PD. The laminin α2 expressed by dystrophic dy(2J) mice is defective in its ability to form polymers. Curr Biol 1999; 9(22):1327-1330.

52. Xu H, Christmas P, Wu XR et al. Defective muscle basement membrane and lack of M-laminin in the dystrophic dy/dy mouse. Proc Natl Acad Sci USA 1994; 91(12):5572-5576.

53. Matsuda R, Nishikawa A, Tanaka H. Visualization of dystrophic muscle fibers in mdx mouse by vital staining with Evans blue: evidence of apoptosis in dystrophin-deficient muscle. J Biochem 1995; 118(5):959-964.

54. Straub V, Rafael JA, Chamberlain JS et al. Animal models for muscular dystrophy show different patterns of sarcolemmal disruption. J Cell Biol 1997; 139:375-385.

55. Mauro A. Satellite cell of skeletal muscle fibers. J Biophys Biochem Cytol 1961; 9:493-495.

56. Campion DR. The muscle satellite cell: a review. Int Rev Cytol 1984; 87:225-251.

57. Sicinski P, Geng Y, Ryder-Cook AS et al. The molecular basis of muscular dystrophy in the mdx mouse: a point mutation. Science (Wash DC) 1989; 244(4912):1578-1580.

58. Vachon PH, Loechel F, Xu H et al. Merosin and laminin in myogenesis; specific requirement for merosin in myotube stability and survival. J Cell Biol 1996; 134(6):1483-1497.

59. Kuang W, Xu H, Vilquin JT et al. Activation of the lama2 gene in muscle regeneration: abortive regeneration in laminin alpha2-deficiency. Lab Invest 1999; 79(12):1601-1613.

60. Grady RM, Grange RW, Lau KS et al. Role for alpha-dystrobrevin in the pathogenesis of dystrophin-dependent muscular dystrophies. Nat Cell Biol 1999; 1(4):215-220.

61. Kameya S, Miyagoe Y, Nonaka I et al. alpha1-syntrophin gene disruption results in the absence of neuronal- type nitric-oxide synthase at the sarcolemma but does not induce muscle degeneration. J Biol Chem 1999; 274(4):2193-2200.

62. Adams ME, Kramarcy N, Krall SP et al. Absence of alpha-syntrophin leads to structurally aberrant neuromuscular synapses deficient in utrophin. J Cell Biol 2000; 150(6):1385-1398.

63. Crosbie RH, Straub V, Yun HY et al. mdx muscle pathology is independent of nNOS perturbation. Hum Mol Genet 1998; 7(5):823-829.

64. Rando TA. Role of nitric oxide in the pathogenesis of muscular dystrophies: a "two hit" hypothesis of the cause of muscle necrosis. Microsc Re Tech 2001; 55(4):223-235.

65. Wehling M, Spencer MJ, Tidball JG. A nitric oxide synthase transgene ameliorates muscular dystrophy in mdx mice. J Cell Biol 2001; 155(1):123-131.

66. Oak SA, Zhou YW, Jarrett HW. Skeletal muscle signaling pathway through the dystrophin glycoprotein complex and Rac1. J Biol Chem 2003; 278(41):39287-39295.

67. Brakebusch C, Fassler R. The integrin-actin connection, an eternal love affair. EMBO J 2003; 22(10):2324-2333.

68. Miranti CK, Brugge JS. Sensing the environment: a historical perspective on integrin signal transduction. Nat Cell Biol 2002; 4(4):E83-90.

69. Graus-Porta D, Blaess S, Senften M et al. Beta1-class integrins regulate the development of laminae and folia in the cerebral and cerebellar cortex. Neuron 2001; 31(3):367-379.

70. Moore SA, Saito F, Chen J et al. Deletion of brain dystroglycan recapitulates aspects of congenital muscular dystrophy. Nature 2002; 418(6896):422-425.
71. Costell M, Gustafsson E, Aszodi A et al. Perlecan maintains the integrity of cartilage and some basement membranes. J Cell Biol 1999; 147(5):1109-1122.
72. Beggs HE, Schahin-Reed D, Zang K et al. FAK deficiency in cells contributing to the basal lamina results in cortical abnormalities resembling congenital muscular dystrophies. Neuron 2003; 40(3):501-514.
73. Kuang W, Xu H, Vachon PH et al. Disruption of the lama2 gene in embryonic stem cells: laminin alpha 2 is necessary for sustenance of mature muscle cells. Exp Cell Res 1998; 241(1):117-125.
74. Stirling CA. Abnormalities in Schwann cell sheaths in spinal nerve roots of dystrophic mice. J Anat 1975; 119(1):169-180.
75. Bradley WG, Jaros E, Jenkison M. The nodes of Ranvier in the nerves of mice with muscular dystrophy. J Neuropathol Exp Neurol 1977; 36(5):797-806.
76. Nakagawa M, Miyagoe-Suzuki Y, Ikezoe K et al. Schwann cell myelination occurred without basal lamina formation in laminin alpha2 chain-null mutant (dy3K/dy3K) mice. Glia 2001; 35(2):101-110.
77. Previtali SC, Nodari A, Taveggia C et al. Expression of laminin receptors in schwann cell differentiation: evidence for distinct roles. J Neurosci 2003; 23(13):5520-5530.
78. Previtali SC, Dina G, Nodari A et al. Schwann cells synthesize alpha7beta1 integrin which is dispensable for peripheral nerve development and myelination. Mol Cell Neurosci 2003; 23(2):210-218.
79. Feltri ML, Graus Porta D, Previtali SC et al. Conditional disruption of beta 1 integrin in Schwann cells impedes interactions with axons. J Cell Biol 2002; 156(1):199-209.
80. Saito F, Moore SA, Barresi R et al. Unique role of dystroglycan in peripheral nerve myelination, nodal structure, and sodium channel stabilization. Neuron 2003; 38(5):747-758.
81. Tinsley JM, Potter AC, Phelps SR et al. Amelioration of the dystrophic phenotype of mdx mice using a truncated utrophin transgene. Nature 1996; 384:349-353.
82. Tinsley J, Deconinck N, Fisher R et al. Expression of full-length utrophin prevents muscular dystrophy in mdx mice. Nat Med 1998; 4(12):1441-1444.
83. Perkins KJ, Davies KE. The role of utrophin in the potential therapy of Duchenne muscular dystrophy. Neuromuscul Disord 2002; 12(Suppl 1):S78-89.
84. Burkin DJ, Wallace GQ, Nicol KJ et al. Enhanced expression of the alpha 7 beta 1 integrin reduces muscular dystrophy and restores viability in dystrophic mice. J Cell Biol 2001; 152(6):1207-1218.
85. Mayer U. Integrins: redundant or important players in skeletal muscle? J Biol Chem 2003; 278(17):14587-14590.
86. Eusebio A, Oliveri F, Barzaghi P et al. Expression of mouse agrin in normal, denervated and dystrophic muscle. Neuromuscul Disord 2003; 13:408-415.
87. Ruegg MA, Bixby JL. Agrin orchestrates synaptic differentiation at the vertebrate neuromuscular junction. Trends Neurosci 1998; 21:22-27.
88. Sanes JR, Lichtman JW. Induction, assembly, maturation and maintenance of a postsynaptic apparatus. Nat Rev Neurosci 2001; 2(11):791-805.
89. Bezakova G, Ruegg MA. New insights into the roles of agrin. Nat Rev Mol Cell Biol 2003; 4(4):295-308.
90. Gesemann M, Brancaccio A, Schumacher B et al. Agrin is a high-affinity binding protein of dystroglycan in non-muscle tissue. J Biol Chem 1998; 273(1):600-605.
91. Talts JF, Andac Z, Gohring W et al. Binding of the G domains of laminin alpha1 and alpha2 chains and perlecan to heparin, sulfatides, alpha-dystroglycan and several extracellular matrix proteins. EMBO J 1999; 18(4):863-870.
92. Denzer AJ, Schulthess T, Fauser C et al. Electron microscopic structure of agrin and mapping of its binding site in laminin-1. EMBO J 1998; 17:335-343.
93. Kammerer RA, Schulthess T, Landwehr R et al. Interaction of agrin with laminin requires a coiled-coil conformation of the agrin-binding site within the laminin γ1 chain. EMBO J 1999; 18(23):6762-6770.
94. Mascarenhas JB, Ruegg MA, Winzen U et al. Mapping of the laminin-binding site of the N-terminal agrin domain (NtA). EMBO J 2003; 22(3):529-536.
95. Fox JW, Mayer U, Nischt R et al. Recombinant nidogen consists of three globular domains and mediates binding of laminin to collagen type IV. EMBO J 1991; 10(11):3137-3146.
96. Nicole S, Davoine CS, Topaloglu H et al. Perlecan, the major proteoglycan of basement membranes, is altered in patients with Schwartz-Jampel syndrome (chondrodystrophic myotonia). Nature Genet 2000; 26(4):480-483.
97. Arikawa-Hirasawa E, Rossi SG, Rotundo RL et al. Absence of acetylcholinesterase at the neuromuscular junctions of perlecan-null mice. Nat Neurosci 2002; 5(2):119-123.

CHAPTER 7

Glycosylation and Muscular Dystrophy

Susan C. Brown and Francesco Muntoni

Abstract

The congenital muscular dystrophies (CMD) are a heterogeneous group of autosomal recessive disorders a number of which have recently been shown to be associated with mutations in the genes encoding for either known or putative glycosyltransferases. These include Fukuyama CMD, Muscle-Eye-Brain disease and Walker-Warburg syndrome, which are associated with eye abnormalities and neuronal migration defects, due to mutations in *fukutin, POMGnT1* and *POMT1* respectively, while mutations in the *fukutin-related protein* (*FKRP*) gene causes congenital muscular dystrophy 1C a form that typically lacks brain involvement. In addition another putative glycosyltransferase, *Large*, is mutated in the myodystrophy mouse and the human homologue of this gene was recently shown to be mutated in a patient with a form of congenital muscular dystrophy named MDC1D. All of these conditions are associated with the hypoglycosylation of α-dystroglycan thus identifying a new pathogenetic mechanism in the muscular dystrophies.

Congenital Muscular Dystrophy

Congenital muscular dystrophy (CMD) presents either at birth or within the first 6 months of life. It is among the most frequent autosomal recessively inherited form of neuromuscular disease and is characterised by hypotonia, muscle weakness, limitations of joint movement and dystrophic changes on skeletal muscle biopsy.[1] Recently it has become evident that a number of forms of CMD are associated with mutations in genes encoding for proteins that are either putative or determined glycosyltransferases.[2-4] These include three severe forms with eye abnormalities and neuronal migration defects namely Fukuyama congenital muscular dystrophy (FCMD), Muscle-Eye-Brain disease (MEB) and Walker-Warburg syndrome (WWS), a severe form of congenital muscular dystrophy (MDC1C [OMIM 606612]) and a milder form of limb girdle muscular dystrophy (LGMD2I [OMIM 607115]). In addition mutations in the gene encoding for the putative glycosyltransferase *LARGE* have been identified in the *myd* mouse and and a patient with a form of congenital muscular dystrophy (MDC1D).[5,6] Overall these entities suggest a new mechanism of pathogenesis in the muscular dystrophies namely that of the post-translational modification of proteins.

Glycosylation and Dystroglycan

Glycosylation takes place in the ER and Golgi compartments and involves a complex series of reactions catalysed by membrane bound glycosyltransferases and glycosidases. There are two main forms of protein glycosylation namely N-linked glycosylation in which the oligosaccharide is added onto an asparagine residue via the amide group and O-linked glycosylation where the oligosaccharide is attached to a hydroxyl group of a serine or threonine residue. In eukaryotic N-glycans, N-acetylglucosamine (GlcNAc) is found as the reducing terminal carbohydrate residue. By contrast O linked glycans feature a variety of reducing terminal sugar residues which include N acetylgalactosamine (GalNAc), fucose (FUc), glucose (GLc), GLcNAc, xylose (Xyl), galactose (Gal), arabinose (Ara) and O-linked mannose (Man).[7,8] Congenital disorders of glycosylation (CDGs) are the first group of conditions in which a defect in the process of glycosylation were recognized and almost invariably they are caused by defective N-linked glycosylation.[9] At least twelve syndromes are currently recognized and almost invariably result in multisystemic defects due to alterations in the modification of a wide range of proteins. Four O-glycosylation defects are also known and comprise of two O-xylosylglycan defects that cause a progeroid variant of Ehlers-Danlos syndrome and the multiple exostoses syndrome and two O-mannosylglycan defects which include Walker-Warburg syndrome (WWS) and Muscle-Eye-Brain disease, that latter of which form the main focus for the present review.

The first indication that aberrant glycosylation might underlie forms of congenital muscular dystrophy arose as a consequence of work carried out on the Drosophila mutant named rotated abdomen (rt). Mutations at the rt locus of Drosphila cause a clockwise helical rotation of the body. Whilst null alleles are viable they exhibit defects in embryonic muscle development, rotation of the whole larval body, and a helical staggering of cuticular patterns in the abdominal segments. The rt gene was noted at this time to have a high homology to the yeast mannosyl-transferases (Pmts)[10] as did the human homologue POMT1 on chromosome 9q34.1.[11] On the basis of the rt phenotype Jurado et al, 1999 suggested that POMT1 could be a candidate for uncharacterized genetic disorders of the muscular system, such as some forms of muscular dystrophy. O-mannosylation was nonetheless thought to be a rare type of glycosylation in mammals, occurring in only a limited number of brain, nerve and skeletal muscle glycoproteins[8] and a specific link between dystroglycan processing and muscular dystrophy was not initially recognised.

Dystroglycan was originally identified as the central component of an assembly of protein complexes (DAPC) that associate with dystrophin.[12] α and β-dystroglycan are encoded by a single gene *DAG1*, the transcript from which is post-translationally cleaved to give rise to two glycoproteins[13] which are tightly associated via noncovalent interactions. The primary sequence of α-dystroglycan predicts a molecular mass of 72 kDa. However, the mass of α-dystroglycan in mammalian skeletal and cardiac muscle is 156 kDa and 140 kDa respectively and in brain and peripheral nerve 120 kDa reflecting the extensive tissue specific patterns of post-translational modification which this protein undergoes. Whilst dystroglycan contains four potential N-linked glycosylation sites, (three of which are in α-dystroglycan and one in β-dystroglycan), it is the O-linked glycosylation that makes the major contribution to the observed molecular weight.[14-16] Multiple O-linked glycosylation sites are located in the serine-theronine- rich 'mucin' domain of the protein and these modifications mediate the interaction with perlecan, laminin-α2 and agrin[17-19] although the specific glycoconjugates involved have not yet been identified.[20] β-dystroglycan interacts with the C terminal region of α-dystroglycan at the membrane periphery[21] and dystrophin, utrophin, caveolin, actin and Grb2 in the cytoplasm,[22-24] thereby linking the extracellular matrix with cytoplasmic and signaling components of the muscle fibre. This arrangement is essential not only for membrane integrity but also proper neuromuscular junction organisation and, as will be explained below, neural cell migration in the CNS.

Null mutations in the dystroglycan gene (*DAG1*) are lethal at an early embryonic stage in mice[25] and may be the reason why no patients have yet been described with mutations in the dystroglycan gene itself. However, a hypoglycosylated form of α-dystroglycan is a feature of the skeletal muscle of WWS, FCMD, MEB and MDC1C patients. This has been primarily determined using three antibodies IIH6, V1A4-1 and a polyclonal antibody directed towards the core peptide. The epitope recognised by each is unknown although the binding of both V1A4-1 and IIH6 is known to be dependent on glycosylation and IIH6 functionally blocks laminin binding in vitro.[26,27] The loss of the epitopes recognised by V1A4-1 and IIH6 in these forms of CMD together with recent immunoblot analysis of α-dystroglycan with the anti-peptide polyclonal antibody, shows that the molecular weight of α-dystroglycan is typically reduced by more than ~60kDa in FCMD, MEB and the myodystrophy (myd^{LARGE}) mouse. This lower molecular weight form of α-dystroglycan also fails to bind laminin, agrin or neurexin on overlay assays providing strong evidence that the O-linked glycans implicated in ligand binding are

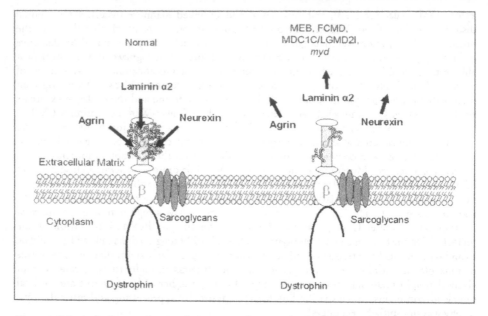

Figure 1. Schematic diagram showing the interaction between the dystroglycan complex and various extracellular matrix components in normal muscle (A). B) shows the disruption of these linkages when the glycosylation of the serine –rich region of α dystroglycan is perturbed.

missing in these disorders[28] (Fig. 1). All these disorders show a variable reduction in laminin α2 in skeletal muscle but apparently normal levels of β -dystroglycan. This review summarises the clinical phenotypes associated with each of these forms of CMD within the context of the recent genetic and biochemical findings.

Walker-Warburg Syndrome

Walker-Warburg syndrome (WWS [MIM 236670]) is the more severe of these group of disorders and associated with a life expectancy of less than 3 years. It is a recessive disorder characterised by severe brain malformations (type II lissencephaly), muscular dystrophy, and structural eye abnormalities.[29] Characteristic features include almost complete lack of brain gyri (agyria or lissencephaly of the cobblestone variant, see below) with agenesis of the corpus callosum, cerebellar hypoplasia, hydrocephaly, and sometimes encephalocele. Cobblestone lissencephaly is associated with neural over-migration during neocortex lamination, which is thought to occur secondarily to a breach of the pial glial limitans. Whilst WWS is known to be genetically heterogenous; a recent study showed that of the cases examined only approximately 20% had a mutation in the O-mannosyltransferase 1 (*POMT1*), of these two were found to be nonsense, two frameshift, and one a missense mutation.[30] The autosomal recessive pattern of inheritance and the type of mutations identified therefore suggest that the syndrome is caused by a loss-of-function mechanism that results in the reduced or absent O-mannosylation of target proteins. POMT1 is thought, but not proven, to catalyse the first step in the O-mannosyl glycan synthesis. Another candidate gene for this disorder is POMT2 which shows an expression pattern in adults that overlaps with that of POMT1[7] however, no patients have as yet been identified with mutations in this gene.

Immunohistochemical analysis of skeletal muscle from patients with POMT1 mutations characteristically shows an absence of glycosylated α-dystroglycan.[30,31] Moreover, a disruption of the α-dystroglycan-laminin axis appears to be a feature of cases proven not to be caused by mutations in POMT1 suggesting that this may be a unifying feature of WWS.[32] Recent work

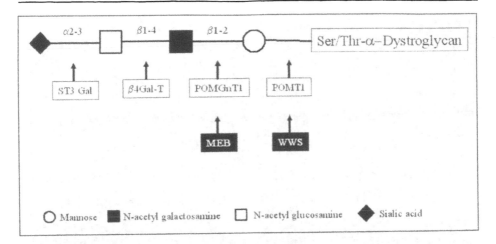

Figure 2. Mutations in POMGnT1, POMT1, fukutin, FKRP and Large are all associated with the production of a hypo-glycosylated from of α dystroglycan. The substrates of these enzymes, with the exception POMGnT1 are as yet unknown. POMT1 is however, thought to initiate the biosynthesis of O-mannosyl glycan. The pathway(s) in which fukutin, FKRP and Large operate remains to be determined.

on a mouse with a brain-specific disruption of α-dystroglycan showed that an absence of dystroglycan is associated with neuronal over-migration and the fusion of cerebral hemispheres.[33] These brain abnormalities are similar to those observed in WWS and so support the concept that α-dystroglycan is one of the targets in this disorder and this may give rise to both the skeletal muscle and CNS abnormalities.

Muscle Eye-Brain Disease

MEB is a relatively milder form of CMD relative to WWS. It is an autosomal recessive disorder characterised by congenital muscular dystrophy, severe congenital myopia, congenital glaucoma, pallor of the optic discs, retinal hypoplasia, mental retardation, hydrocephalus, abnormal electroencephalograms and myoclonic jerks. Patients with MEB are floppy from birth with generalised muscle weakness, including facial and neck muscles. Brain MRIs reveal an abnormal structure of the neuronal cortical folding with thickened cortex (pachygyria) and hypoplastic brainstem and cerebellum. MEB is caused by loss of function mutations in the gene encoding protein O-linked mannose β1,2-N-acetylglucosaminyltransferase 1 (*POMGnT1* (MEB [OMIM 253280]).[34] The human gene is widely expressed and the primary disease target in MEB remains unproven. However, α-dystroglycan is strongly implicated by the fact that sialyl O-mannosyl glycan is known to be a laminin-binding ligand of α-dystroglycan[35] and POMGnT1 catalyzes the transfer of N-acetylglucosamine to O-mannose of glycoproteins (Fig. 2). In addition both α-dystroglycan (Fig. 3) and laminin α2 chain expression are reduced in the muscles of MEB patients.[34,36] MEB is prevalent in Finland however, a number of novel POMGnT1 mutations have recently been reported in other ethnic groups suggesting that this disorder has a more world wide distribution than initially appreciated. A possible genotype – phenotype correlation has been suggested, with missense, nonsense and frameshifting mutations towards the 5' end of the gene tending to associate with more severe structural brain involvement. These latter cases showed some clinical overlap with WWS (see below) thus broadening the clinical spectrum of this disorder.[36] A recent report showed a highly significant decrease in POMGnT1 activity in four MEB patient muscle extracts relative to controls suggesting that POMGnT1 activity provides a rapid and relatively simple diagnostic test for this disease. Since MEB, WWS and FCMD are heterogenous with respect to both clinical presentation and radiological examination, assays of POMGnT1 would seem to be a useful screening procedure in all CMD patients associated with brain malformations.[37]

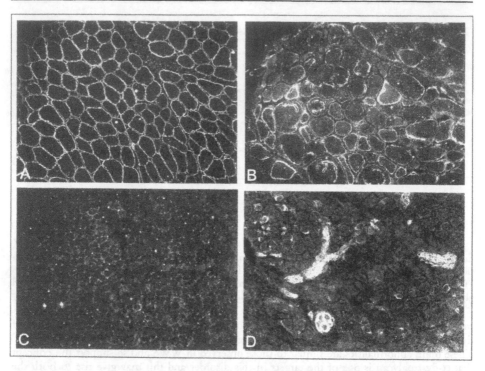

Figure 3. Immunolabelling of α dystroglycan using an antibody to a glycosylated epitope in muscle from a normal control (A), LGMD2I (B), MEB (C) and MDC1C (D) patient. Note the near absence of this epitope in the MEB and MDC1C patient and less severe reduction in LGMD2I relative to the control. The section of the MDC1C patient shows also nerves which retain α dystroglycan staining consistent with the absence of nervous system involvement in this patient.

Fukuyama Muscular Dystrophy

Fukuyama-type congenital muscular dystrophy (FCMD) is one of the most common autosomal-recessive disorders in Japan and is almost entirely confined to this country. The main clinical features include weakness of facial and limb muscles, hypotonia and an inability to walk. Patients usually become bedridden prior to 10 years of age due to generalized muscle atrophy and joint contracture and death usually occurs by 20 years of age. All cases display severe mental retardation with IQ scores between 30 and 50 and seizures occur in nearly half of the cases. Overall the severity of brain involvement is milder than WWS and MEB. The brain malformation takes the form of polymicrogyria (also part of the type II lissencephaly spectrum, but milder than that observed in MEB and WWS) and patients appear to have a disruption in both radial and tangential neuronal migration in addition to altered neuronal migration in the cerebral and cerebellar cortex.[38] Ophthalmologic findings such as peripheral abnormalities of the retina or abnormal eye movements are also often observed in FCMD patients.

Most FCMD patients (87%) carry an ancestral mutation that arose as a consequence of the integration of a 3 kb retrotransposon element into the 3' untranslated region of the *fukutin* gene on chromosome 9q31 which leads to a reduction in fukutin mRNA levels. In addition to these findings, two point mutations[39] and four other mutations[40] have been identified in rare FCMD alleles. There are also several other alleles which show no changes within the coding region and are probably attributable to mutations in the regulatory regions such as promoter sequences or intronic sequences critical for alternative splicing.[41] Patients homozygous for the founder mutation are less severely affected than compound heterozygotes with a point mutation on the other allele presumably because of a higher residual activity of fukutin.[42] Until recently this disorder was not seen outside Japan however, two patients of Turkish origin homozygous for truncating mutations have now been reported. Interestingly these patients had a WWS phenotype rather than that typically associated with FCMD with severe brain and eye anomalies.[43,44]

Targeted homozygous null mutations of the fukutin gene in mice leads to embryonic lethality at day 6.5-7.5, prior to the development of skeletal muscle, cardiac muscle, or mature neurons (H. Kurahashi, S. Takeda, C. Meno, M. Horie, M. Taniguchi, H. Otani, H. Hamada, T. Toda, unpublished data). In view of this mice chimeric for fukutin deficiency have been generated.[45] Animals with less than a 50% contribution of heterozygous ES cells show a typical muscular dystrophy and a marked dis-organisation of the laminar structures of both the cerebral and cerebellar cortices and hippocampus. α-dystroglycan immunolabelling is reduced as is its laminin-binding activity. In addition these mice showed abnormalities of the lens, loss of laminar structure in the retina, and retinal detachment. Observations which confirm a role for fukutin in the maintenance of muscle integrity, cortical histogenesis and ocular development.

Sequence analysis predicts fukutin to be an enzyme that modifies cell-surface glycoproteins or glycolipids with some homology to the *fringe*-like family of enzymes.[46] Initial transfection experiments suggested that fukutin was secreted after passing through the Golgi,[47] however, others have since suggested that it is a Golgi resident protein.[48] The major sites of expression of fukutin appear to be heart, brain, skeletal muscle and pancreas.[47]

The skeletal muscle of patients with FCMD show an immunohistochemical reduction of both α- dystroglycan (Hayashi et al, 1999) and laminin α2 chain (laminin-2 and -4) and electron microscopy confirms a disruption of the muscle fiber basal lamina.[49,50] The brains of FCMD fetuses characteristically show breaches in the glia limitans–basal lamina.[51-53] The glia limitans is formed by the endfeet of astrocytes, and in situ hybridisation and immunohistochemical analyses suggest that fukutin is normally expressed in both fetal and adult glial cells (some of which are astrocytes) in addition to neurons where it is developmentally regulated.[54]

MDC1C and LGMD2I

Mutations in the *FKRP* gene give rise to two main clinical entities namely MDC1C and LGMD2I.[55,56] The main features of both include muscle hypertrophy that is more prominent in the legs, markedly elevated serum creatine kinase, normal intelligence and normal brain MRI. Cardiomyopathy, respiratory insufficiency, and macroglossia become evident at a later stage. Patients with MDC1C present at birth and subsequently develop significant leg hypertrophy associated with wasting of the deltoid and pectoralis muscle. The weakness in the arms is more pronounced than in the legs, and independent ambulation is not usually achieved. Patients with LGMD2I, also show some significant muscle hypertrophy with shoulder girdle weakness and may be divided into a Duchenne-like group, who have early onset and loss of independent ambulation in the teens, and a milder group with later onset and preserved ambulation after the second decade. In a recent study more than half of LGMD2I patients had evidence of dilated cardiomyopathy, which was not consistently associated with an early or more severe presentation. Respiratory failure affected a third of the patients with a Duchenne-like phenotype after the age of 16 years.[57]

Initially neither disorder was thought to be associated with brain involvement however, there is now evidence of more severe *FKRP* mutations that can result in structural cerebellar changes in a proportion of patients with MDC1C.[58,59] In addition we have recently identified rare, severe mutations which result in more extensive structural brain and eye involvement, leading to a phenotype resembling MEB and WWS.[44] Specifically we have identified a homozygous T919A missense mutation in a patient with a clinical diagnosis of MEB who presented with severe myopia, cortical pachygyria and cerebellar dysplasia, and a G953T homozygous missense mutation in a patient with a clinical diagnosis of WWS who presented with severe cobblestone lissencephaly, cerebellar dysplasia and microphthalmia and died in infancy. These observations underscore the wide spectrum of conditions caused by mutations in the

FKRP gene to include both MEB and WWS, in which muscular dystrophy and structural brain and eye involvement coexist. The cobblestone lissencephaly and the structural eye involvement characterises the severe end of the FKRP phenotypic spectrum. This range therefore includes the catastrophic and early fatal outcome seen in WWS cases to patients who remain ambulant in their seventh decade of life with minimal physical disability. Significant phenotypic variability has also been described for patients with mutations in the *fukutin* gene. In addition to this the clinical spectrum arising from mutations in *POMGnT1* is much wider than was originally appreciated.[36]

FKRP is comprised of 4 exons that encodes for a 495 amino acid protein. It is ubiquitously expressed in human tissues with the highest levels apparent in skeletal and cardiac muscle. Patients with MDC1C are compound heterozygous between a nonsense and a missense mutation or carry two missense mutations.[57] To date no cases with two nonsense mutations have been observed, suggesting that a total loss of protein may be embryonic lethal. None of the patients with MDC1C had the Leu276Ileu change, which in the population examined was invariably found in patients with LGMD2I.[57] This mutation has been calculated to occur at a heterozygous frequency of 1:400 in the UK.[60] The second allelic mutation in patients with the Leu276Ileu change determines the severity of the LGMD2I phenotype although the degree of intra-familial clinical variability suggests that additional factors may also play a significant role.[57]

Skeletal muscle from both MDC1C and LGMD2I patients displays a variable depletion of α- but not β-dystroglycan at the muscle fibre membrane (Fig. 3). This reduction in α-dystroglycan immunolabelling is typically more severe in MDC1C and at the severe end of the LGMD2I spectrum.[61] Many patients show a reduction in laminin α2 chain expression although this is more subtle than the changes observed with α- dystroglycan.[61] There are as yet no animal models for these conditions. Muscle hypertrophy is a clinical feature of patients with FKRP mutations but as yet its origin has not been ascertained. Interestingly true muscle fibre hypertrophy is a feature of mice in which a muscle specific conditional knock-out of dystroglycan has been introduced.[62]

The enzymatic activity of FKRP has yet to be proven although over-expression of FKRP alters dystroglycan processing in vitro.[48] Patients with MDC1C characteristically show a marked reduction in the average molecular mass of α-dystroglycan relative to normal muscle suggesting that O-linked glycosylation may be impaired.[63] The relatively narrow band pattern seen in some patients with mutations in the *FKRP* gene are comparable to that observed at around 10 weeks of gestation in the human fetus.[61] These observations suggest that defects in FKRP may act to alter the expression of the higher molecular mass glycoforms of α-dystroglycan characteristic of mature muscle.

MDC1D and the Myodystrophy Mouse

The myodystrophy (*myd*; now renamed Large^myd) mouse carries a loss of function mutation in the *LARGE* gene encoding for a putative bifunctional glycosyltransferase. Homozygous Large^myd mice display a severe, progressive muscular dystrophy and a mild cardiomyopathy in addition to retinal, peripheral and central nervous system involvement.[3,64,65] Abnormalities in neuronal migration are observed in the brain particularly the cortex and cerebellum which is similar to that seen in fukutin-deficient mice. A profound loss of muscle α- dystroglycan has also been observed.

LARGE encodes a putative, bifunctional glycosyltransferase, the substrate for which is at present unclear. More recently one patient with a G1525A (Glu509Lys) missense mutation and a 1 bp insertion, 1999insT has been reported and this disorder named MDC1D. This 17-year-old girl presented with congenital muscular dystrophy, profound mental retardation, white matter changes and subtle structural abnormalities on brain MRI. Her skeletal muscle biopsy showed reduced immunolabelling of α-dystroglycan and immunoblotting demonstrated a reduced molecular weight form of α-dystroglycan that did however, retain some laminin binding activity.[6]

Dystroglycan and the Pathogenesis of CMD

It is well recognised that a disruption of the dystroglycan complex is associated with a number of forms of muscular dystrophy.[66] It was nonetheless unclear until relatively recently as to whether defects in dystroglycan could give rise to the structural and functional brain abnormalities seen in FCMD and MEB. The generation of mice that have a brain-selective deletion of dystroglycan[33] suggests that this is in fact the case since these animals display a disarray of cerebral cortical layering, fusion of cerebral hemispheres and cerebellar folia, an aberrant migration of granule cells and discontinuities in the pial surface basal lamina (glia limitans) which data from other sources shows often results from aberrant neuronal migration.[33] Whilst these animals also lack β-dystroglycan which is normally expressed at the muscle fibre membrane of patients with MEB, FCMD and WWS, these observations do nonetheless support the hypothesis that interactions between α dystroglycan and its ligands are essential for normal brain development.

The precise cellular processes that are disrupted in FCMD, WWS, MEB, MDC1C and MDC1D remain unclear. The six-layered neocortex is formed by sequential waves of postmitotic neurons that migrate radially from the ventricular zone (VZ) to the pial surface and is generated in an 'inside-out' manner with early and later migrating neurons forming the deep and superficial cortical plate layers respectively.[67] These processes are in part reliant on the formation of a basement membrane by the meningeal cells which then associates tightly with radial glia processes to form the pial-glial barrier that is crucial for appropriate neural migration and layer formation. Some of the abnormalities of the neocortex in WWS namely, neuronal over-migration and invasion of the subarachnoid space by neurons is similar to the situation seen in mice with disturbances in the structural components of the basal lamina for example mice lacking the gene for myristoylated alanine-rich C kinase substrate (MARCKS), β1 or α6 integrins (reviewed in ref. 68). In view of the previously mentioned basal lamina defects in FCMD muscle and brain and the well established association between disruption of the dystroglycan complex and muscular dystrophy, it seems likely that disruptions in the formation and/or structural integrity of the basement membrane may account for many of the features typical of the cobblestone lissencephalics. One question that is still open is whether additional mechanisms may contribute to the neuronal migration disorder observed in this group of patients. Defects in the ability of neurons to migrate and alterations in the relay of signals from the extracellular matrix have both been shown to alter neocortical layering in either patients with lissencephaly or mouse mutants.[69] For example mutations in either LIS-1 (*LIS1*) or double cortin (*DCX*) lead to cytoskeletal defects that interfere with the ability of neuronal cells to migrate and mice with a graded reduction in Lis1 show a dose-dependent disorganisation of the cortical layers. Mice carrying mutations in either reelin (*Reln*, an extracellular matrix protein) or its receptors *Vldlr* (very-low-density lipoprotein receptor) or *Lrp8* (apolipoprotein E receptor), or *Dab1* (mouse homologue of the fly protein disabled), all show defects in neocortical layering that has been proposed to be due to an alteration in signaling between the extracellular matrix and the cell interior.[67] The genes encoding for POMGnT1, fukutin and dystroglycan are broadly expressed in late embryonic and early postnatal cerebellar neurons, including premigratory granule neurons of the external granule cell layer. Expression of POMGnT1 and fukutin was also observed in neurons of the internal granule cell layer after migration, whereas DG mRNA was largely downregulated. Immunolabelling shows that α- and β-DG are expressed on the Bergmann glial scaffolds used by granule cells during early postnatal radial migration, and in situ hybridization confirmed that these cells also express POMGnT1 and fukutin. These observations therefore raise the possibility that abnormal glycosylation of α-dystroglycan on glial scaffolds and neurons and their processes could affect interactions with ligands expressed by migrating granule cells. This would be in addition to the defects in the pial glial basal lamina and suggests that a disruption in several cellular processes may act to disrupt neuronal migration.

Conclusions

Recent work now shows that mutations in the genes encoding for POMT1, FKRP and fukutin can all result in a WWS phenotype, implying that these conditions are part of a single disease spectrum. Whilst both *POMT1* and *POMGnT1*, encode enzymes potentially involved in the biosynthesis of specific glycans associated with dystroglycan, the biochemical activities of fukutin, FKRP and Large are as yet unknown. However, all these disorders result in similar immunocytochemical and Western blot abnormalities of expression of α-dystroglycan, implying that the post-translational modification of this molecule may play a central role in the pathogenesis of this group of disorders. In view of the overlapping clinical spectrum identified so far it is possible that these putative glycosyltransferases operate along a similar pathway.

Abnormalities in α-dystroglycan expression occur in a number of other as yet unclassified congenital muscular dystrophies only some of which have central nervous system involvement.[70] It is therefore likely that the number of diseases associated with the abnormal post translational modification of proteins such as dystroglycan will increase. Moreover, it is also possible that a number of other protein targets for these enzymes exist although their identity and role in the pathogenesis of these disorders is at present unclear. Tenascin-R is one potential candidate since it has a relatively high relative abundance of O-linked sialyated glycans.[71] Moreover, tenascin has previously been shown to promote or inhibit cell adhesion and neurite outgrowth in specific assays and is known to be prominent in development of the nervous system. Neural cell adhesion (NCAM) is another glycoprotein of interest since in muscle tissue it contains a muscle specific domain (MSD) to which mucin type O-glycans are attached and mutation of the potential O-glycosylation sites within the MSD diminishes the ability of NCAM to facilitate myoblast fusion.[72] Integrins link the extracellular matrix with the cell interior and have a well documented role in cell signalling and so may also represent important targets. Future works will seeks to identify the precise targets for some of the putative enzymes mutated in these disorders will be crucial both for our future understanding of the essential pathways involved in neuronal migration and in explaining the reasons for the spectrum of changes that are observed.

References

1. Dubowitz V. Congenital muscular dystrophy: An expanding clinical syndrome. Ann Neurol 2000; 47(2):143-144.
2. Muntoni F, Brockington M, Blake DJ et al. Defective glycosylation in muscular dystrophy. Lancet 2002; 360(9343):1419-1421.
3. Grewal PK, Hewitt JE. Glycosylation defects: A new mechanism for muscular dystrophy? Hum Mol Genet 2003.
4. Martin PT, Freeze HH. Glycobiology of neuromuscular disorders. Glycobiology 2003.
5. Grewal PK, Holzfeind PJ, Bittner RE et al. Mutant glycosyltransferase and altered glycosylation of alpha-dystroglycan in the myodystrophy mouse. Nat Genet 2001; 28(2):151-154.
6. Longman C, Brockington M, Torelli S et al. Mutations in the human LARGE gene cause MDC1D, a novel form of congenital muscular dystrophy with severe mental retardation and abnormal glycosylation of {alpha}-dystroglycan. Hum Mol Genet 2003.
7. Willer T, Valero MC, Tanner W et al. O-mannosyl glycans: From yeast to novel associations with human disease. Curr Opin Struct Biol 2003; 13(5):621-630.
8. Endo T. O-mannosyl glycans in mammals. Biochim Biophys Acta 1999; 1473(1):237-246.
9. Jaeken J. Komrower Lecture. Congenital disorders of glycosylation (CDG): It's all in it! J Inherit Metab Dis 2003; 26(2-3):99-118.
10. Martin-Blanco E, Garcia-Bellido A. Mutations in the rotated abdomen locus affect muscle development and reveal an intrinsic asymmetry in Drosophila. Proc Natl Acad Sci USA 1996; 93(12):6048-6052.
11. Jurado LA, Coloma A, Cruces J. Identification of a human homolog of the Drosophila rotated abdomen gene (POMT1) encoding a putative protein O-mannosyl-transferase, and assignment to human chromosome 9q34.1. Genomics 1999; 58(2):171-180.
12. Blake DJ, Weir A, Newey SE et al. Function and genetics of dystrophin and dystrophin-related proteins in muscle. Physiol Rev 2002; 82(2):291-329.

13. Ibraghimov-Beskrovnaya O, Milatovich A, Ozcelik T et al. Human dystroglycan: Skeletal muscle cDNA, genomic structure, origin of tissue specific isoforms and chromosomal localization. Hum Mol Genet 1993; 2(10):1651-1657.

14. Ibraghimov-Beskrovnaya O, Ervasti JM, Leveille CJ et al. Primary structure of dystrophin-associated glycoproteins linking dystrophin to the extracellular matrix. Nature 1992; 355(6362):696-702.

15. Ervasti JM, Campbell KP. Membrane organization of the dystrophin-glycoprotein complex. Cell 1991; 66(6):1121-1131.

16. Holt KH, Crosbie RH, Venzke DP et al. Biosynthesis of dystroglycan: Processing of a precursor propeptide. FEBS Lett 2000; 468(1):79-83.

17. Henry MD, Campbell KP. Dystroglycan: An extracellular matrix receptor linked to the cytoskeleton. Curr Opin Cell Biol 1996; 8(5):625-631.

18. Henry MD, Campbell KP. Dystroglycan inside and out. Curr Opin Cell Biol 1999; 11(5):602-607.

19. Moll J, Barzaghi P, Lin S et al. An agrin minigene rescues dystrophic symptoms in a mouse model for congenital muscular dystrophy. Nature 2001; 413(6853):302-307.

20. Martin PT. Dystroglycan glycosylation and its role in matrix binding in skeletal muscle. Glycobiology 2003.

21. Bozzi M, Veglia G, Paci M et al. A synthetic peptide corresponding to the 550-585 region of alpha-dystroglycan binds beta-dystroglycan as revealed by NMR spectroscopy. FEBS Lett 2001; 499(3):210-214.

22. Yang B, Jung D, Motto D et al. SH3 domain-mediated interaction of dystroglycan and Grb2. J Biol Chem 1995; 270(20):11711-11714.

23. Ilsley JL, Sudol M, Winder SJ. The interaction of dystrophin with beta-dystroglycan is regulated by tyrosine phosphorylation. Cell Signal 2001; 13(9):625-632.

24. Chen YJ, Spence HJ, Cameron JM et al. Direct interaction of beta-dystroglycan with F-actin. Biochem J 2003; (Pt).

25. Williamson RA, Henry MD, Daniels KJ et al. Dystroglycan is essential for early embryonic development: Disruption of Reichert's membrane in Dag1-null mice. Hum Mol Genet 1997; 6(6):831-841.

26. Brown SC, Fassati A, Popplewell L et al. Dystrophic phenotype induced in vitro by antibody blockade of muscle alpha-dystroglycan-laminin interaction. J Cell Sci 1999; 112(Pt 2):209-216.

27. Durbeej M, Larsson E, Ibraghimov-Beskrovnaya O et al. Nonmuscle alpha-dystroglycan is involved in epithelial development. J Cell Biol 1995; 130(1):79-91.

28. Michele DE, Campbell KP. Dystrophin-glycoprotein complex: Post-translational processing and dystroglycan function. J Biol Chem 2003.

29. Dobyns WB, Pagon RA, Armstrong D et al. Diagnostic criteria for Walker-Warburg syndrome. Am J Med Genet 1989; 32(2):195-210.

30. Beltran-Valero DB, Currier S, Steinbrecher A et al. Mutations in the O-Mannosyltransferase gene POMT1 give rise to the severe neuronal migration disorder Walker-Warburg syndrome. Am J Hum Genet 2002; 71(5):1033-1043.

31. Sabatelli P, Columbaro M, Mura I et al. Extracellular matrix and nuclear abnormalities in skeletal muscle of a patient with Walker-Warburg syndrome caused by POMT1 mutation. Biochim Biophys Acta 2003; 1638(1):57-62.

32. Jimenez-Mallebrera C, Torelli S, Brown SC et al. Profound skeletal muscle depletion of alpha-dystroglycan in Walker-Warburg syndrome. Eur J Paediatr Neurol 2003; 7(3):129-137.

33. Moore SA, Saito F, Chen J et al. Deletion of brain dystroglycan recapitulates aspects of congenital muscular dystrophy. Nature 2002; 418(6896):422-425.

34. Yoshida A, Kobayashi K, Manya H et al. Muscular dystrophy and neuronal migration disorder caused by mutations in a glycosyltransferase, POMGnT1. Dev Cell 2001; 1(5):717-724.

35. Chiba A, Matsumura K, Yamada H et al. Structures of sialylated O-linked oligosaccharides of bovine peripheral nerve alpha-dystroglycan. The role of a novel O-mannosyl-type oligosaccharide in the binding of alpha-dystroglycan with laminin. J Biol Chem 1997; 272(4):2156-2162.

36. Taniguchi K, Kobayashi K, Saito K et al. Worldwide distribution and broader clinical spectrum of muscle-eye-brain disease. Hum Mol Genet 2003; 12(5):527-534.

37. Zhang W, Vajsar J, Cao P et al. Enzymatic diagnostic test for Muscle-Eye-Brain type congenital muscular dystrophy using commercially available reagents. Clin Biochem 2003; 36(5):339-344.

38. Saito Y, Kobayashi M, Itoh M et al. Aberrant neuronal migration in the brainstem of fukuyama-type congenital muscular dystrophy. J Neuropathol Exp Neurol 2003; 62(5):497-508.

39. Kobayashi K, Nakahori Y, Mizuno K et al. Founder-haplotype analysis in Fukuyama-type congenital muscular dystrophy (FCMD). Hum Genet 1998; 103(3):323-327.

40. Kondo-Iida E, Saito K, Tanaka H et al. Molecular genetic evidence of clinical heterogeneity in Fukuyama-type congenital muscular dystrophy. Hum Genet 1997; 99(4):427-432.

41. Kobayashi K, Sasaki J, Kondo-Iida E et al. Structural organization, complete genomic sequences and mutational analyses of the Fukuyama-type congenital muscular dystrophy gene, fukutin. FEBS Lett 2001; 489(2-3):192-196.
42. Toda T, Kobayashi K, Kondo-Iida E et al. The Fukuyama congenital muscular dystrophy story. Neuromuscul Disord 2000; 10(3):153-159.
43. Silan F, Yoshioka M, Kobayashi K et al. A new mutation of the fukutin gene in a nonJapanese patient. Ann Neurol 2003; 53(3):392-396.
44. Beltrán Valero de b, Voit T, Longman C et al. Mutations in the FKRP gene can cause Muscle-Eye-Brain disease and Walker-Warburg syndrome. J Med Genet2004; 41:e61.
45. Takeda S, Kondo M, Sasaki J et al. Fukutin is required for maintenance of muscle integrity, cortical histiogenesis and normal eye development. Hum Mol Genet 2003; 12(12):1449-1459.
46. Aravind L, Koonin EV. The fukutin protein family—predicted enzymes modifying cell-surface molecules. Curr Biol 1999; 9(22):R836-R837.
47. Kobayashi K, Nakahori Y, Miyake M et al. An ancient retrotransposal insertion causes Fukuyama-type congenital muscular dystrophy. Nature 1998; 394(6691):388-392.
48. Esapa CT, Benson MA, Schroder JE et al. Functional requirements for fukutin-related protein in the Golgi apparatus. Hum Mol Genet 2002; 11(26):3319-3331.
49. Ishii H, Hayashi YK, Nonaka I et al. Electron microscopic examination of basal lamina in Fukuyama congenital muscular dystrophy. Neuromuscul Disord 1997; 7(3):191-197.
50. Matsubara S, Mizuno Y, Kitaguchi T et al. Fukuyama-type congenital muscular dystrophy: Close relation between changes in the muscle basal lamina and plasma membrane. Neuromuscul Disord 1999; 9(6-7):388-398.
51. Nakano I, Funahashi M, Takada K et al. Are breaches in the glia limitans the primary cause of the micropolygyria in Fukuyama-type congenital muscular dystrophy (FCMD)? Pathological study of the cerebral cortex of an FCMD fetus. Acta Neuropathol (Berl) 1996; 91(3):313-321.
52. Yamamoto T, Toyoda C, Kobayashi M et al. Pial-glial barrier abnormalities in fetuses with Fukuyama congenital muscular dystrophy. Brain Dev 1997; 19(1):35-42.
53. Saito Y, Murayama S, Kawai M et al. Breached cerebral glia limitans-basal lamina complex in Fukuyama-type congenital muscular dystrophy. Acta Neuropathol (Berl) 1999; 98(4):330-336.
54. Yamamoto T, Kato Y, Karita M et al. Fukutin expression in glial cells and neurons: Implication in the brain lesions of Fukuyama congenital muscular dystrophy. Acta Neuropathol (Berl) 2002; 104(3):217-224.
55. Brockington M, Yuva Y, Prandini P et al. Mutations in the fukutin-related protein gene (FKRP) identify limb girdle muscular dystrophy 2I as a milder allelic variant of congenital muscular dystrophy MDC1C. Hum Mol Genet 2001; 10(25):2851-2859.
56. Brockington M, Blake DJ, Brown SC et al. The gene for a novel glycosyltransferase is mutated in congenital muscular dystrophy MDC1C and limb girdle muscular dystrophy 2I. Neuromuscul Disord 2002; 12(3):233-234.
57. Mercuri E, Brockington M, Straub V et al. Phenotypic spectrum associated with mutations in the fukutin-related protein gene. Ann Neurol 2003; 53(4):537-542.
58. Topaloglu H, Brockington M, Yuva Y et al. FKRP gene mutations cause congenital muscular dystrophy, mental retardation, and cerebellar cysts. Neurology 2003; 60(6):988-992.
59. Louhichi N, Triki C, Quijano-Roy S et al. New FKRP mutations causing congenital muscular dystrophy associated with mental retardation and central nervous system abnormalities. Identification of a founder mutation in Tunisian families. Neurogenetics 2003.
60. Poppe M, Cree L, Bourke J et al. The phenotype of limb-girdle muscular dystrophy type 2I. Neurology 2003; 60(8):1246-1251.
61. Brown SC, Torelli S, Brockington M et al. Abnormalities in α dystroglycan expression in MDC1C and LGMD2I muscular dystrophies. American Journal of Pathology In press.
62. Cohn RD, Henry MD, Michele DE et al. Disruption of DAG1 in differentiated skeletal muscle reveals a role for dystroglycan in muscle regeneration. Cell 2002; 110(5):639-648.
63. Brockington M, Blake DJ, Prandini P et al. Mutations in the fukutin-related protein gene (FKRP) cause a form of congenital muscular dystrophy with secondary laminin alpha2 deficiency and abnormal glycosylation of alpha-dystroglycan. Am J Hum Genet 2001; 69(6):1198-1209.
64. Rayburn HB, Peterson AC. Naked axons in myodystrophic mice. Brain Res 1978; 146(2):380-384.
65. Holzfeind PJ, Grewal PK, Reitsamer HA et al. Skeletal, cardiac and tongue muscle pathology, defective retinal transmission, and neuronal migration defects in the Large(myd) mouse defines a natural model for glycosylation-deficient muscle - eye - brain disorders. Hum Mol Genet 2002; 11(21):2673-2687.
66. Campbell KP. Three muscular dystrophies: Loss of cytoskeleton-extracellular matrix linkage. Cell 1995; 80(5):675-679.

67. Gupta A, Tsai LH, Wynshaw-Boris A. Life is a journey: A genetic look at neocortical development. Nat Rev Genet 2002; 3(5):342-355.
68. Magdaleno SM, Curran T. Brain development: Integrins and the Reelin pathway. Curr Biol 2001; 11(24):R1032-R1035.
69. Henion TR, Qu Q, Smith FI. Expression of dystroglycan, fukutin and POMGnT1 during mouse cerebellar development. Brain Res Mol Brain Res 2003; 112(1-2):177-181.
70. Muntoni F, Valero dB, Bittner R et al. 114th ENMC international workshop on congenital muscular dystrophy (CMD) 17-19 january 2003, naarden, the Netherlands: (8th Workshop of the International Consortium on CMD; 3rd Workshop of the MYO-CLUSTER project GENRE). Neuromuscul Disord 2003; 13(7-8):579-588.
71. Zamze S, Harvey DJ, Pesheva P et al. Glycosylation of a CNS-specific extracellular matrix glycoprotein, tenascin-R, is dominated by O-linked sialylated glycans and "brain-type" neutral N-glycans. Glycobiology 1999; 9(8):823-831.
72. Suzuki M, Angata K, Nakayama J et al. Polysialic acid and mucin-type O-glycansonthe neural cell adhesion molecule (NCAM) differentially regulatemyoblast fusion. J Biol Chem 2003.

CHAPTER 8

Overview of the Limb-Girdle Muscular Dystrophies and Dysferlinopathy

Kate M.D. Bushby and Steven H. Laval

The Current Status of the Limb-Girdle Muscular Dystrophies

The limb-girdle muscular dystrophies (LGMD) comprise a group of disorders now known to reflect the whole spectrum of molecular pathogenesis of muscle disease. Initially identified as a heterogeneous set of muscular dystrophies predominantly affecting the pelvic and shoulder girdle musculature, the underlying molecular pathology has now been determined for three forms of autosomal dominant LGMD and ten forms of autosomal recessive disease (Table 1).[1,2] Three further forms of autosomal dominant disease have been identified by linkage analysis but the genes remain to be elucidated. While the enhanced diagnostic techniques available through these discoveries mean that for many patients with LGMD a precise diagnosis is now available, with concomitant benefits for management and genetic counselling, the existence of some patients in whom a diagnosis cannot yet be achieved suggests that more forms of LGMD remain to be characterized.

As the majority of forms of LGMD were defined initially by linkage analysis, recent classifications are based on genetically mapped disease loci, designated LGMD1 A, B, C etc chronologically in the order of the locus description for the dominant disorders and LGMD2A, B, C etc for the recessive disorders.[3] Clarification of the genes and proteins underlying these diseases has defined a new level of heterogeneity in that proteins from various compartments of the muscle fibre can be affected. This includes components of the muscle fibre membrane, enzymes involved in specific functions within the muscle fibre, components of the sarcomere and the nuclear membrane (see Fig. 1).[1] Therefore, although some of the individual forms of LGMD are very rare, they do provide an insight into the breadth of molecular mechanisms which lead to the final common pathway of a dystrophic phenotype.

LGMD as a Disorder of the Muscle Fibre Membrane

The crucial role and composition of the muscle fibre membrane in maintaining muscle fibre integrity is indicated by the range of muscular dystrophies which result from disruption of this structure. For example, mutations in the components of the sarcoglycan complex in the muscle fibre membrane underlie LGMD2C, D, E and F (reviewed in Chapter 10).[29] These muscular dystrophies clinically are similar to the disorders caused by mutations in the dystrophin gene (Duchenne and Becker muscular dystrophy) reflecting a similar spectrum of disease severity from rapidly progressive childhood disease to a milder adult condition. As with dystrophinopathy, patients frequently develop cardiomyopathy, and hypertrophy of muscle groups, especially the calf musculature, is common.[30,31] Altered glycosylation of α-dystroglycan, a peripheral membrane component of the dystrophin associated glycoprotein complex, is thought to be the mechanism by which defects in FKRP (a putative glycosyltransferase) cause muscular dystrophy (reviewed in Chapter 7). FKRP mutations are

Table 1. *Current state of knowledge on the genes and proteins involved in LGMD*

	Gene Location	Gene Product	Sub-Cellular Localisation	Proposed Function	Frequency	Key Clinical Correlates	Typical Protein Findings	Key Refs.
LGMD1A	5q22-q34	Myotilin	Sarcomere (Z-line)	Binds α-actinin, γ-filamin and bundles F-actin.	Very rare, two families only	Dysarthria, tight achilles tendons	Secondary deficiency of laminin γ-1.	4,5 Ch. 11
LGMD1B	1q11-21	Lamin A/C	Nuclear Membrane	Structural component of nuclear lamina	Relatively common and worldwide. New dominant mutations common	Consider especially if presence of cardiac complications or contractures	Lamin A/C labelling is not usually abnormal in these autosomal dominant cases. May be a secondary reduction in laminin β-1.	6 Ch. 12
LGMD1C	3p25	Caveolin-3	Sarcolemma	Concentration of signalling molecules/biogenesis of T-tubules	Uncommon but worldwide	Calf hypertrophy, hyperCKaemia	Loss of caveolin-3 immunolabelling in muscle sections	7 Ch. 9
LGMD1D	6q23	?	?	?	Single family only	Cardiac involvement		8
LGMD1E	7q	?	?	?	Rare			9
LGMD1F	7q	?	?	?	Single family only	Possible anticipation		10
LGMD2A	15q15.1-q21.1	Calpain-3	Cytosolic	Regulation of NF-κB/IκBα in protection from apoptosis. Also binds titin. Cleaves γ-filamin.	Relatively common	Atrophic disease, frequently also contractures, calf hypertrophy rare	Diagnostic calpain-3 antibodies work only on immunoblotting. Patients with calpain-3 mutations will usually have abnormal calpain-3 on immunoblot, but in some patients may be normal. Secondary calpain-3 deficiency has been described in dysferlinopathy	11-13 Ch. 11
LGMD2B	2p13	Dysferlin	Sarcolemma/Cytosolic	Membrane repair	Worldwide	May present with distal weakness as well as proximal, typically onset in late teens/early twenties	Loss of dysferlin immunolabelling on sections and immunoblotting. May be secondary calpain-3 deficiency and alteration in caveolin-3 labelling pattern	14,15

Table continued on next page

Table 1. Continued

	Gene Location	Gene Product	Sub-Cellular Localisation	Proposed Function	Frequency	Key Clinical Correlates	Typical Protein Findings	Key Refs.
LGMD2C	13q12	γ-sarcoglycan	Sarcolemma	Stabilizes DGC at the sarcolemma, binds γ-filamin	Variable proportions of different sarcoglycan mutations worldwide	Can be variable in severity, from DMD to BMD like. Cardiac and respiratory complications common	Typically see total loss of the sarcoglycan primarily involved and partial loss of the other members of the complex. With β- and δ-sarcoglycan mutations whole complex is typically lost.	16,17 Ch. 10
LGMD2D	17q12-q21.33	α-sarcoglycan	Sarcolemma	Stabilizes DGC at the sarcolemma				18,19 Ch. 10
LGMD2E	4q12	β-sarcoglycan	Sarcolemma	Stabilizes DGC at the sarcolemma				20,21 Ch. 10
LGMD2F	5q33-q34	δ-sarcoglycan	Sarcolemma	Stabilizes DGC at the sarcolemma, binds γ-filamin				22,23 Ch. 10
LGMD2G	17q11-q12	Telethonin	Sarcomere (Z-line)	Substrate for Titin kinase	Brazil only		Loss of telethonin labelling	24 Ch. 11
LGMD2H	9q31-q34.1	TRIM32	Cytosolic	E3 ubiquitin ligase involved in targeting proteins to the proteasome	Canadian Hutterite population only		Specific antibodies not widely available	25,26 Ch. 11
LGMD2I	19q13.3	Fukutin Related Protein	Golgi Apparatus	Glycosylation of α-dystroglycan, the extra-cellular component of the DGC	Common, worldwide	Calf hypertrophy, common cardiac and respiratory involvement	Various secondary protein abnormalities including reduction in laminins (laminin α-2 reduction may be seen on immunoblotting only) and α-dystroglycan	27 Ch. 7
LGMD2J	2q	Titin	Sarcomere	Molecular ruler protein specifying sarcomeric structure. Has an intrinsic kinase activity.	Reported only in rare homozygous cases in families with dominant tibial muscular dystrophy in Finland.		Secondary calpain-3 deficiency	28 Ch. 11

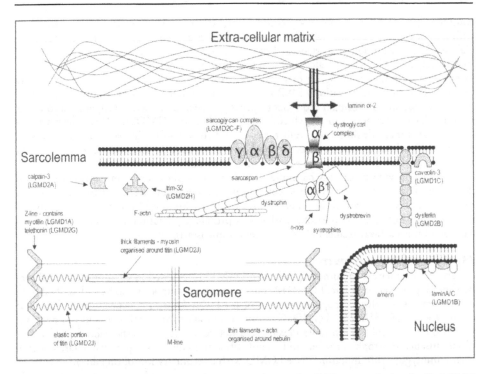

Figure 1. Organization of LGMD associated proteins in the muscle cell. Those proteins involved in LGMD are shaded in grey. The dystrophin glycoprotein complex at the sarcolemma links the extra-cellular matrix to the F-actin cytoskeleton via laminin α-2, the dystroglycan complex and dystrophin. The sarcoglycans form a distinct sub-complex liked to dystroglycan via sarcospan. Other LGMD associated proteins are located elsewhere in the sarcolemma (dysferlin, caveolin-3), the cytosol (calpain-3, TRIM-32), the nuclear lamina (laminA/C) or the sarcomere (titin, myotilin, telethonin).

responsible for muscular dystrophies with a wide range of disease severity, from congenital muscular dystrophy with brain involvement to a more common and much milder form of LGMD, LGMD2I. Patients with this disorder typically have muscle hypertrophy and are at risk of cardiac and respiratory failure.[27,32,33]

Also within the muscle fibre membrane but not part of the dystrophin associated complex, mutations in caveolin-3 (Chapter 9) cause LGMD1C (a rare, typically fairly mild form of LGMD with muscle hypertrophy) as well as a variety of other muscle phenotypes including rippling muscle disease, distal myopathy, hyperCKaemia and myalgia.[34-37] Dysferlin mutations, reviewed in more detail later in this chapter, are responsible for a form of LGMD, LGMD2B, as well as forms of distal myopathy (Miyoshi myopathy and distal myopathy with anterior tibial involvement).[38-40] Localisation of all of these various proteins within the muscle fibre membrane does not necessarily suggest a shared mechanism for disease causation. Disruption of the dystrophin associated complex is most likely to cause muscular dystrophy through a disruption of the structural link between the intracellular and extracellular environments provided by the dystrophin associated complex,[41] while mutations in dysferlin are likely to result in impairment of muscle membrane repair and maintenance.[15,42]

LGMD as Enzymopathy

Mutations in the gene encoding calpain-3, a muscle specific protease, in LGMD2A provided the first indication that enzyme defects as well as disruption of structural proteins might be involved in the causation of a muscular dystrophy phenotype.[11] LGMD2A is a relatively common form of LGMD, and is a predominantly atrophic disease characterized frequently with loss of muscle bulk and variable joint contractures as well as progressive weakness.[43,44] Calpain-3 shares a number of structural characteristics with the other ubiquitous or tissue specific calpains, which act as calcium dependent proteases. The natural substrate of calpain-3 in muscle is not known though it may regulate proteolysis of a variety of important muscle proteins. Proteolytic inactivation of IκBα has been suggested as a role for calpain-3, and an increase in apoptosis in calpain-3 deficient muscle has been described.[12] Calpain-3 also binds titin, the large sarcomeric protein which is itself involved in LGMD2J.[28,45] Recent work has demonstrated that calpain-3 specifically binds to and cleaves γ-filamin using pull-down assays and expression in a cell culture system,[13] implicating calpain-3 in the processes of cytoskeletal remodelling which accompany myoblast fusion and repair.

A second enzymatic pathway has more recently been implicated in a rare form of LGMD, described to date only in the Hutterite population of Canada. LGMD2H is associated with a mutation in TRIM 32 which is a member of the "tripartite motif" protein family probably involved in the ubiquitin-proteasome pathway of protein degradation. It is likely that the disruption of this pathway results in loss of protein homeostasis.[25,26]

LGMD and the Sarcomere

The involvement of sarcomeric proteins in LGMD is reviewed in Chapter 11. The disorders in this group are very rare. LGMD1A, due to myotilin mutations, has been described only in two families worldwide. It appears to be associated with dysarthria, which is not a feature of other types of LGMD.[4,46,47] LGMD2G is due to mutations in telethonin and has been reported only in Brazil.[24] Titin mutations present heterozygously are responsible for a form of distal muscular dystrophy prevalent in Finland (tibial muscular dystrophy) while rare patients with homozygous titin mutations have LGMD with a secondary loss of calpain-3.[45]

LGMD and the Nuclear Envelope

Finally, mutations in lamin A/C rarely cause an LGMD phenotype (LGMD1B).[6,48] Lamin A/C and emerin are components of the nuclear lamina and their roles in muscular dystrophy are reviewed in chapter 12. Mutations in lamin A/C are responsible for an extraordinarily wide range of different phenotypes. A shared feature of the muscle related laminopathies is a high risk of cardiac conduction defects and sudden death.[49]

Dysferlinopathy

Clinical Considerations

The dysferlin gene was cloned simultaneously by two groups working on apparently two different muscle diseases, LGMD2B and Miyoshi myopathy. LGMD2B was defined by linkage to chromosome 2p13 in a number of large families with a generally slowly progressive proximal muscle disease with onset in the late teens or early twenties.[39,50,51] Miyoshi myopathy, a disorder first described in Japan, is characterized by inability to walk on tiptoes reflecting early weakness of the gastrocnemius muscle.[38,52,53] Onset in Miyoshi myopathy is also typically towards the end of the second, or beginning of the third decade. An overlap between LGMD2B and MM was initially proposed following the description of large families linked to chromosome 2 segregating both proximal and distal phenotypes.[54-56] Following the cloning of the dysferlin gene these families were indeed shown to have variable disease in association with the same homozygous dysferlin mutations. A third phenotype associated with dysferlinopathy, distal muscular dystrophy with anterior tibial involvement, has also been described.[40] As the disease progresses, weakness typically becomes global and it is no longer possible to determine the mode of onset.[57] Muscle imaging may indicate preclinical involvement of the distal musculature with a proximal presentation and vice versa, so that although patients can often be assigned to a clear proximal or distal onset group, the distinctions of these disorders become blurred with time. Patients with dysferlinopathy of whatever presentation typically share a number of features including a tight range of age at onset in the late teens and early twenties and a very high serum creatine kinase at presentation. Interestingly, and in contrast to other forms of muscular dystrophies, patients with dysferlinopathy are often reported to have had good sporting prowess prior to the onset of weakness, and there are a couple of reports of dysferlinopathy patients having had normal creatine kinase presymptomatically. These features, together with the relatively common finding of inflammatory features on muscle biopsy, mean that patients with dysferlinopathy have frequently been misdiagnosed as having inflammatory muscle disease, which is unresponsive to steroid treatment. The inflammatory infiltrate seen in dysferlinopathy is predominantly T-cell and macrophage in origin with few B cells. Up-regulation of MHC class I molecules and inflammatory infiltrates are restricted to necrotic fibres and perivascular regions. Overall, it appears that the inflammatory response is directed against degenerating myofibres and is therefore a secondary event.[58] What remains unclear is why such a presentation appears to be relatively specific for dysferlinopathy, as inflammation to this degree is not usually seen in other types of muscular dystrophy, with the exception of facioscapulohumeral muscular dystrophy.

The early pathological hallmarks of dysferlinopathy have been described in detail by electron microscopy. There are small discontinuities of the plasma membrane and in some cases replacement of the sarcolemma with a layer of vesicles. Other changes include thickening and/or duplication of the basal lamina, papillary projections and small subsarcolemmal vacuoles.[59,60] This may suggest that dysferlin plays a role in maintaining the structural integrity of the sarcolemma and is consistent with dysferlin being involved in membrane repair (see below).

Progression of muscle weakness in dysferlinopathy is variable both within and between families, although it is not clear whether this is due to environmental or genetic factors. There does not appear to be a major involvement of the cardiac or respiratory systems and the condition seems compatible with normal fertility and in most cases with normal life expectancy.

The *dysferlin* gene is very large (55 exons) and there do not appear to be major hot-spots for mutations. Missense and nonsense mutations are both described in all areas of the gene, with no clear genotype-phenotype correlations. Founder mutations, both associated with variable phenotypes, have been described in the aboriginal population of Canada and in the Libyan Jewish population.[56,57]

The Ferlin Gene Family

Dysferlin was the first human member of the ferlin family of proteins to have been identified, named after the *C.elegans* fertility factor *fer-1* to which they are homologous.[61] Four members of this protein family have been identified in man; dysferlin, myoferlin, otoferlin and ferlL4.[62-65] All four contain multiple C2 domains and a C terminal transmembrane domain. In *C.elegans fer-1* mutants are infertile due to the failure of membranous organelles to fuse with the plasma membrane during spermatogenesis, suggesting a role for these proteins in membrane fusion or vesicle trafficking events. All of the ferlins are widely expressed, including dysferlin, however, mutations in these genes rise to a highly tissue specific phenotypes, with dysferlinopathy being skeletal muscle restricted and mutations in otoferlin causing a form of sensorineural deafness.[63] No phenotypes have been ascribed to deficiency of myoferlin which is also expressed in muscle and is upregulated in DMD.[62,64] Fer1-L4 is widely expressed and has alternative splicing.[65] Experiments in muscle development have shown that myoferlin is expressed in prefusion myoblasts, earlier than dysferlin, which is expressed in post fusion myotubes. Functional studies on the first C2 domain (C2A) of myoferlin suggest that it is able to bind tritiated phospholipids in the presence of calcium, further emphasizing the lipid binding properties of this family of proteins.[66] Ferlin C2A domain function may be homologous to the role of C2 domains in synaptotagmin which also have a calcium sensitive binding to phospholipids which is thought to mediate regulation of membrane fusion. However, the functions of the different C2 domains in dysferlin have not all been studied yet in detail. The presence of so many C2 domains in a single protein is highly unusual and is likely to be functionally significant.

Dysferlin Protein Localisation and Interactions

Initial studies indicated that dysferlin localized exclusively to the sarcolemma,[67] although with other antibodies some perinuclear staining was also observed.[68] However, use of techiniques to amplify the dysferlin signal revealed distinct intracellular staining reminiscent of vesicles and a patchy accumulation of dysferlin in the cytoplasm of fast fibres in patients with mutations in DGC proteins and LGMD of unknown aetiology.[60] Dysferlin staining is also patchy or cytoplasmic in LGMD1C and rippling muscle disease, both caused by caveolin-3 mutations.[69] Caveolin-3 has also been shown to interact with dysferlin, the gene product deficient in LGMD2B and Miyoshi myopathy by coimmunoprecipitation, although the significance of this observation is unclear.[70] An interaction with caveolin-3 is consistent with the hypothesis that dysferlin acts in the process by which the sarcolemma is repaired following disease-dependent injury.

The Putative Function of Dysferlin

All cells require a rapid and efficient means to reseal the plasma membrane following injury, however this process is more critical to survival in highly active tissues such as striated muscle.[65] Membrane repair is an active process allowing internal membranes to be recruited to the plasma membrane to form a patch over the membrane disruption. This process is triggered by calcium and effected by a protein complex analogous to the synaptic vesicle fusion machinery, containing C2-domain proteins.[71] Using myofibres from a mouse model of dysferlinopathy, a calcium-dependent membrane resealing activity has been shown to be perturbed in dysferlin-negative fibres using high intensity laser irradiation and a membrane-impermeant fluorescent dye.[15] Damaged myofibres displayed patches of dysferlin positive membrane which excluded other sarcolemmal markers such as caveolin-3, indicating that they were derived from internal, dysferlin positive vesicles. This process may involve an interaction between dysferlin and the annexins, as dysferlin appears to be stably associated with annexin A2, whilst an association between dysferlin and annexin A1 is disrupted upon membrane wounding.[42]

Conclusions

The suggested role of dysferlin in membrane repair defines a new mechanism for the production of a muscular dystrophy phenotype. Skeletal muscle fibres are highly adapted to the mechanical stress under which they perform their very specialised function. Dysferlin's role in membrane repair cannot be an "all or nothing" effect, as total absence of any ability to repair skeletal muscle damage could be devastating for the survival of the muscle fibres, and indeed in the absence of dysferlin a certain amount of membrane repair can still occur.[15] Amongst the various types of LGMD, various underlying pathogenic mechanisms have been invoked to lead to the production of a muscular dystrophy phenotype. It remains to be seen what a major role disruption of membrane repair may play in other types of muscle disease, and indeed in "normal" processes such as ageing.

References

1. Bushby K, Beckmann JS. The 105th ENMC Sponsored workshop: Pathogenesis in the nonsarcoglycan limb-girdle muscular dystrophies, Naarden, 2002. Neuromusc Disord 2003; 13(1):80-90.
2. Bushby K. Making sense of the limb-girdle muscular dystrophies. Brain 1999; 122:1403-1420.
3. Bushby KMD, Beckmann JS. Report of the 30th and 31st ENMC international workshop- the limb-girdle muscular dystrophies, and proposal for a new nomenclature. Neuromusc Disord 1995; 5:337-344.
4. Hauser MA, Horrigan SK, Salmikangas P et al. Myotilin is mutated in limb-girdle muscular dystrophy 1A. Hum Mol Genet 2000; 9(14):2141-2147.
5. Salmikangas P, van der Ven PF, Lalowski M et al. Myotilin, the limb-girdle muscular dystrophy 1A (LGMD1A) protein, cross-links actin filaments and controls sarcomere assembly. Hum Mol Genet 2003; 12(2):189-203.

6. Muchir A, Bonne G, van der Kooi AJ et al. Identification of mutations in the gene encoding lamins A/C in autosomal dominant limb-girdle muscular dystrophy with atrioventricular conduction disturbances (LGMD1B). Hum Mol Genet 2000; 9(9):1453-1459.
7. Minetti C, Sotgia F, Bruno C et al. Mutations in the caveolin-3 gene cause autosomal dominant limb-girdle muscular dystrophy. Nature Genetics 1998; 18:365-368.
8. Messina DN, Speer MC, Pericak-Vance MA et al. Linkage of familial dilated cardiomyopathy with conduction defect and muscular dystrophy to chromosome 6q23. Am J Hum Genet 1997; 61(4):909-917.
9. Speer MC, Vance JM, Grubber JM et al. Identification of a new autosomal dominant limb-girdle muscular dystrophy locus on chromosome 7. Am J Hum Genet 1999; 64:556-562.
10. Palenzuela L, Andreu AL, Gamez J et al. A novel autosomal dominant limb-girdle muscular dystrophy (LGMD1F) maps to 7q32.1-32.2. Neurology 2003; 61:404-406.
11. Richard I, Broux O, Allamand V et al. Mutations in the proteolytic enzyme, calpain 3, cause limb girdle muscular dystrophy type 2A. Cell 1995; 81:27-40.
12. Baghdiguian S, Martin M, Richard I et al. Calpain 3 deficiency is associated with myonuclear apoptosis and profound perturbation of the IκBα/NF-κB pathway in limb-girdle muscular dystrophy type 2A. Nature Medicine 1999; 5:503-511.
13. Guyon JR, Kudryashova E, Potts A et al. Calpain 3 cleaves filamin C and regulates its ability to interact with gamma- and delta-sarcoglycans. Muscle Nerve 2003; 28(4):472-483.
14. Bashir R, Britton S, Strachan T et al. A gene related to caenorhabditis elegans spermatogenesis factor fer-1 is mutated in limb-girdle muscular dystrophy type 2B. Nature Genetics 1998; 20:37-42.
15. Bansal D, Miyake K, Vogel SS et al. Defective membrane repair in dysferlin-deficient muscular dystrophy. Nature 2003; 423(6936):168-172.
16. Noguchi S, McNally EM, Ben Othmane K et al. Mutations in the dystrophin-associated protein gamma-sarcoglycan in chromosome 13 muscular dystrophy. Science 1995; 270(5237):819-822.
17. McNally EM, Duggan DJ, Gorospe JR et al. Mutations that disrupt the carboxyl-terminus of y-sarcoglycan cause muscular dystrophy. Hum Mol Gen 1996; 5(11):1841-1847.
18. Roberds S, Leturcq F, Allamand V et al. Missense mutations in the adhalin gene linked to autosomal recessive muscular dystrophy. Cell 1994; 78:625-633.
19. Piccolo F, Roberds SL, Jeanpierre M et al. Primary adhalinopathy: A common cause of autosomal recessive muscular dystrophy of variable severity. Nature Genetics 1995; 10:243-245.
20. Lim LE, Duclos F, Broux O et al. β-sarcoglycan: Characterisation and role in limb-girdle muscular dystrophy linked to 4q12. Nature Genetics 1995; 11:257-265.
21. Bonnemann CG, Modi R, Noguchi S et al. β-sarcoglycan (A3b) mutations cause autosomal recessive muscular dystrophy with loss of the sarcoglycan complex. Nature Genetics 1995; 11:266-273.
22. Passos-Bueno MR, Moreira ES, Vainzof M et al. Linkage analysis in autosomal recessive limb-girdle muscular dystrophy (AR LGMD) maps a sixth form to 5q33-34 (LGMDF) and indicates that there is at least one more subtype of AR LGMD. Hum Mol Genet 1996; 5(6):815-820.
23. Nigro V, Moreira ES, Piluso G et al. Autosomal recessive limb-girdle muscular dystrophy, LGMD2F, is caused by a mutation in the δ-sarcoglycan gene. Nature Genetics 1996; 14:195-198.
24. Moreira ES, Wiltshire TJ, Faulkner G et al. Limb-girdle muscular dystrophy type 2G is caused by mutations in the gene encoding the sarcomeric protein telethonin. Nature Genetics 2000; 24:163-166.
25. Weiler T, Greenberg CR, Zelinski T et al. A gene for autosomal recessive limb-girdle muscular dystrophy in Manitoba Hutterites maps to chromosome region 9q31-33: Evidence for another limb-girdle muscular dystrophy locus. Am J Hum Genet 1998; 63:140-147.
26. Frosk P, Weiler T, Nylen E et al. Limb-girdle muscular dystrophy type 2H associated with mutation in TRIM32, a putative E3-ubiquitin-ligase gene. Am J Hum Genet 2002; 70:663-672.
27. Brockington M, Yuva Y, Prandini P et al. Mutations in the fukutin-related protein gene (FKRP) identify limb girdle muscular dystrophy 2I as a milder allelic variant of congenital muscular dystrophy MDC1C. Hum Mol Genet 2001; 10(25):2851-2859.
28. Hackman P, Vihola A, Haravuori H et al. Tibial muscular dystrophy is a titinopathy caused by mutations in TTN, the gene encoding the giant skeletal-muscle protein titin. Am J Hum Genet 2002; 71:492-500.
29. Angelini C, Fanin M, Freda MP et al. The clinical spectrum of sarcoglycanopathies. Neurology 1999; 52:176-179.
30. Politano L, Nigro V, Passamano L et al. Evaluation of cardiac and respiratory involvement in sarcoglycanopathies. Neuromusc Disord 2001; 11:178-185.
31. Calvo F, Teijeira S, Fernandez JM et al. Evaluation of heart involvement in gamma-sarcoglycanopathy (LGMD2C). A study of ten patients. Neuromusc Disord 2000; 10:560-566.

32. Brockington M, Blake DJ, Prandini P et al. Mutations in the fukutin-related protein gene (FKRP) cause a form of congenital muscular dystrophy with secondary laminin alpha2 deficiency and abnormal glycosylation of alpha-dystroglycan. Am J Hum Genet 2001; 69(6):1198-1209.

33. Poppe M, Cree L, Bourke J et al. The phenotype of limb-girdle muscular dystrophy type 2I. Neurology 2003; 60(8):1246-1251.

34. McNally EM, de Sa Moreira E, Duggan DJ et al. Caveolin-3 in muscular dystrophy. Hum Mol Gen 1998; 7(5):871-877.

35. Merlini L, Carbone I, Capanni C et al. Familial isolated hyperCKaemia associated with a new mutation in the caveolin-3 (CAV-3) gene. J Neurol Neurosurg Psychiatry 2002; 73(1):65-67.

36. Tateyama M, Aoki M, Nishino I et al. Mutation in the caveolin-3 gene causes a peculiar form of distal myopathy. Neurology 2002; 58(2):323-325.

37. Kubisch C, Schoser BG, von During M et al. Homozygous mutations in caveolin-3 cause a severe form of rippling muscle disease. Ann Neurol 2003; 53(4):512-520.

38. Liu J, Aoki M, Illa I et al. Dysferlin, a novel skeletal muscle gene, is mutated in Miyoshi myopathy and limb girdle muscular dystrophy. Nature Genetics 1998; 21(1):31-36.

39. Bashir R, Britton S, Strachan T et al. A gene related to Caenorhabditis elegans spermatogenesis factor fer-1 is mutated in limb-girdle muscular dystrophy type 2B. Nature Genetics 1998; 20(1):37-42.

40. Illa I, Serrano-Munuera C, Gallardo E et al. Distal anterior compartment myopathy: A dysferlin mutation causing a new muscular dystrophy phenotype. Ann Neurol 2001; 49:130-134.

41. Straub V, Campbell KP. Muscular dystrophies and the dystrophin-glycoprotein complex. Current Opinion in Neurology 1997; 10:168-175.

42. Lennon NJ, Kho A, Bacskai BJ et al. Dysferlin interacts with annexins A1 and A2 and mediates sarcolemmal wound-healing. J Biol Chem 2003; 278:50466-50473.

43. Sorimachi H, Kinbara K, Kimura S et al. Muscle-specific Calpain, p94, responsible for limb girdle muscular dystrophy type 2A, Associates with Connectin through IS2, a p94-specific Sequence. Biological Chemistry 1995; 270(52):31155-31162.

44. Herasse M, Ono Y, Fougerousse F et al. Expression and functional characteristics of calpain 3 isoforms generated through tissue-specific transcriptional and posttranscriptional events. Mol Cell Biol 1999; 19(6):4047-4055.

45. Haravuori H, Vihola A, Straub V et al. Secondary calpain3 deficiency in 2q-linked muscular dystrophy: Titin is the candidate gene. Neurology 2001; 56(7):869-877.

46. Speer MC, Gilchrist JM, Stajich JM et al. Anticipation in autosomal dominant limb-girdle muscular dystrophy (LGMD1A). Am J Hum Genet 1994; 55:A7.

47. Hauser MA, Conde CB, Kowaljow V et al. myotilin mutation found in second pedigree with LGMD1A. Am J Hum Genet 2002; 71(6):1428-1432.

48. van der Kooi AJ, van Meegen M, Ledderhof TM et al. Genetic localisation of a newly recognised autosomal dominant limb-girdle muscular dystrophy with cardiac involvement (LGMD1B) to chromosome 1q11-21. Am J Hum Genet 1997; 60:891-895.

49. Bonne G, Mercuri E, Muchir A et al. Clinical and molecular genetic spectrum of autosomal dominant Emery-dreifuss muscular dystrophy due to mutations of the lamin A/C gene. Ann Neurol 2000; 48:170-180.

50. Bashir R, Strachan T, Keers S et al. A gene for autosomal recessive limb-girdle muscular dystrophy maps to chromosome 2p. Hum Mol Gen 1994; 3:455-457.

51. Passos-Bueno MR, Bashir R, Moreira ES et al. Confirmation of the 2p locus for late-onset autosomal recessive limb-girdle muscular dystrophy and refinement of the candidate region. Genomics 1995; 27:192-195.

52. Miyoshi K, Kawai H, Iwasa M et al. Autosomal recessive distal muscular dystrophy as a new type of progressive muscular dystrophy. Brain 1986; 109:31-54.

53. Linssen WHJP, Notermans NC, van der Graaf Y et al. Miyoshi-type distal muscular dystrophy. Clinical spectrum in 24 Dutch patients. Brain 1997; 120:1989-1996.

54. Weiler T, Greenberg CR, Nylen E et al. Limb-girdle muscular dystrophy and Miyoshi myopathy in an Aboriginal Canadian kindred map to LGMD2B and segregate with the same haplotype. Am J Hum Genet 1996; 59:872-878.

55. Illarioshkin SN, Ivanova-Smolenskaya IA, Tanaka H et al. Clinical and molecular analysis of a large family with three distinct phenotypes of progressive muscular dystrophy. Brain 1996; 119(6):1895-1909.

56. Weiler T, Bashir R, Anderson LVB et al. Identical mutation in patients with limb girdle muscular dystrophy type 2B or Miyoshi myopathy suggests a role for modifier gene(s). Hum Mol Gen 1999; 8(5):871-877.

57. Argov Z, Sadeh M, Mazor K et al. Muscular dystrophy due to dysferlin deficiency in Libyan Jews. Clinical and genetic features. Brain 2000; 123:1229-1237.

58. Illa I, Serrano-Munuera C, Gallardo E et al. Immunohistochemical study of inflammation on muscle biopsies from patients with dysferlinopathy. Neurology 2001; 56:A210.
59. Selcen D, Stilling G, Engel AE. The earliest pathologic alterations in dysferlinopathy. Neurology 2001; 56:1472-1481.
60. Piccolo F, Moore SA, Ford GC et al. Intracellular accumulation and reduced sarcolemmal expression of dysferlin in limb-girdle muscular dystrophies. Annals of Neurology 2000; 48:902-912.
61. Achanzar WE, Ward S. A nematode gene required for sperm vesicle fusion. Cell Science 1997; 110:1073-1081.
62. Britton S, Freeman T, Keers S et al. The third human FER-1 like protein is highly homologous to dysferlin. Genomics 2000; 68:313-321.
63. Yasunaga S, Grati M, Cohen-Salmon M et al. A mutation in OTOF, encoding otoferlin, a FER-1-like protein, causes DFNB9, a nonsyndromic form of deafness. Nat Genet 1999; 21(4):363-369.
64. Belt Davis D, Delmonte AJ, Ly CT et al. Myoferlin, a candidate gene and potential modifier of muscular dystrophy. Hum Mol Gen 2000; 9(2):217-226.
65. Doherty KR, McNally EM. Repairing the tears: Dysferlin in muscle membrane repair. Trends Mol Med 2003; 9(8):327-330.
66. Davis DB, Doherty KR, Delmonte AJ et al. Calcium-sensitive phospholipid binding properties of normal and mutant ferlin C2 domains. J Biol Chem 2002; 277(25):22883-22888.
67. Anderson LVB, Davison K, Moss JA et al. Dysferlin is a plasma membrane protein and is expressed early in human development. Hum Mol Gen 1999; 8(5):855-861.
68. Matsuda C, Aoki M, Hayashi YK et al. Dysferlin is a surface membrane-associated protein that is absent in Miyoshi myopathy. Neurology 1999; 53:1119-1122.
69. Prelle A, Sciacco M, Tancredi L et al. Clinical, morphological and immunological evaluation of six patients with dysferlin deficiency. Acta Neuropathol (Berl) 2003; 105(6):537-542.
70. Matsuda C, Hayashi YK, Ogawa M et al. The sarcolemmal proteins dysferlin and caveolin-3 interact in skeletal muscle. Hum Mol Genet 2001; 10:1761-1766.
71. McNeil PL, Terasaki M. Coping with the inevitable: How cells repair a torn surface membrane. Nat Cell Biol 2001; 3(5):E124-E129.

CHAPTER 9

Caveolin-3 and Limb-Girdle Muscular Dystrophy

Ferruccio Galbiati and Michael P. Lisanti

Abstract

Caveolin-3 is the principal structural protein component of caveolae membrane domains in skeletal muscle cells. Caveolae are plasma membrane invaginations implicated in the regulation of signal transduction events. The roles that caveolin-3 plays in skeletal muscle cell physiology are becoming more apparent. Several mutations within the human caveolin-3 gene have been identified over the last few years. These mutations are responsible for different forms of muscle diseases, including limb-girdle muscular dystrophy type 1C (LGMD-1C), hyperCKemia (HCK), distal myopathy (DM), and rippling muscle disease (RMD). In this chapter, we will discuss the functional significance of these caveolin-3 mutations in humans.

Caveolae and Caveolins

Caveolae are 50-100 nm flask-shaped invaginations of the plasma membrane (Fig. 1). Caveolin is the principal protein component of caveolae membranes in vivo.[1-3] The caveolin gene family consists of three members, namely caveolin-1, -2, and -3. Caveolin-1 and caveolin-2 contribute to the formation of caveolae in many cell types. Caveolin-3 is the only caveolin family member that is expressed in striated muscle cell types (i.e., skeletal muscle and heart). Several independent investigations have demonstrated that caveolins play a crucial role in cell signaling—by acting as scaffolding proteins to concentrate, organize, and functionally modulate signaling molecules.[4-7] These molecules include H-Ras, Src-like tyrosine kinases, nitric oxide synthase (NOS) isoforms, hetero-trimeric G-proteins, protein kinase A (PKA), protein kinase C (PKC), and components of the p42-44 mitogen activated protein (MAP) kinase pathway.[8-24]

Functional Roles of Caveolin-3 in Skeletal Muscle Fibers

Caveolin-3 is not expressed in undifferentiated myoblasts, but only in differentiated multinucleated myotubes in vitro.[25] In myotubes, caveolin-3 is localized at the plasma membrane[25] (Fig. 2). Consistent with this data, caveolin-3 was found at the sarcolemma (skeletal muscle cell plasma membrane) in mature skeletal muscle tissue in vivo[26,27] (Fig. 2). Interestingly, caveolin-3 was observed in T-tubules during the differentiation of myoblasts in culture and during skeletal muscle development in mice.[28] These results suggest that caveolin-3 may have a role in the formation of the T-tubule system during muscle development. In support of this hypothesis, Lisanti, Galbiati and colleagues have demonstrated that caveolin-3 (-/-) null mice, which do not express caveolin-3, show T-tubule abnormalities.[27] Thus, it appears that caveolin-3 localization changes during muscle formation: caveolin-3 is localized mainly within T-tubules in developing muscle, while it is found primarily at the sarcolemma in adult mature skeletal muscle.

Molecular Mechanisms of Muscular Dystrophies, edited by Steve J. Winder. ©2006 Eurekah.com.

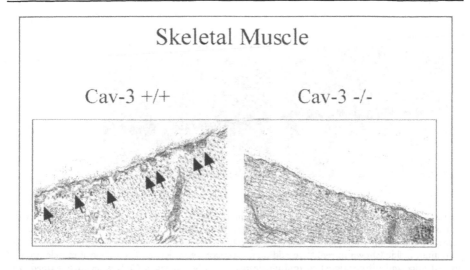

Figure 1. Caveolae in skeletal muscle fibers. Transmission electron micrographs of skeletal muscle tissue from normal (Cav-3 +/+) and caveolin-3 null (Cav-3 -/-) mice. Note the presence of flask-shaped invaginations at the plasma membrane of normal skeletal muscle (left panel).[27] In contrast, note that a lack of caveolin-3 results in loss of caveolae at the sarcolemma (right panel).[27] These results indicates that caveolin-3 is clearly responsible for the formation of caveolae membrane invaginations in skeletal muscle cells.

Caveolin-3 interacts with a number of signaling molecules at the plasma membrane of skeletal muscle cells. For example, c-Src, Src-like kinases (Lyn), and hetero-trimeric G-proteins were found to be associated with caveolae membranes in C2C12 cells, a myoblast cell line,[29] suggesting that caveolin-3 may functionally modulate Src- and G-protein-mediated signaling pathways.

There are a number of lines of evidence suggesting that caveolin-3 may also play a role in the regulation of energy metabolism in skeletal muscle fibers. In fact, caveolin-3 was shown to interact with phosphofructokinase-M (PFK-M; M, muscle-specific isoform) in C2C12 cells.[30] PFK-M is a key regulatory enzyme in the glycolytic pathway within skeletal muscle. In addition, the interaction between caveolin-3 and PFK-M was positively modulated by extracellular glucose and stabilized by activators of PFK-M.[30]

Glucose uptake is essential for normal muscle functioning. Interestingly, caveolin-3 was able to stimulate the phosphorylation of IRS-1, a downstream regulator of the insulin receptor, if coexpressed with the insulin receptor itself.[31] In support of a possible role for caveolae in mediating insulin signaling, the ligand-bound insulin receptor was localized to plasma membrane caveolae by electron microscopy in adipocytes.[32]

Neuronal nitric oxide synthase (nNOS) mediates nitric oxide production in skeletal muscle cells. Caveolin-3 has been demonstrated to directly interact with nNOS in vitro.[33,34] Interaction with caveolin-3 resulted in the inhibition of nNOS enzymatic activity.[33,34] Thus, it appears that caveolin-3 may functionally regulate nitric oxide-dependent functions in muscle tissue. Consistent with these data, Galbiati and colleagues have shown that the transgenic over-expression of caveolin-3 in the heart results in the inhibition of NOS activity in vivo.[35] These authors demonstrated that the increased interaction between caveolin-3 and both endothelial NOS (eNOS), and nNOS, in the hearts of caveolin-3 transgenic mice, represents a valid molecular explanation for the described NOS inhibition.[35]

Dystrophin and its associated glycoproteins, such as dystroglycans and sarcoglycans, form the "dystrophin complex" in skeletal muscle cells (Fig. 3). The major role of this complex is to act as a link between the extracellular matrix and intracellular cytoskeletal elements. Several

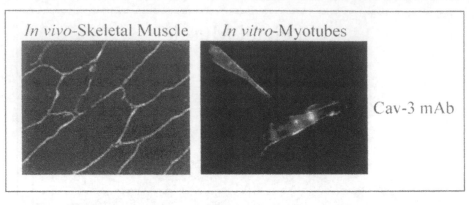

Figure 2. Caveolin-3 is expressed at the sarcolemma of skeletal muscle cells, both in vivo and in vitro. Caveolin-3 localization was assessed in skeletal muscle tissue from normal mice by immunolfuorescence analysis using an antibody probe specific for caveolin-3 (left panel). Nuclei were counterstained with propidium iodide (shown in red).[26] In addition, myoblasts were derived from skeletal muscle tissue of normal mice and differentiated to multinucleated myotubes in vitro. Cells were then stained with anti-caveolin-3 IgGs (right panel).[25] Note that caveolin-3 (shown in green) is localized at the plasma membrane of skeletal muscle cells both in vivo and in vitro. To view color versions of figures, go to http://www.eurekah.com/chapter.php?chapid=1616&bookid=133&catid=20.

independent investigations have shown that caveolin-3 participates in the organization of the dystrophin complex in muscle. For example, dystrophin, α-sarcoglycan, and β-dystroglycan were found within caveolae membranes in differentiated C2C12 cells.[29] Consistent with this data, dystrophin coimmunoprecipitated and coimmunolocalized with caveolin-3 in differentiated C2C12 cells.[29] In addition, dystrophin was localized to caveolae membranes in smooth muscle cells by immunoelectron microscopy.[36] These data indicate that caveolin-3 is dystrophin-associated, although the biogenesis of the dystrophin complex does not absolutely require caveolin-3. In fact, caveolin-3 can be physically separated from the dystrophin complex under certain conditions.[37]

Interestingly, dystrophin and its associated glycoproteins are dramatically reduced in transgenic mice over-expressing caveolin-3, which show a Duchenne-like muscular dystrophy phenotype.[26] The virtual absence of the dystrophin complex from skeletal muscle tissue of caveolin-3 transgenic mice may be explained by considering that the WW-like domain of caveolin-3 binds the PPXY sequence at the C-terminus of β-dystroglycan, which is the same binding site recognized by the WW domain of dystrophin.[38](Fig. 3). As dystrophin is anchored to the plasma membrane through direct binding to β-dystroglycan, over-expression of caveolin-3 may competitively displace dystrophin from the plasma membrane and promote, as a consequence, its degradation.

Caveolinopathies

Caveolinopathies are a class of muscle diseases associated with mutations in the human caveolin-3 gene. They include hyperCKemia (HCK), distal myopathy (DM), rippling muscle disease (RMD), and limb-girdle muscular dystrophy type 1C (LGMD-1C) (Table 1).

HyperCKemia

Sporadic

Creatine kinase (CK) is a cytosolic enzyme. Upon muscle injury (either lysis or necrosis), the enzyme is released into the blood. As a consequence, elevated CK levels in the blood are an indication of myopathy. Minetti, Lisanti and colleagues have reported a mutation in the

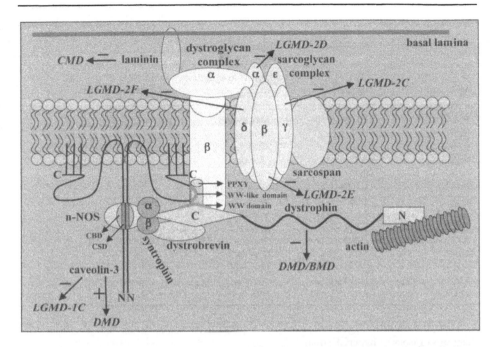

Figure 3. The dystrophin glycoprotein complex (DGC) and associated proteins at the muscle cell plasma membrane. The dystrophin glycoprotein complex links the extracellular matrix to cytoskeletal elements. β-dystroglycan is the only component of the DGC that directly interacts with dystrophin. The WW domain within dystrophin and the WW-like domain within caveolin-3 both bind to the PPXY motif at the C-terminus of β-dystroglycan. Up-regulation of caveolin-3 (as seen in Duchenne muscular dystrophy patients, in mdx mice, and in caveolin-3 over-expressing transgenic mice) can displace dystrophin from the plasma membrane, inducing destabilization and, eventually, degradation of the protein. Arrows indicate diseases caused by the absence (-) or up-regulation (+) of the indicated protein. DMD: Duchenne muscular dystrophy; BMD: Becker muscular dystrophy; CMD: congenital muscular dystrophy; LGMD-1C, -2C, -2D, -2E, -2F: Limb-girdle muscular dystrophy type 1C, -2C, -2D, -2E, -2F; nNOS, neuronal form of nitric oxide synthase; CSD, caveolin scaffolding domain; CBD, caveolin binding domain. Modified from ref. 56.

caveolin-3 gene in two unrelated children with persistent elevated blood levels of creatine kinase.[39] These patients did not show muscle weakness or other symptoms of myopathy. Sequence analysis of the caveolin-3 gene indicated a substitution of a glutamine for an arginine at amino acid position 26 (R26Q). The expression of caveolin-3 was only partially reduced in the two children. Heterologous expression of the R26Q caveolin-3 mutant (Cav-3 R26Q) in culture indicated that Cav-3 R26Q is expressed at significantly lower levels as compared with wild-type caveolin-3, and is retained at the level of the Golgi complex.[40] However, Cav-3 R26Q did not behave in a dominant negative fashion when coexpressed with wild-type caveolin-3. These results may explain the partial reduction of caveolin-3 expression in the muscle biopsies of these two patients.

Familial

More recently, a novel mutation in the caveolin-3 gene was described in two patients (mother and son) with isolated elevated CK levels, but without muscle symptoms.[41] Genetic analysis revealed a proline to leucine substitution at amino acid position 28 (P28L), the first caveolin-3 mutation associated with isolated familial hyperCKemia. Muscle biopsies from these two patients indicated a reduced caveolin-3 membrane association. This phenotype appears similar to that observed in patients with the caveolin-3 R26Q mutation.

Table 1. Caveolin-3 mutations in caveolinopathies

Mutation	Disease	Down-Regulation of Caveolin-3	References
R26Q	LGMD-1C	Moderate	51,52
R26Q	HCK	Moderate	39,40
R26Q	DM	Moderate	42
R26Q	RMD	Moderate	43-45
P28L	HCK	Moderate	41
A45T	LGMD-1C	Severe	50
A45T	RMD	Severe	43
A45V	RMD	Severe	43
T63P	LGMD-1C	Severe	51
ΔTFT (63-65)	LGMD-1C	Severe	47-49
L86P	RMD	Severe	46
A92T	RMD	Severe	46
P104L	LGMD-1C	Severe	47-49
P104L	RMD	Severe	43

Thus, these results suggest that caveolin-3 expression should be evaluated in the differential diagnosis of isolated hyperCKemia.

Distal Myopathy

The same substitution of a glutamine for an arginine at amino acid position 26 (R26Q) in the caveolin-3 gene was described in one patient with sporadic distal myopathy. In this patient, muscle atrophy was restricted to the small muscles of the hands and feet.[42] Caveolin-3 expression was reduced in these muscle fibers, as demonstrated by immunohistochemistry and Western blotting analysis. These data are indicative of the clinical heterogeneity of myopathies with mutations in the caveolin-3 gene.

Rippling Muscle Disease

Rippling muscle disease is a muscle disorder characterized by involuntary, mechanically induced contractions of skeletal muscle. Interestingly, three different groups reported the caveolin-3 R26Q mutation in patients with rippling muscle disease.[43-45] These patients are representative of both autosomal dominant and sporadic cases. Caveolin-3 expression is reduced in the skeletal muscle of these patients. Interestingly, Schroder and colleagues have shown that the Cav-3 R26Q mutation gives rise to 3 different disease phenotypes—rippling muscle disease, distal myopathy, and limb girdle muscular dystrophy—all within the same family [44]

The caveolin-3 R26Q mutation is not the only one associated with rippling muscle disease. Additional mutations within the human caveolin-3 gene have been shown to cause RMD by Kubisch and colleagues. More precisely, the caveolin-3 A45T and P104L mutations were found in three families with rippling muscle disease.[43] Moreover, the same authors described the caveolin-3 A45V and A45T mutations in two additional families with RMD. All these mutants were expressed at lower levels at the plasma membrane, compared to wild type caveolin-3, if transfected into C2C12 cells.

Vorgered and colleagues have reported two novel homozygous mutations in the caveolin-3 gene (L86P and A92T) in two unrelated patients with an unusually severe form of RMD.[46] Immunohistochemical and immunoblot analysis of muscle biopsies from these patients revealed a severe reduction of caveolin-3 expression. Caveolae were almost completely absent from the sarcolemma, as demonstrated by electron microscopy.

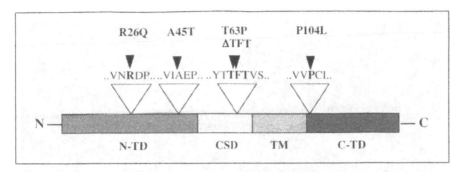

Figure 4. Schematic diagram summarizing Cav-3 gene mutations associated with Limb-girdle Muscular Dystrophy in humans. Four point mutations (R26Q, A45T, T63P, and P104L) and one deletion (ΔTFT, deletion of amino acids 63-65) have been described in LGMD-1C in humans. The point mutations R26Q and A45T occur within the N-terminal domain (N-TD) of caveolin-3. The point mutation T63P and the microdeletion TFT (amino acids 63-65) are located within the caveolin scaffolding domain (CSD). The point mutation P104L occurs just between the transmembrane domain (TM) and the C-terminal domain (C-TD). The identity of twelve amino acid residues is conserved in all three human caveolins, as well as the two C. elegans caveolins.[47,57] Interestingly, R26, F64, and P104 represent three of the twelve conserved residues. Thus, these LGMD-1C mutations provide genetic evidence that these residues are important in vivo.

Caveolin-3 Mutations in LGMD-1C

Limb-girdle muscular dystrophies are a genetically heterogeneous group of disorders. They are characterized by weakness affecting the pelvic and shoulder girdle musculature. The diagnostic criteria include elevated CK, proximal muscular dystrophy, and, in many cases, cardiomyopathy. LGMDs range from severe forms with early onset and rapid progression, to milder forms with later onset and slower progression (see Chapter 8).

The inheritance of LGMDs may be autosomal dominant (LGMD-1) or recessive (LGMD-2). To date, at least sixteen LGMD genes have been mapped. Six of them are autosomal dominant (LGMD-1A→F), and ten autosomal recessive (LGMD-2A→J). Many of the LGMD genes have been identified and cloned (see Chapter 8). Among them, LGMD-1C is caused by a number of mutations in the human caveolin-3 gene. These mutations include the following amino acid substitutions: P104L, ΔTFT/63-65 (deletion of amino acids 63 to 65), A45T, T63P, and R26Q (Fig. 4 and Table 1).

P104L and ΔTFT/63-65

The first mutations in the caveolin-3 gene were reported by Minetti, Sotgia, Lisanti, and colleagues in 1998.[47] They demonstrated that autosomal dominant limb-girdle muscular dystrophy (termed LGMD-1C) was due to (i) a missense mutation within the membrane spanning domain (P104L), and (ii) a 9-base pair microdeletion that removes three amino acids within the caveolin scaffolding domain (ΔTFT/63-65). Patients with LGMD-1C displayed ~95% reduction of caveolin-3 protein expression. In contrast, the expression of dystrophin and dystrophin-associated glycoproteins was not affected by the loss of caveolin-3 expression. Calf hypertrophy and mild-to-moderate proximal muscle weakness were the clinical features of these patients.

A molecular characterization was then performed in vitro by Lisanti, Galbiati and colleagues. They over-expressed caveolin-3 P104L (Cav-3 P/L) and ΔTFT/63-65 (Cav-3 ΔTFT) in NIH 3T3 cells and demonstrated that these mutants are expressed at lower levels, as compared with wild-type caveolin-3, and undergo degradation through a proteasome-dependent pathway.[48,49] In addition, they showed that Cav-3 P/L and Cav-3 ΔTFT are retained at the level of the Golgi complex (Fig. 5) and behave in a dominant negative fashion, causing the degradation, and retention at the Golgi complex, of wild-type caveolin-3 as well.[48,49] Interestingly,

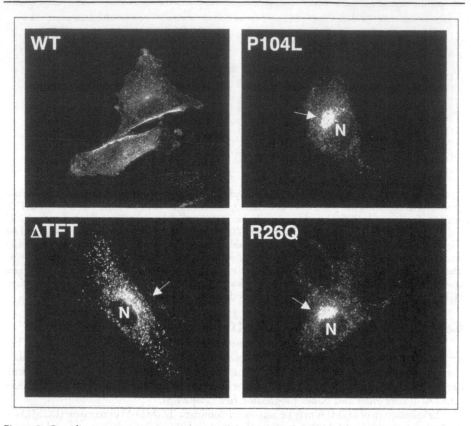

Figure 5. Caveolin-3 mutants are retained intracellularly in L6 skeletal muscle myoblasts. L6 cells were transiently transfected with Cav-3 (WT), Cav-3 (R26Q), Cav-3 (P104L), or Cav-3 (ΔTFT [63-65]). Thirty-six hours after transfection, cells were immunostained with a monoclonal antibody directed against Cav-3. Cav-3 WT localizes to the plasma membrane. In contrast, Cav-3 (R26Q), Cav-3 (P104L), and Cav-3 (ΔTFT) are all retained intracellularly in a peri-nuclear (Golgi) compartment. Reproduced from ref. 40.

treatment with proteosomal inhibitors prevented the degradation of LGMD-1C mutants, although they did not reach the plasma membrane. Importantly, proteasomal inhibitor treatment rescued wild-type caveolin-3 when coexpressed with LGMD-1C mutants of caveolin-3. In fact, wild-type caveolin-3 was not degraded and reached the plasma membrane.[48,49] Thus, these results have direct clinical implications for treatment of patients with LGMD-1C.

A45T

Voit and colleagues have shown a heterozygous 136G→A substitution in the caveolin-3 gene in a 4-year-old girl with limb girdle muscular dystrophy type 1C.[50] The patient presented with myalgia and muscle cramps in her dystrophic skeletal muscle. The sporadic missense mutation resulted in an alanine to threonine substitution at amino acid position 45 (A45T). Caveolin-3 expression at the sarcolemma was dramatically reduced in this LGMD-1C patient, suggesting a dominant-negative effect of the mutant on wild-type caveolin-3.[50]

As caveolin-3 participates in the organization of the dystrophin complex, the authors examined the expression of dystrophin and its associated glycoproteins. They demonstrated that the expression of α-dystroglycan was almost completely lost in the skeletal muscle of the LGMD-1C patient.[50] In contrast, the expression of β-dystroglycan, α-, β-, γ-, δ-sarcoglycan, dystrophin, and α$_2$-laminin was normal.[50]

nNOS has been shown to directly interact with caveolin-3 in vitro.[33,34] Thus, these authors asked whether the expression of nNOS was affected by the near complete loss of caveolin-3 in the LGMD-1C patient. They demonstrated that nNOS expression is dramatically reduced in the Cav-3 A45T-expressing muscle, suggesting that alterations of nitric oxide production may represent a possible molecular explanation for the muscle disease observed in this patient.[50]

T63P

The novel missense mutation T63P has been described by Brown and colleagues in an eleven-year-old Japanese girl.[51] The patient had proximal muscle weakness, exercise-induced myalgia, and mildly elevated CK levels. In addition, muscle biopsy revealed a marked variation in the diameter of muscle fibers, necrotic and degenerating fibers.[51] In addition, they observed an increased number of fibers with central nuclei, which is indicative of muscle regeneration.[51] Caveolin-3 expression was significantly reduced, but not absent, in this patient. Interestingly, the authors reported that dysferlin, a surface membrane protein whose deficiency results in LGMD-2B, is mislocalized in the skeletal muscle tissue of the Cav-3 T63P patient. More precisely, although the total protein expression is unaffected, dysferlin displayed a patchy staining with some cytoplasmic localization.[51] In contrast, dystrophin and α-sarcoglycan immunostaining was normal. Interestingly, they also demonstrated that caveolin-3 coimmunoprecipitated with dysferlin in normal skeletal muscle tissue.[51] Thus, these data suggest that caveolin-3 may play a role in the anchoring of dysferlin to the plasma membrane, and that reduced caveolin-3 expression may result in a weaker association of dysferlin to the sarcolemma (see Chapter 8).

R26Q

As discussed above, the R26Q substitution was reported in patients with hyperCKemia, distal myopathy, and rippling muscle disease. However, the Cav-3 R26Q mutation was also described in patients with limb-girdle muscular dystrophy. Pellissier and colleagues reported a 71-year-old woman with LGMD associated with a R26Q mutation in the caveolin-3 gene.[52] Caveolin-3 expression was reduced at the plasma membrane of skeletal muscle tissue from this patient. A similar reduction of caveolin-3 expression was reported by Brown and colleagues in skeletal muscle tissue from a LGMD patient with an R26Q substitution.[51] Thus, it appears that the arginine to glutamine substitution at amino acid position 26 within the human caveolin-3 gene can lead to various clinical phenotypes, including HCK, DM, RMD, and LGMD-1C. As described earlier in this chapter, the R26Q mutation was also found in HCK, DM, RMD, and LGMD-1C patients—all within the same family.[44]

Mouse Models of LGMD-1C

Caveolin-3 P104L Transgenic Mice

The first mouse model of LGMD-1C was generated by Shimizu and colleagues (Table 2). These authors expressed the Cav-3 P104L mutant in mouse skeletal muscle tissue as a transgene.[53] As previously discussed, the P104L mutation is one of the first two mutations described within the caveolin-3 gene in LGMD patients. Caveolin-3 is virtually absent from the sarcolemma of Cav-3 P104L transgenic mice.[53] This is consistent with the autosominal dominant form of genetic transmission of the Cav-3 P104L mutation in humans,[47] and the dominant-negative effect that this mutant form exercises on wild-type caveolin-3.[48] Transgenic mice expressing mutant caveolin-3 showed a myopathic phenotype resembling LGMD-1C in humans.[53] However, the muscle damage was much more severe in caveolin-3 P104L transgenic mice, as compared with LGMD-1C patients. Given the fact that the Cav-3 P104L mutant promotes the degradation of wild-type caveolin-3, this discrepancy may be explained considering that additional signaling molecules that directly interact with caveolin-3 may undergo degradation in the skeletal muscle of Cav-3 P104L transgenic mice.

Table 2. Mouse models of LGMD-1C

Genotype	Phenotype
Cav-3 P104L Tg Mice	• Dramatic reduction of caveolin-3 expression; • Smaller than normal littermates; • Severe myopathic features: variation in fiber size, atrophic fibers, central nuclei, increased endomysium; • Normal expression of dystrophin and β–dystroglycan; and • Increased nNOS activity
Cav-3 (-/-) Null Mice	• Loss of caveolin-3 and caveolae at the sarcolemma; • Mild-to-moderate myopathic changes: variability in muscle fiber size and presence of necrotic fibers; • Normal expression of dystrophin and its associated glycoproteins; • Exclusion of the dystrophin-glycoprotein complex from lipid raft domains; and • T-tubule system abnormalities: dilated and longitudinally oriented T-tubules

Tg, transgenic; Null, knock-out or targeted gene disruption

The authors also evaluated the effect of the Cav-3 P104L mutant on the dystrophin complex. They showed that both dystrophin and β-dystroglycan protein expression in skeletal muscle is unchanged in these transgenic mice, as assessed by Western blotting analysis.[53] This result is consistent with the observations of Minetti, Lisanti and colleagues in LGMD patients with the P104L mutation.[47] Interestingly, nNOS activity was significantly increased in skeletal muscle tissue expressing the caveolin-3 mutant.[53] This data is supported by the previously reported ability of caveolin-3 to inhibit nNOS activity in vitro.[33,34] Thus, it is possible to speculate that elevated nNOS activity may explain, at least in part, the myopathic phenotype observed in the muscle of Cav-3 P104L transgenic mice.

Caveolin-3 (-/-) Null Mice

LGMD-1C mutations within the caveolin-3 gene, with the exception of the R26Q substitution, induce a dramatic reduction of total caveolin-3 protein expression in the skeletal muscles of LGMD-1C patients. For this reason, caveolin-3 (-/-) null mice, which do not express caveolin-3 in skeletal muscle tissue, represent a valid mouse model for studying this form of muscular dystrophy. Lisanti, Galbiati and colleagues have generated caveolin-3 null mice and demonstrated that a lack of caveolin-3 induces a mild muscle myopathy[27](Table 2). Interestingly, this data is consistent with the relative weak dystrophic phenotype observed in patients with LGMD-1C.[47] In addition, as observed in LGMD-1C patients,[47] the expression of dystrophin, α-sarcoglycan, and β-dystroglycan is not affected by the loss of caveolin-3,[27] However, these proteins are excluded from cholesterol-enriched plasma membrane microdomains, termed lipid rafts.[27] It is possible to speculate that such mislocalization may contribute to the dystrophic phenotype observed in caveolin-3 null mice.

As discussed earlier, caveolin-3 has been proposed to mediate the development of the T-tubule system. A number of experimental observations indicate that the T-tubule system is altered in caveolin-3 null mice. Dihydropyridine receptor-1α and ryanodine receptor, two markers of the T-tubules system, were mislocalized in caveolin-3 null mice, as demonstrated by immunofluorescence analysis.[27] In addition, T-tubules were specifically stained with potassium ferrocyanate and visualized by electron microscopy. T-tubules appeared dilated, swollen, and ran in irregular directions in skeletal muscle tissue of caveolin-3 null mice, in contrast to an orderly transverse orientation in normal control wild-type muscle.[27]

Consistent with this data, disorganization of the T-tubule system was reported in skeletal muscle biopsies from LGMD-1C patients.[54] T-tubules play a key role in skeletal muscle functioning, as they participate in the signaling events that lead to muscle contraction. Thus, the alteration of the T-tubule system may partially explain the muscle weakness reported in LGMD-1C patients.[47]

The dystrophic phenotype of caveolin-3 null mice observed by Lisanti, Galbiati and colleagues is consistent with data published by Kikuchi et al which have also developed caveolin-3-deficient mice.[55] Lack of caveolin-3 resulted in the loss of caveolae at the sarcolemma, without affecting the level of expression of dystrophin and its associated glycoproteins.[55] The authors also demonstrated muscle degeneration in the soleus and diaphragm of caveolin-3 null mice.[55]

Conclusions

As discussed throughout this chapter, nine different mutations within the human caveolin-3 gene have been associated with a variety of disease phenotypes, including hyperCKemia, distal myopathy, rippling muscle disease, and limb-girdle muscular dystrophy type 1C. The existence of distinct phenotypes in patients with the same mutation, even within the same family, would suggest that specific polymorphisms within modified genes may influence the individual phenotype. One of the many challenges ahead of us will be the identification of possible modifying factors and/or genes in the individual genetic background of affected patients that may contribute to the four different clinical phenotypes observed in caveolinopathies.

Acknowledgments

We thank Dr. R. Campos-Gonzalez (BD-Pharmingen/Transduction Laboratories) for donating mAbs directed against caveolin-3. M.P.L. was supported by grants from the National Institutes of Health (NIH), and the Susan G. Komen Breast Cancer Foundation, as well as a Hirschl/Weil-Caulier Career Scientist Award. F.G. was supported by a grant from the American Heart Association (AHA) and start-up funds from the Department of Pharmacology at the University of Pittsburgh.

References

1. Glenney JR. Tyrosine phosphorylation of a 22 kD protein is correlated with transformation with Rous sarcoma virus. J Biol Chem 1989; 264:20163-20166.
2. Glenney JR, Soppet D. Sequence and expression of caveolin, a protein component of caveolae plasma membrane domains phosphorylated on tyrosine in RSV-transformed fibroblasts. Proc Natl Acad Sci USA 1992; 89:10517-10521.
3. Rothberg KG, Heuser JE, Donzell WC et al. Caveolin, a protein component of caveolae membrane coats. Cell. 1992; 68:673-682.
4. Lisanti MP, Scherer P, Tang Z-L et al. Caveolae, caveolin and caveolin-rich membrane domains: A signalling hypothesis. Trends In Cell Biology. 1994; 4:231-235.
5. Couet J, Li S, Okamoto T et al. Molecular and cellular biology of caveolae: Paradoxes and Plasticities. Trends in Cardiovascular Medicine 1997; 7:103-110.
6. Okamoto T, Schlegel A, Scherer PE et al. Caveolins, A family of scaffolding proteins for organizing "preassembled signaling complexes" at the plasma membrane. J Biol Chem (Mini-review) 1998; 273:5419-5422.
7. Sargiacomo M, Scherer PE, Tang Z-L et al. Oligomeric structure of caveolin: Implications for caveolae membrane organization. Proc Natl Acad Sci USA 1995; 92:9407-9411.
8. Song KS, Li S, Okamoto T et al. Copurification and direct interaction of Ras with caveolin, an integral membrane protein of caveolae microdomains. Detergent free purification of caveolae membranes. J Biol Chem 1996; 271:9690-9697.
9. Li S, Couet J, Lisanti MP. Src tyrosine kinases, G alpha subunits and H-Ras share a common membrane-anchored scaffolding protein, Caveolin. Caveolin binding negatively regulates the auto-activation of Src tyrosine kinases. J Biol Chem 1996; 271:29182-29190.

10. Garcia-Cardena G, Oh P, Liu J et al. Targeting of nitric oxide synthase to endothelilal cell caveolae via palmitoylation: Implications for caveolae localization. Proc Natl Acad Sci USA 1996; 93:6448-6453.

11. Smart E, Ying Y-S, Conrad P et al. Caveolin moves from caveolae to the Golgi apparatus in response to cholesterol oxidation. J Cell Biol 1994; 127:1185-1197.

12. Moldovan N, Heltianu C, Simionescu N et al. Ultrastructural evidence of differential solubility in Triton X-100 of endothelial vesicles and plasma membrane. Exp Cell Res 1995; 219:309-313.

13. Mineo C, James GL, Smart EJ et al. Localization of EGF-stimulated Ras/Raf-1 interaction to caveolae membrane. J Biol Chem 1996; 271:11930-11935.

14. Liu P, Ying YS, Anderson RGW. PDGF activates MAP kinase in isolated caveolae. Proc Natl Acad Sci USA 1997; 94:13666-13670.

15. Shenoy-Scaria AM, Dietzen DJ, Kwong J et al. Cysteine-3 of Src family tyrosine kinases determines palmitoylation and localization in caveolae. Journal of Cell Biology 1994; 126:353-363.

16. Smart EJ, Foster D, Ying Y-S et al. Protein kinase C activators inhibit receptor-mediated potocytosis by preventing internalization of caveolae. J Cell Biol 1993; 124:307-313.

17. Schnitzer JE, Oh P, Jacobson BS et al. Caveolae from luminal plasmalemma of rat lung endothelium: Microdomains enriched in caveolin, Ca2+-ATPase, and inositol triphosphate receptor. Proc Natl Acad Sci USA 1995; 92:1759-1763.

18. Robbins S, Quintrell N, Bishop JM. Differential palmitoylation of the two isoforms of the Src family kinase, HCK, affects their localization to caveolae. J Cellular Biochem 1995; (Supplement 19A):27 (abstr.).

19. Chang W-J, Ying Y, Rothberg KG et al. Purification and characterization of smooth muscle cell caveolae. Journal of Cell Biology 1994; 126:127-138.

20. Couet J, Sargiacomo M, Lisanti MP. Interaction of a receptor tyrosine kinase, EGF-R, with caveolins. Caveolin binding negatively regulates tyrosine and serine/threonine kinase activities. J Biol Chem 1997; 272:30429-30438.

21. Ju H, Zou R, Venema VJ et al. Direct interaction of endothelial nitric-oxide synthase and caveolin-1 inhibits synthase activity. J Biol Chem 1997; 272(30):18522-18525.

22. Feron O, Belhassen L, Kobzik L et al. Endothelial nitric oxide synthase targeting to caveolae. Specific interactions with caveolin isoforms in cardiac myocytes and endothelial cells. J Biol Chem 1996; 271(37):22810-22814.

23. Segal SS, Brett SE, Sessa WC. Codistribution of NOS and caveolin throughout peripheral vasculature and skeletal muscle of hamsters. Am J Physiol 1999; 277:H1167-1177.

24. Garcia-Cardena G, Fan R, Stern D et al. Endothelial nitric oxide synthase is regulated by tyrsosine phosphorylation and interacts with caveolin-1. J Biol Chem 1996; 271:27237-27240.

25. Volonte D, Peoples AJ, Galbiati F. Modulation of Myoblast Fusion by Caveolin-3 in Dystrophic Skeletal Muscle Cells: Implications for Duchenne Muscular Dystrophy and Limb-Girdle Muscular Dystrophy-1C. Mol Biol Cell 2003; 14:4075-4088.

26. Galbiati F, Volonte D, Chu JB et al. Transgenic overexpression of caveolin-3 in skeletal muscle fibers induces a Duchenne-like muscular dystrophy phenotype. Proc Natl Acad Sci USA 2000; 97:9689-9694.

27. Galbiati F, Engelman JA, Volonte D et al. Caveolin-3 null mice show a loss of caveolae, changes in the microdomain distribution of the dystrophin-glycoprotein complex, and T-tubule abnormalities. J Biol Chem 2001; 276:21425-21433.

28. Parton RG, Way M, Zorzi N et al. Caveolin-3 associates with developing T-tubules during muscle differentiation. J Cell Biol 1997; 136:137-154.

29. Song KS, Scherer PE, Tang Z-L et al. Expression of caveolin-3 in skeletal, cardiac, and smooth muscle cells. Caveolin-3 is a component of the sarcolemma and cofractionates with dystrophin and dystrophin-associated glycoproteins. J Biol Chem 1996; 271:15160-15165.

30. Scherer PE, Lisanti MP. Association of phosphofructokinase-M with caveolin-3 in differentiated skeletal myotubes: Dynamic regulation by extracellular glucose and intracellular metabolites. J Biol Chem 1997; 272:20698-20705.

31. Yamamoto M, Toya Y, Schwencke C et al. Caveolin is an activator of insulin receptor signaling. J Biol Chem 1998; 273:26962-26968.

32. Gustavsson J, Parpal S, Karlsson M et al. Localization of the insulin receptor in caveolae of adipocyte plasma membrane. Faseb J 1999; 13:1961-1971.

33. Garcia-Cardena G, Martasek P, Masters BS et al. Dissecting the interaction between nitric oxide synthase (NOS) and caveolin. Functional significance of the nos caveolin binding domain in vivo. J Biol Chem 1997; 272(41):25437-25440.

34. Venema VJ, Ju H, Zou R et al. Interaction of neuronal nitric-oxide synthase with caveolin-3 in skeletal muscle. Identification of a novel caveolin scaffolding/inhibitory domain. J Biol Chem 1997; 272:28187-28190.
35. Aravamudan B, Volonte D, Ramani R et al. Transgenic overexpression of caveolin-3 in the heart induces a cardiomyopathic phenotype. Hum Mol Genet 2003; 12:2777-2788.
36. North A, Galazkiewicz B, Byers T et al. Complementary distribution of vinculin and dystrophin define two distinct sarcolemma domains in smooth muscle. J Cell Biol 1993; 120:1159-1167.
37. Crosbie RH, Yamada H, Venzke DP et al. Caveolin-3 is not an essential component of the dystrophin glycoprotein complex. FEBS Lett 1998; 427:279-282.
38. Sotgia F, Lee JK, Das K et al. Caveolin-3 Directly Interacts with the C-terminal Tail of beta - Dystroglycan. Identification of a central ww-like domain within caveolin family members. Biol Chem 2000; 275:38048-38058.
39. Carbone I, Bruno C, Sotgia F et al. Mutation in the CAV3 gene causes partial caveolin-3 deficiency and hyperCKemia. Neurology 2000; 54:1373-1376.
40. Sotgia F, Woodman SE, Bonuccelli G et al. Phenotypic behavior of caveolin-3 R26Q, a mutant associated with hyperCKemia, distal myopathy, and rippling muscle disease. Am J Physiol Cell Physiol 2003; 285:C1150-1160.
41. Merlini L, Carbone I, Capanni C et al. Familial isolated hyperCKaemia associated with a new mutation in the caveolin-3 (CAV-3) gene. J Neurol Neurosurg Psychiatry 2002; 73:65-67.
42. Tateyama M, Aoki M, Nishino I et al. Mutation in the caveolin-3 gene causes a peculiar form of distal myopathy. Neurology 2002; 58:323-325.
43. Betz RC, Schoser BG, Kasper D et al. Mutations in CAV3 cause mechanical hyperirritability of skeletal muscle in rippling muscle disease. Nat Genet 2001; 28:218-219.
44. Fischer D, Schroers A, Blumcke I et al. Consequences of a novel caveolin-3 mutation in a large German family. Ann Neurol 2003; 53:233-241.
45. Vorgerd M, Ricker K, Ziemssen F et al. A sporadic case of rippling muscle disease caused by a de novo caveolin-3 mutation. Neurology 2001; 57:2273-2277.
46. Kubisch C, Schoser BG, von During M et al. Homozygous mutations in caveolin-3 cause a severe form of rippling muscle disease. Ann Neurol 2003; 53:512-520.
47. Minetti C, Sotogia F, Bruno C et al. Mutations in the caveolin-3 gene cause autosomal dominant limb-girdle muscular dystrophy. Nature Genetics 1998; 18:365-368.
48. Galbiati F, Volonte D, Minetti C et al. Phenotypic behavior of caveolin-3 mutations that cause autosomal dominant limb girdle muscular dystrophy (LGMD-1C). Retention of LGMD-1C caveolin-3 mutants within the Golgi complex. J Biol Chem 1999; 274:25632-25641.
49. Galbiati F, Volonte D, Minetti C et al. Limb-girdle muscular dystrophy (LGMD-1C) mutants of caveolin-3 Undergo Ubiquitination and Proteasomal Degradation. Treatment with proteasomal inhibitors blocks the dominant negative effect of LGMD-1C mutants and rescues wild-type caveolin-3. J Biol Chem 2000; 275:37702-37711.
50. Herrmann R, Straub V, Blank M et al. Dissociation of the dystroglycan complex in caveolin-3-deficient limb girdle muscular dystrophy. Hum Mol Gene 2000; 9:2335-2340.
51. Matsuda C, Hayashi YK, Ogawa M et al. The sarcolemmal proteins dysferlin and caveolin-3 interact in skeletal muscle. Hum Mol Genet 2001; 10:1761-1766.
52. Figarella-Branger D, Pouget J, Bernard R et al. Limb-girdle muscular dystrophy in a 71-year-old woman with an R27Q mutation in the CAV3 gene. Neurology 2003; 61:562-564.
53. Sunada Y, Ohi H, Hase A et al. Transgenic mice expressing mutant caveolin-3 show severe myopathy associated with increased nNOS activity. Hum Mol Genet 2001; 10:173-178.
54. Minetti C, Bado M, Broda P et al. Impairment of caveolae formation and T-system disorganization in human muscular dystrophy with caveolin-3 deficiency. Am J Pathol 2002; 160:265-270.
55. Hagiwara Y, Sasaoka T, Araishi K et al. Caveolin-3 deficiency causes muscle degeneration in mice. Hum Mol Genet 2000; 9:3047-3054.
56. Galbiati F, Razani B, Lisanti MP. Caveolae and caveolin-3 in muscular dystrophy. Trends Mol Med 2001; 7:435-441.
57. Tang Z, Okamoto T, Boontrakulpoontawee P et al. Identification, sequence, and expression of an invertebrate caveolin gene family from the nematode Caenorhabditis elegans : Implications for the molecular evolution of mammalian caveolin genes. J Biol Chem 1997; 272:2437-2445.
58. Mathews KD, Morris SA. Limb-Girdle Muscular Dystrophy. Current Neurology and Neuroscience Reports 2003; 3:78-85.

CHAPTER 10

The Sarcoglycans

Elizabeth M. McNally

Abstract

The sarcoglycans are transmembrane proteins found as a plasma membrane-associated complex. First characterized as a subunit of the dystrophin glycoprotein complex in skeletal muscle, the sarcoglycan complex is secondary disrupted and destabilized from the plasma membrane when dystrophin is mutated, as in Duchenne Muscular Dystrophy. Autosomal recessive mutations in several sarcoglycan genes, α, β, γ and δ, also lead to disruption of the sarcoglycan complex yielding a similar phenotype to what is seen in Duchenne Muscular Dystrophy and referred to as the type 2 Limb Girdle Muscular Dystrophies. Mutations in δ-sarcoglycan have been reported in human cardiomyopathy patients. The exact function of the sarcoglycan complex is not known, but it appears to have both mechanical and nonmechanical roles in stabilizing the plasma membrane in cardiac and skeletal muscle. There is a variant of the sarcoglycan complex in vascular smooth muscle and in some nonmuscle cell and tissue types. Specifically, ε-sarcoglycan is highly expressed in the central and peripheral nervous systems. Dominant mutations in ε-sarcoglycan lead to an inherited movement disorder. Thus, the sarcoglycan complex has important roles in both muscle and nonmuscle tissues.

Historical Perspective

The dystrophin glycoprotein complex (DGC) is a multimeric, multifaceted collection of proteins found at the plasma membrane of muscle cells (Fig. 1). The DGC has cytoplasmic elements, dystrophin, dystrobrevins and syntrophins, and nitric oxide synthase, and transmembrane elements, dystroglycan, the sarcoglycans and sarcospan. The entire DGC can be purified as a macromolecular complex from detergent solubilized membranes.[1-3] Using an additional detergent solubilization technique, the sarcoglycan complex can be separated from the remainder of the DGC.[4] Mutations in dystrophin lead to Duchenne and Becker muscular dystrophy in humans and in the mdx mouse.[5,6] Mutations that disrupt dystrophin cause instability of the remainder of the DGC including the sarcoglycans.[1,3] Mutations that target the carboxyl-terminus of dystrophin lead to a severe muscular dystrophy phenotype; it is in this region that dystrophin interacts with dystroglycan and anchors the DGC.

Genetics

The characteristic phenotype in DMD, including early onset muscle degeneration and regeneration, calf hypertrophy and elevated creatine kinase (CK) is linked to the X chromosome and is present only in males. In regions of the world with higher rates of consanguinity, a phenotype similar to DMD was noted in both males and females and was termed Severe Childhood Autosomal Recessive Muscular Dystrophy (SCARMD).[7] Genetic linkage analysis on families with SCARMD demonstrated that the genetic defect was located on chromosome 13q12, although some SCARMD families excluded involvement of the chromosome 13 locus, indicating genetic heterogeneity of this disorder.[8-11]

Molecular Mechanisms of Muscular Dystrophies, edited by Steve J. Winder. ©2006 Eurekah.com.

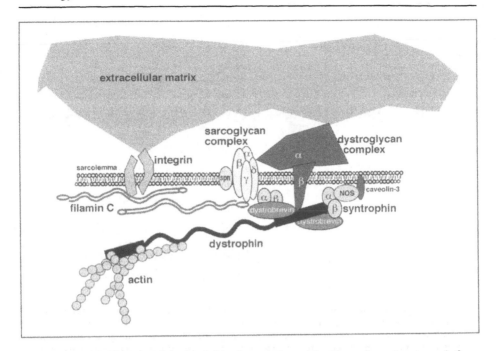

Figure 1. Shown is a schematic of the dystrophin glycoprotein complex. Dystroglycan, is composed of two subunits and interacts with dystrophin in the intracellular, cytoplasmic domain while the extracellular domain binds laminin-α2 in the extracellular matrix. The cytoplasmic proteins of the DGC include the syntrophins, nitric oxide synthase (NOS) and the dystrobrevins. The transmembrane components include β-dystroglycan, the sarcoglycans and sarcospan (spn). Filamin C is a cytoplasmic, actin binding protein of muscle that interacts with the cytoplasmic tails of γ sarcoglycan and δ sarcoglycan and also with the integrin complex.

Biochemistry and Cell Biology

The first biochemical analyses to identify the sarcoglycan complex underestimated the full complexity of this subcomplex. Only through human genetic studies, biochemical analyses and the additional analysis of the human genome sequence has it become clear that there are six sarcoglycan sequences, α, β, γ, δ, ε and ζ. Antibodies were generated to a number of sarcoglycan proteins, initially referred to as 50 kDa dystrophin associated glycoprotein (DAG), 43 kDa DAG and the 35 kDa DAG.[2,3] It was these three elements that were purified from the main DGC and, together were termed the sarcoglycan complex.[4] It was subsequently shown that both the 50 kDa and 35 kDa proteins are heterogeneous in that they are produced from multiple genes (see below). Antibodies directed to the sarcoglycan components and the DGC were used to demonstrate the secondary reduction and degeneration of the DGC that occurs in the muscle of DMD subjects and the mdx mouse.[1,12] Interestingly, SCARMD subjects also showed a reduction of sarcoglycan proteins as did the well-studied small animal model of cardiomyopathy and muscular dystrophy, the BIO 14.6 Syrian hamster model (see below).[13-16] Genetic and mutational analyses in Duchenne-like muscular dystrophy patients revealed that mutations in the genes encoding α-sarcoglycan, β-sarcoglycan, γ-sarcoglycan and δ-sarcoglycan lead to human forms of muscular dystrophy.[16-20]

α-Sarcoglycan

α-sarcoglycan is a type I transmembrane protein that migrates as a 50 kDa protein (Fig. 2). The primary amino acid sequence predicts a protein of approximately 43 kDa with the difference in size related to N-linked glycosylation.[21] The gene encoding human α-sarcoglycan maps

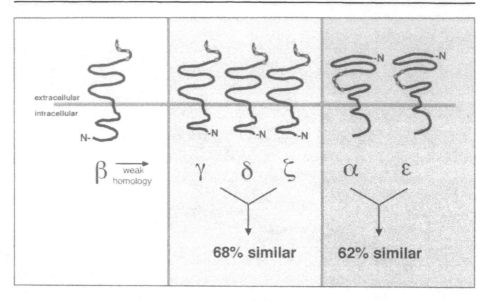

Figure 2. Shown is a schematic of the sarcoglycan proteins. β-, γ-, δ, and ζ-sarcoglycan are type II transmembrane proteins. β-sarcoglycan is only very weakly homologous to γ-, δ- and ζ-sarcoglycan. γ-, δ- and ζ-sarcoglycan have an identical gene structure and likely arose from multiple gene duplications. In nonvertebrates, there is only a single sarcoglycan sequence in this class. α-sarcoglycan and ε-sarcoglycan are type I transmembrane proteins that also appear to have arisen from a gene duplication event as nonvertebrates have only a single one of these sequences. Mutations in α-, β-, γ- and δ-sarcoglycan produce limb girdle muscular dystrophy while mutations in ε-sarcoglycan produce a nonprogressive movement disorder, myoclonus dystonia. The circles represent conserved cysteine residues necessary for intramolecular interaction.

to chromosome 17p21, and autosomal recessive mutations in the α-sarcoglycan have now been described from muscular dystrophy patients from all over the world.[22,23] The majority of mutations in the α-sarcoglycan gene are point mutations that encode missense changes. Compound heterozygous mutations have been noted. These deleterious polymorphisms may affect cysteine residues that are important for secondary structure and potentially, interactions within the sarcoglycan complex. Like the other sarcoglycan gene mutations, in most carriers, heterozygous α-sarcoglycan mutations typically are not associated with any muscle phenotype. Mice engineered to lack α-sarcoglycan develop a progressive muscular dystrophy similar to what has been noted in the mdx mouse that lacks dystrophin.[24,25] Mice lacking α-sarcoglycan do not develop cardiomyopathy, although cardiomyopathy has been noted rarely in human subjects with α-sarcoglycan mutations.[24,26] The α-sarcoglycan protein has homology to ecto ATPases and biochemical studies have suggested that this may be a feature of α-sarcoglycan.[27]

β-Sarcoglycan

β-sarcoglycan is a 43 kDa type II transmembrane protein (Fig. 2).[16,19] There are three potential N-linked glycosylation sites in the extracellular domain and potential phosphorylation sites in the cytoplasmic portion. The gene encoding human β-sarcoglycan is on chromosome 4q12, and recessive mutations in the β-sarcoglycan gene lead to muscular dystrophy and cardiomyopathy. Mice lacking β-sarcoglycan have been generated and develop cardiomyopathy and muscular dystrophy similar to their human counterparts.[28,29] Notably, the smooth muscle sarcoglycan complex is disrupted in these mice since β-sarcoglycan is expressed normally in smooth muscle including vascular smooth muscle.[28]

γ-Sarcoglycan

The gene encoding γ-sarcoglycan maps to human chromosome 13q12, and recessive mutations in the γ-sarcoglycan gene lead to cardiomyopathy and muscular dystrophy.[18] There is a single common mutation, Δ521-T, that is relatively prevalent. This frameshift mutation encodes a series of novel amino acids leading to a stop codon and is thought to produce an unstable, truncated protein. Generally, as a result of this Δ521-T mutation, there is little sarcoglycan expressed at the surface of muscle, although exceptions have been described.[30,31] γ-sarcoglycan is a 35 kDa type II transmembrane protein that has a single N-linked glycosylation site and predicted phosphorylation sites in the cytoplasmic domain (Fig. 2). Mouse models have been engineered that harbor null alleles for γ-sarcoglycan.[32,33] These mice display a phenotype similar to what is seen in β-sarcoglycan null mice (above) and δ-sarcoglycan null mice (see below).

δ-Sarcoglycan

δ-sarcoglycan is highly related to γ-sarcoglycan in that they share nearly 70 percent amino acid similarity (Fig. 2).[34] Moreover, the gene structure for δ-sarcoglycan on human chromosome 5q31 is similar to the gene structure for γ-sarcoglycan. The identical placement of intron and exons in these genes strongly suggest that they arose from a gene duplication event. Recessive mutations in the δ-sarcoglycan gene produce LGMD type 2F, a disorder similar to DMD and BMD.[20] Rare cases of familial dilated cardiomyopathy have also been attributed in mutations in the δ-sarcoglycan gene, although in this case the pattern of inheritance of dilated cardiomyopathy was thought to be autosomal dominant.[35] Like γ-sarcoglycan, δ-sarcoglycan is a type II transmembrane 35 kDa protein with a single N-linked glycosylation site. There is an alternative splice form that alters the very carboxyl-terminus of the protein.[36] This region contains a cluster of conserved cysteine residues that have weak homology to the epidermal growth factor like cysteine residues.[37] The cysteine cluster is found in β-, γ-, δ- and ζ-(see below) sarcoglycan, and may serve as a receptor site for an as yet unknown ligand. The alternative splice form of δ-sarcoglycan lacks these cysteine residues and is widely expressed by mRNA expression analysis. Like β-sarcoglycan, δ-sarcoglycan is expressed in both striated and smooth muscle, including vascular smooth muscle.

There are several different animal models of δ-sarcoglycan mutations. The first, the BIO4.6 Syrian Hamster model is a recessive model of cardiomyopathy and muscular dystrophy that was first described 40 years ago.[38,39] This genetic model of cardiomyopathy displays pathology very similar to what is seen in humans with dystrophin or sarcoglycan gene mutations. That is, focal degeneration occurs in both heart and skeletal in response to dystrophin and sarcoglycan gene mutations. In the BIO 14.6 hamster, the focal nature of tissue damage was initially suggested to relate to vascular spasm since, in the heart, the appearance resembled focal micro-infarcts.[40,41] A mouse model was generated with a null allele of δ-sarcoglycan, and these mice displayed the same focal nature of tissue damage.[42] In these mice, it was noted that the sarcoglycan complex was disrupted in vascular smooth muscle, and it was reasoned that disruption of the vascular smooth muscle sarcoglycan complex was responsible for vascular spasm and tissue damage. This concept was further supported by studies of α-sarcoglycan mutant mice that displayed little cardiomyopathy and had an intact vascular smooth muscle sarcoglycan complex.[24]

An alternative explanation is that the defect is one that arises in striated muscle and that vascular defects arise as a nonvascular smooth muscle cell-autonomous defect. Supporting this idea is the cardiomyopathic findings in mice lacking γ-sarcoglycan. Mice lacking γ-sarcoglycan and δ-sarcoglycan have an identical phenotype with focal degeneration, but the vascular smooth muscle sarcoglycan complex is intact in these γ-sarcoglycan null animals.[43] Abnormal vascular reactivity arises from degeneration in cardiomyocytes and furthers the course of cardiomyopathy progression. Vascular spasm mediated through this vascular smooth muscle-cell xtrinsic process is a target for therapeutic intervention.

ε-Sarcoglycan

ε-sarcoglycan is highly related to α-sarcoglycan and is encoded by a gene on human chromosome 7p21.[44,45] Like α-sarcoglycan, ε-sarcoglycan is type I transmembrane protein (Fig. 2), but unlike α-sarcoglycan, ε-sarcoglycan is expressed in tissues outside of muscle.[44,45] ε-sarcoglycan is expressed highly in the developing nervous system (EMM, unpublished results), and ε-sarcoglycan is expressed in striated and smooth muscle. Interestingly, mutations in the ε-sarcoglycan gene lead to an unusual phenotype of myoclonus dystonia.[46] In this syndrome, ε-sarcoglycan gene mutations lead to a dominant, nonprogressive movement disorder. The genetics of myoclonus dystonia are complicated by variable penetrance related to parent-of-origin effects. This phenomenon is explained by imprinting of the maternal ε-sarcoglycan allele, first noted in mice.[47] Thus, as the maternal allele is silenced, the gene defect, in effect is inherited from the parental allele.[48,49]

The high degree of sequence similarity between α-sarcoglycan and ε-sarcoglycan has been noted.[44,45] It has been suggested that the mild to absent cardiac phenotype associated with α-sarcoglycan mutations may relate to upregulation and compensation by ε-sarcoglycan.[25] In vascular smooth muscle, it is clear that ε-sarcoglycan can substitute for α-sarcoglycan.[50]

ζ-Sarcoglycan

The most recently described sarcoglycan sequence is ζ-sarcoglycan.[51] ζ-sarcoglycan is highly related to both γ-sarcoglycan and δ-sarcoglycan and is similarly a type II 35 kDa protein (Fig. 2). The intron and exon structure of the ζ-sarcoglycan gene suggests a gene triplication event leading to the relationship between γ-, δ- and ζ-sarcoglycan. ζ-sarcoglycan is encoded by a gene on human chromosome 8p22. ζ-sarcoglycan can be purified from muscle microsomal membranes and through immunoprecipitation experiments, was shown to interact with dystrophin and β-sarcoglycans.[51] Moreover, ζ-sarcoglycan expression is reduced in microsomal membranes from muscle with sarcoglycan gene mutations. Genetic mutations have not been described in ζ-sarcoglycan gene so it is unclear whether genetic defects in this gene can lead to muscular dystrophy. In addition to being expressed in striated muscle, ζ-sarcoglycan is highly expressed in smooth muscle sources including vascular smooth muscle (Fig. 3).

Sarcospan

Sarcospan is a member of the tetraspanin family and, as it name implies, contains four transmembrane domains. Initially characterized as the 25 kDa component of the DGC, the identity of sarcospan was clarified by microsequencing.[52] Mice deficient for sarcospan have been generated and display no obvious phenotype.[53] This apparent lack of phenotype may arise from the expression of additional tetraspanins in muscle and potential compensation by these alternative tetraspanins. Biochemical and genetic studies have confirmed that the sarcoglycan complex is required for proper targeting of sarcospan.[54] Sarcospan is intimately associated with the sarcoglycan complex, but the detergent approach used to define the sarcoglycan complex does not include sarcospan.[55] Thus, it appears that sarcospan is not, by strict criteria, a member of the sarcoglycan complex although its close association with the sarcoglycan complex is important. In other tissues, tetraspanin proteins interact with integrins to mediate interaction with other membrane complexes.[56]

Heterogeneity of the Sarcoglycan Complex

In mammals, the six sarcoglycan sequences, α, β, γ, δ, ε, and ζ, appear to constitute the complete sarcoglycan family. The major sarcoglycan complex found in striated muscle includes α-sarcoglycan, β-sarcoglycan, γ-sarcoglycan and δ-sarcoglycan. As originally characterized through biochemical purification, the sarcoglycan complex contained a 50 kDa subunit (α-sarcoglycan), a 43 kDa subunit (β-sarcoglycan) and a 35 kDa subunit with a ratio of 1:1:2.[2,57] The heterogeneity of the 35 kDa component was not originally appreciated, and it was through

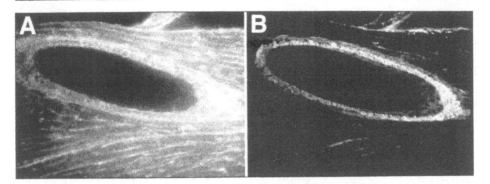

Figure 3. Shown is expression of ζ-sarcoglycan in striated and vascular smooth muscle. An anti-ζ-sarcoglycan antibody was raised against a peptide specific to ζ-sarcoglycan.[51] Sections were prepared from mouse cardiac muscle through coronary arteries. A) The anti-ζ-sarcoglycan antibody reacts with the vascular smooth muscle layer surrounding the coronary artery as well as the plasma membrane of the surrounding cardiomyocytes. B) Staining with anti-smooth muscle actin confirms the identity of the vascular smooth muscle layer.

genomic and subsequent biochemical analysis that it became clear that the 35 kDa sarcoglycan subunit in striated muscle contains both γ-sarcoglycan and δ-sarcoglycan in a 1:1 ratio.[2,57] Further complicating this issue is the recent identification of ζ-sarcoglycan.[51] γ-sarcoglycan, δ-sarcoglycan and ζ-sarcoglycan are very similar in amino acid composition, so the specificity of antibodies should be considered when reexamining earlier studies. Cross reactivity of antibodies between γ-sarcoglycan, δ-sarcoglycan and ζ-sarcoglycan may confound some of the earlier analyses examining expression patterns and tissue specificity.

ε-sarcoglycan can be detected at the plasma membrane of skeletal muscle, but by immunoblotting and immunostaining, the amount of this protein is less that the related protein α-sarcoglycan.[32] However, the apparent lower expression of ε-sarcoglycan may arise from a lower affinity anti-ε-sarcoglycan antibody.[32] That said, the original biochemical characterization of the sarcoglycan complex from skeletal muscle supports that the 50 kDa component is mainly comprised by α-sarcoglycan. The pl of each of α-sarcoglycan and ε-sarcoglycan is similar (5.44 for α-sarcoglycan versus 6.10 for ε-sarcoglycan) but the molecular mass of these two species differs slightly with ε-sarcoglycan migrating slightly slower on SDS-PAGE analysis. Furthermore, using an anti-ε-sarcoglycan antibody, ε-sarcoglycan appears to be more highly expressed in vascular smooth muscle tissue when examining tissue sections where both striated and vascular smooth muscle are found in the same sections.[50] Like ε-sarcoglycan, ζ-sarcoglycan appears to be more highly expressed in vascular smooth muscle than in striated muscle, although expression can also be detected in striated muscle.[51] In vascular smooth muscle, a complex of ζ-sarcoglycan, δ-sarcoglycan and β-sarcoglycan constitutes the vascular smooth muscle sarcoglycan complex (Fig. 4).[50] Whether this same complex is also present in striated muscle is not known. The major sarcoglycan complex of striated muscle is α-sarcoglycan, β-sarcoglycan, γ-sarcoglycan and δ-sarcoglycan.

Sarcoglycans Are Conserved in Invertebrates

Genomic information from both C. elegans and Drosophila melanogaster support the presence of at least three distinct sarcoglycan genes in each of these genetic model systems.[58,59] In both species, there is a single α/ε-sarcoglycan-like sequence, a β-sarcoglycan-like sequence and a single γ/δ/ζ-sarcoglycan-like sequence. In mammals, gene duplication produced α-sarcoglycan and ε-sarcoglycan while there is even greater heterogeneity for γ-, δ- and ζ-sarcoglycan. Analysis of mice with null mutations in sarcoglycan genes suggests that each sarcoglycan protein possesses unique function and that sarcoglycan subunits cannot substitute for one and other,

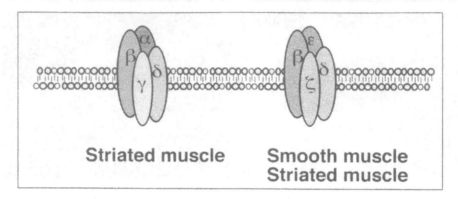

Striated muscle   Smooth muscle
                                   Striated muscle

Figure 4. There are at least two different sarcoglycan complexes. The first complex shown to the left is the most abundant in striated muscle and consists of α-, β-, γ- and δ-sarcoglycan. The vascular smooth muscle sarcoglycan complex contains ε-, β-, δ- and ζ-sarcoglycan. This complex may also be present at much lower abundance in striated muscle.

except potentially in the case of α-sarcoglycan.[25] There is upregulation of ε-sarcoglycan in α-sarcoglycan null hearts and little to no cardiomyopathy in these mice, suggesting that ε-sarcoglycan can potentially assume some of the function of α-sarcoglycan.[25]

Sequences identified as sarcoglycans are noted in the Danio rerio and Fugu databases (unpublished results). Antibodies raised against mammalian sarcoglycan epitopes demonstrate crossreactivity to muscle membrane proteins from lower vertebrates confirming the presence of a sarcoglycan complex in leech and torpedo.[60-62] Genetic studies have been initiated in *C. elegans* where genetic mutants were studied using an RNA interference technique.[58] Worms with a combination of dystrophin, dystroglycan and γ/δ/ζ "knock-downs" displayed a head-bending phenotype, and the significance of this finding is still under investigation. Whether an invertebrate model of sarcoglycan gene mutations may be useful for larger scale genetic screens, including suppressor screens, remains to be established. Of note, in examining the worm and fly databases, orthologues for sarcospan were not identified.[58,59]

Assembly of the Sarcoglycan Complex

The assembly of the sarcoglycan complex is of relevance since so many sarcoglycan mutations result in the dissolution and/or aberrant assembly of the sarcoglycan complex (Fig. 5). Sarcoglycan assembly has been studied in cultured cells, including the C2C12 mouse muscle cell line and primary myoblasts cultures, in mouse models of sarcoglycan mutations and inferred from staining patterns of residual sarcoglycans in muscle biopsies from human patients with sarcoglycan mutations.[63-65] Taking these data together, in striated muscle, δ-sarcoglycan and β-sarcoglycan form a core unit to which α-sarcoglycan or ε-sarcoglycan can bind. γ-sarcoglycan is added last to this complex. The complex assembles in the endoplasmic reticulum where it is also coassembles with β-dystroglycan, although this interaction is less well studied.[64]

Interactions with the Remainder of the DGC and Other Cytoskeletal Proteins

Recent studies using a differential ionic strength to dissolve interactions within the DGC have suggested that the sarcoglycan complex associates with dystrophin by way of an interaction with dystrobrevin.[55] α-dystrobrevin is a splice form highly expressed in muscle that has homology to the carboxyl-terminus of dystrophin and, like dystrophin, can bind to the syntrophins.[66,67] The syntrophin-dystrobrevin interaction is tight and there is evidence that this complex preferentially interacts with the sarcoglycan complex.[67]

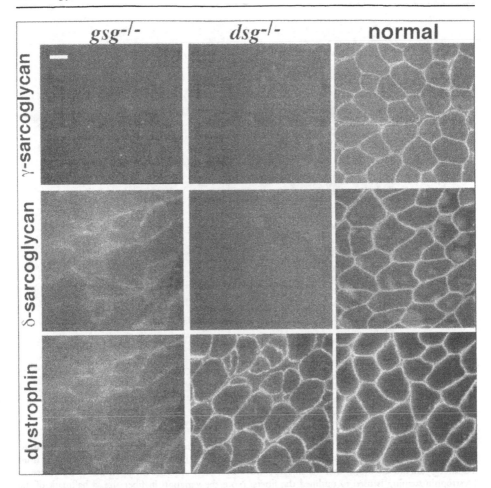

Figure 5. The sarcoglycan complex is disrupted as a result of sarcoglycan and dystrophin gene mutations. Mutations in a γ-sarcoglycan (*gsg*⁻/⁻) result in the loss of γ-sarcoglycan and a secondary reduction in δ-sarcoglycan. Mutations in δ-sarcoglycan (*dsg*⁻/⁻) produce a secondary decrease in γ-sarcoglycan as well as the loss of δ-sarcoglycan. Dystrophin staining is normal in both these sarcoglycan mutants. The sarcoglycan complex is similarly decreased in response to dystrophin mutations. The antibodies used are listed at the left.

It has also been suggested that the role of the sarcoglycan complex is to stabilize the interaction between α-dystroglycan and β-dystroglycan.[68] The two subunits, α and β, of dystroglycan are produced from a single gene and result from proteolytic cleavage from a single polypeptide precursor (see Chapter 16). The α subunit of dystroglycan is differentially glycosylated in many different tissues types and was originally identified as cranin in the nervous system.[69,70] In muscle, the predominant form of α-dystroglycan is a 156 kDa protein. In the BIO 14.6 Syrian hamster that is mutant for δ-sarcoglycan, biochemical analysis of the residual dystroglycan complex reveals that it no longer associates with the sarcoglycan complex.[68]

There is a muscle specific form of the protein filamin, called filamin C, is a cytoskeletal protein that is found at both the Z band and plasma membrane of muscle.[71,72] Filamins are actin bindings proteins that play a role in reorganizing the cytoskeleton in nonmuscle cells. The precise function of filamin C in striated muscle is not known. It was shown that filamin C can bind directly to the cytoplasmic domains of both γ-sarcoglycan and δ-sarcoglycan (Fig. 1).[71] In addition, the cellular distribution of filamin C changes in response to the loss of the

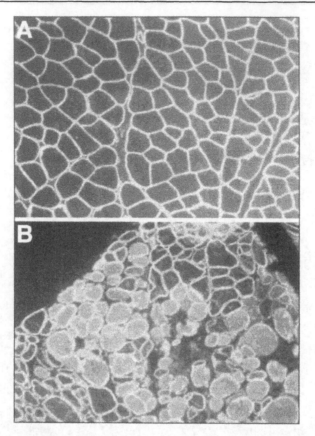

Figure 6. Normal muscle from a mouse that has been injected with Evans blue dye (EBD) is shown in A. Dystrophin staining outlines the myofibers. No EBD uptake is seen in normal muscle. B shows muscle from a δ-sarcoglycan mutant mouse that was injected with EBD. EBD opacifies the fibers due to fluorescence. Dystrophin staining is used to outlined the fibers. Note the variation in fiber size, a hallmark of the dystrophic process in muscle.

sarcoglycan complex where it redistributes from overlying the Z band to the plasma membrane. The genetic defect in human subjects is a dominant form of LGMD that maps to chromosome 7 where the locus for filamin C is. However, mutations in filamin C gene have not been identified.[73,74]

Where sarcoglycan is absent, whether by a primary mutation in sarcoglycan genes or through secondary destabilization in the case of dystrophin gene mutations, the plasma membrane becomes abnormally leaky. This has been best demonstrated with the vital tracer Evans blue dye (EBD) as normal muscle is impermeable to EBD. EBD uptake is a feature of dystrophin deficient muscle but not of muscle with a mutation in α2 chain of laminin-2 (merosin).[75,76] This laminin is the extracellular ligand for dystroglycan and mutations in this gene lead to congenital muscular dystrophy (see Chapter 6).[69] Disruption of the sarcoglycan complex, where the dystrophin protein is still normally found at the plasma membrane, is also associated with EBD uptake.[32] Therefore, aspects of the membrane permeability defects can be attributed specifically to the loss of the sarcoglycan complex (Fig. 6).

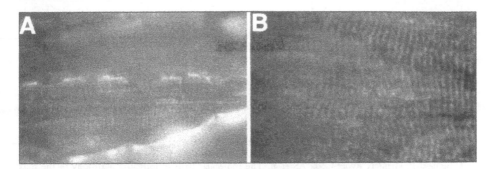

Figure 7. DGC elements are concentrated in costameres at the plasma membrane. Costameres are membrane-associated structures that overlie the Z line. Sarcoglycan proteins are also found in a nonuniform, linear distribution on longitudinal sections of muscle. In cross sections of muscle, this pattern cannot be appreciated. In longitudinal sections, the linear pattern is over the Z line. A) The pattern seen on longitudinal sections from an anti-β-sarcoglycan antibody. B) The pattern seen on longitudinal sections from an anti-ζ-sarcoglycan antibody.

The function of the sarcoglycan complex is not known, but like the DGC itself, the complex appears to have both mechanical and signaling roles. The only domain found within a sarcoglycan sequence to date is the ectoATPase domain in α-sarcoglycan.[27] As for physiological function, the function of the DGC appears to be to stabilize the plasma membrane against some of the forces associated with muscle contraction.[77,78] Mice lacking dystrophin display enhanced damage in response to repeated contraction, especially eccentric contraction that is thought to mimic the forces associated with physiologic exercise. Muscle from mice lacking δ-sarcoglycan display an intermediate phenotype with an intermediate degree of increased damage in response to eccentric contraction.[65] Muscle from mice lacking γ-sarcoglycan do not show enhanced damage in response to eccentric contraction highlighting that the sarcoglycan complex, and the DGC itself, play more than just a simple mechanical stabilizing role.[79] The sarcoglycan complex, like the DGC, is concentrated over costameres (Fig. 7). Costameres are specialized structures of striated muscle that overlie the Z band.[80,81] It is clear that the Z band had additional roles beyond simply anchoring thin filaments; in cardiac muscle, titin, MLP and telethonin play a role in elastic recoil through Z band anchoring.[82,83] As the DGC and sarcoglycan form the specific attachment of the Z band to the membrane, they are also likely to be multifunctional.

Future Directions

Future studies of the sarcoglycan and the DGC complex center on understanding the nonmechanical function of the DGC and identifying the ligand that sarcoglycan binds to in the extracellular matrix. Understanding better the downstream signaling defects that arise from dissolution of the sarcoglycan complex have bearing, in that these pathways may be targets for therapy in treating both the LGMDs and DMD/BMD. Finally, signalling defects that arise form the loss of this complex may be relevant to understand better muscle wasting and degeneration as it occurs in other muscle diseases and with the muscle loss that occurs in aging.

Ackowledgments

EMM is supported by the NIH, the American Heart Association, the Muscular Dystrophy Association and the Burroughs Wellcome Fund.

References

1. Ervasti JM, Ohlendieck K, Kahl SD et al. Deficiency of a glycoprotein component of the dystrophin complex in dystrophic muscle. Nature 1990; 345(6273):315-319.
2. Ervasti JM, Campbell KP. Membrane organization of the dystrophin-glycoprotein complex. Cell 1991; 66(6):1121-1131.
3. Yoshida M, Ozawa E. Glycoprotein complex anchoring dystrophin to sarcolemma. J Biochem (Tokyo) 1990; 108(5):748-752.
4. Yoshida M, Suzuki A, Yamamoto H et al. Dissociation of the complex of dystrophin and its associated proteins into several unique groups by n-octyl beta-D-glucoside. Eur J Biochem 1994; 222(3):1055-1061.
5. Hoffman EP, Brown Jr RH, Kunkel LM. Dystrophin: The protein product of the Duchenne muscular dystrophy locus. Cell 1987; 51(6):919-928.
6. Sicinski P, Geng Y, Ryder-Cook AS et al. The molecular basis of muscular dystrophy in the mdx mouse: A point mutation. Science 1989; 244(4912):1578-1580.
7. Ben Hamida M, Fardeau M, Attia N. Severe childhood muscular dystrophy affecting both sexes and frequent in Tunisia. Muscle Nerve 1983; 6(7):469-480.
8. Ben Othmane K, Ben Hamida M, Pericak-Vance MA et al. Linkage of Tunisian autosomal recessive Duchenne-like muscular dystrophy to the pericentromeric region of chromosome 13q. Nat Genet 1992; 2(4):315-317.
9. Romero NB, Tome FM, Leturcq F et al. Genetic heterogeneity of severe childhood autosomal recessive muscular dystrophy with adhalin (50 kDa dystrophin-associated glycoprotein) deficiency. C R Acad Sci III 1994; 317(1):70-76.
10. Passos-Bueno MR, Oliveira JR, Bakker E et al. Genetic heterogeneity for Duchenne-like muscular dystrophy (DLMD) based on linkage and 50 DAG analysis. Hum Mol Genet 1993; 2(11):1945-1947.
11. Azibi K, Bachner L, Beckmann JS et al. Severe childhood autosomal recessive muscular dystrophy with the deficiency of the 50 kDa dystrophin-associated glycoprotein maps to chromosome 13q12. Hum Mol Genet 1993; 2(9):1423-1428.
12. Ohlendieck K, Campbell KP. Dystrophin-associated proteins are greatly reduced in skeletal muscle from mdx mice. J Cell Biol 1991; 115(6):1685-1694.
13. Matsumura K, Tome FM, Collin H et al. Deficiency of the 50K dystrophin-associated glycoprotein in severe childhood autosomal recessive muscular dystrophy. Nature 1992; 359(6393):320-322.
14. Yamanouchi Y, Mizuno Y, Yamamoto H et al. Selective defect in dystrophin-associated glycoproteins 50DAG (A2) and 35DAG (A4) in the dystrophic hamster: An animal model for severe childhood autosomal recessive muscular dystrophy (SCARMD). Neuromuscul Disord 1994; 4(1):49-54.
15. Roberds SL, Ervasti JM, Anderson RD et al. Disruption of the dystrophin-glycoprotein complex in the cardiomyopathic hamster. J Biol Chem 1993; 268(16):11496-11499.
16. Bonnemann CG, Modi R, Noguchi S et al. Beta-sarcoglycan (A3b) mutations cause autosomal recessive muscular dystrophy with loss of the sarcoglycan complex. Nat Genet 1995; 11(3):266-273.
17. Roberds SL, Leturcq F, Allamand V et al. Missense mutations in the adhalin gene linked to autosomal recessive muscular dystrophy. Cell 1994; 78(4):625-633.
18. Noguchi S, McNally EM, Ben Othmane K et al. Mutations in the dystrophin-associated protein gamma-sarcoglycan in chromosome 13 muscular dystrophy. Science 1995; 270(5237):819-822.
19. Lim LE, Duclos F, Broux O et al. Beta-sarcoglycan: Characterization and role in limb-girdle muscular dystrophy linked to 4q12. Nat Genet 1995; 11(3):257-265.
20. Nigro V, de Sa Moreira E, Piluso G et al. Autosomal recessive limb-girdle muscular dystrophy, LGMD2F, is caused by a mutation in the delta-sarcoglycan gene. Nat Genet 1996; 14(2):195-198.
21. Roberds SL, Anderson RD, Ibraghimov-Beskrovnaya O et al. Primary structure and muscle-specific expression of the 50-kDa dystrophin-associated glycoprotein (adhalin). J Biol Chem 1993; 268(32):23739-23742.
22. McNally EM, Yoshida M, Mizuno Y et al. Human adhalin is alternatively spliced and the gene is located on chromosome 17q21. Proc Natl Acad Sci USA 1994; 91(21):9690-9694.
23. Piccolo F, Roberds SL, Jeanpierre M et al. Primary adhalinopathy: A common cause of autosomal recessive muscular dystrophy of variable severity. Nat Genet 1995; 10(2):243-245.
24. Duclos F, Straub V, Moore SA et al. Progressive muscular dystrophy in alpha-sarcoglycan-deficient mice. J Cell Biol 1998; 142(6):1461-1471.
25. Liu LA, Engvall E. Sarcoglycan isoforms in skeletal muscle. J Biol Chem 1999; 274(53):38171-38176.
26. van der Kooi AJ, de Voogt WG, Barth PG et al. The heart in limb girdle muscular dystrophy. Heart 1998; 79(1):73-77.
27. Betto R, Senter L, Ceoldo S et al. Ecto-ATPase activity of alpha-sarcoglycan (adhalin). J Biol Chem 1999; 274(12):7907-7912.
28. Durbeej M, Cohn RD, Hrstka RF et al. Disruption of the beta-sarcoglycan gene reveals pathogenetic complexity of limb-girdle muscular dystrophy type 2E. Mol Cell 2000; 5(1):141-151.

29. Araishi K, Sasaoka T, Imamura M et al. Loss of the sarcoglycan complex and sarcospan leads to muscular dystrophy in beta-sarcoglycan-deficient mice. Hum Mol Genet 1999; 8(9):1589-1598.
30. Bonnemann CG, Wong J, Jones KJ et al. Primary gamma-sarcoglycanopathy (LGMD 2C): Broadening of the mutational spectrum guided by the immunohistochemical profile. Neuromuscul Disord 2002; 12(3):273-280.
31. Crosbie RH, Lim LE, Moore SA et al. Molecular and genetic characterization of sarcospan: Insights into sarcoglycan-sarcospan interactions. Hum Mol Genet 2000; 9(13):2019-2027.
32. Hack AA, Ly CT, Jiang F et al. Gamma-sarcoglycan deficiency leads to muscle membrane defects and apoptosis independent of dystrophin. J Cell Biol 1998; 142(5):1279-1287.
33. Sasaoka T, Imamura M, Araishi K et al. Pathological analysis of muscle hypertrophy and degeneration in muscular dystrophy in gamma-sarcoglycan-deficient mice. Neuromuscul Disord 2003; 13(3):193-206.
34. Nigro V, Piluso G, Belsito A et al. Identification of a novel sarcoglycan gene at 5q33 encoding a sarcolemmal 35 kDa glycoprotein. Hum Mol Genet 1996; 5(8):1179-1186.
35. Tsubata S, Bowles KR, Vatta M et al. Mutations in the human delta-sarcoglycan gene in familial and sporadic dilated cardiomyopathy. J Clin Invest 2000; 106(5):655-662.
36. Jung D, Duclos F, Apostol B et al. Characterization of delta-sarcoglycan, a novel component of the oligomeric sarcoglycan complex involved in limb-girdle muscular dystrophy. J Biol Chem 1996; 271(50):32321-32329.
37. McNally EM, Duggan D, Gorospe JR et al. Mutations that disrupt the carboxyl-terminus of gamma-sarcoglycan cause muscular dystrophy. Hum Mol Genet 1996; 5(11):1841-1847.
38. Gertz EW. Cardiomyopathic Syrian hamster: A possible model of human disease. Prog Exp Tumor Res 1972; 16:242-260.
39. Bajusz E. Hereditary cardiomyopathy: A new disease model. Am Heart J 1969; 77(5):686-696.
40. Factor SM, Sonnenblick EH. Hypothesis: Is congestive cardiomyopathy caused by a hyperreactive myocardial microcirculation (microvascular spasm)? Am J Cardiol 1982; 50(5):1149-1152.
41. Factor SM, Minase T, Cho S et al. Microvascular spasm in the cardiomyopathic Syrian hamster: A preventable cause of focal myocardial necrosis. Circulation 1982; 66(2):342-354.
42. Coral-Vazquez R, Cohn RD, Moore SA et al. Disruption of the sarcoglycan-sarcospan complex in vascular smooth muscle: A novel mechanism for cardiomyopathy and muscular dystrophy. Cell 1999; 98(4):465-474.
43. Wheeler MT, Allikian MJ, Heydeman A et al. Smooth muscle cell extrinsic vascular spasm arises from cardiomyocyte degeneration in sarcoglycan mutant cardiomyopathy. J Clin Invest 2004; 113:668-675.
44. Ettinger AJ, Feng G, Sanes JR. Epsilon-Sarcoglycan, a broadly expressed homologue of the gene mutated in limb-girdle muscular dystrophy 2D. J Biol Chem 1997; 272(51):32534-32538.
45. McNally EM, Ly CT, Kunkel LM. Human epsilon-sarcoglycan is highly related to alpha-sarcoglycan (adhalin), the limb girdle muscular dystrophy 2D gene. FEBS Lett 1998; 422(1):27-32.
46. Zimprich A, Grabowski M, Asmus F et al. Mutations in the gene encoding epsilon-sarcoglycan cause myoclonus-dystonia syndrome. Nat Genet 2001; 29(1):66-69.
47. Piras G, El Kharroubi A, Kozlov S et al. Zac1 (Lot1), a potential tumor suppressor gene, and the gene for epsilon-sarcoglycan are maternally imprinted genes: Identification by a subtractive screen of novel uniparental fibroblast lines. Mol Cell Biol 2000; 20(9):3308-3315.
48. Grabowski M, Zimprich A, Lorenz-Depiereux B et al. The epsilon-sarcoglycan gene (SGCE), mutated in myoclonus-dystonia syndrome, is maternally imprinted. Eur J Hum Genet 2003; 11(2):138-144.
49. Muller B, Hedrich K, Kock N et al. Evidence that paternal expression of the epsilon-sarcoglycan gene accounts for reduced penetrance in myoclonus-dystonia. Am J Hum Genet 2002; 71(6):1303-1311.
50. Straub V, Ettinger AJ, Durbeej M et al. epsilon-sarcoglycan replaces alpha-sarcoglycan in smooth muscle to form a unique dystrophin-glycoprotein complex. J Biol Chem 1999; 274(39):27989-27996.
51. Wheeler MT, Zarnegar S, McNally EM. zeta-Sarcoglycan, a novel component of the sarcoglycan complex, is reduced in muscular dystrophy. Hum Mol Genet 2002; 11(18):2147-2154.
52. Crosbie RH, Heighway J, Venzke DP et al. Sarcospan, the 25-kDa transmembrane component of the dystrophin-glycoprotein complex. J Biol Chem 1997; 272(50):31221-31224.
53. Lebakken CS, Venzke DP, Hrstka RF et al. Sarcospan-deficient mice maintain normal muscle function. Mol Cell Biol 2000; 20(5):1669-1677.
54. Crosbie RH, Lebakken CS, Holt KH et al. Membrane targeting and stabilization of sarcospan is mediated by the sarcoglycan subcomplex. J Cell Biol 1999; 145(1):153-165.
55. Yoshida M, Hama H, Ishikawa-Sakurai M et al. Biochemical evidence for association of dystrobrevin with the sarcoglycan-sarcospan complex as a basis for understanding sarcoglycanopathy. Hum Mol Genet 2000; 9(7):1033-1040.

56. Hemler ME. Specific tetraspanin functions. J Cell Biol 2001; 155(7):1103-1107.
57. Yamamoto H, Hagiwara Y, Mizuno Y et al. Heterogeneity of dystrophin-associated proteins. J Biochem (Tokyo) 1993; 114(1):132-139.
58. Grisoni K, Martin E, Gieseler K et al. Genetic evidence for a dystrophin-glycoprotein complex (DGC) in Caenorhabditis elegans. Gene 2002; 294(1-2):77-86.
59. Greener MJ, Roberts RG. Conservation of components of the dystrophin complex in Drosophila. FEBS Lett 2000; 482(1-2):13-18.
60. Royuel M, Hugon G, Rivier Fl et al. Dystrophin-associated proteins in obliquely striated muscle of the leech. Histochem J 2001; 33(3):135-139.
61. Royuela M, Paniagua R, Rivier F et al. Presence of invertebrate dystrophin-like products in obliquely striated muscle of the leech, Pontobdella muricata (Annelida, Hirudinea). Histochem J 1999; 31(9):603-608.
62. Royuela M, Hugon G, Rivier F et al. Variations in dystrophin complex in red and white caudal muscles from Torpedo marmorata. J Histochem Cytochem 2001; 49(7):857-865.
63. Chan YM, Bonnemann CG, Lidov HG et al. Molecular organization of sarcoglycan complex in mouse myotubes in culture. J Cell Biol 1998; 143(7):2033-2044.
64. Noguchi S, Wakabayashi E, Imamura M et al. Formation of sarcoglycan complex with differentiation in cultured myocytes. Eur J Biochem 2000; 267(3):640-648.
65. Hack AA, Lam MY, Cordier L et al. Differential requirement for individual sarcoglycans and dystrophin in the assembly and function of the dystrophin-glycoprotein complex. J Cell Sci 2000; 113(Pt 14)(2):2535-2544.
66. Newey SE, Benson MA, Ponting CP et al. Alternative splicing of dystrobrevin regulates the stoichiometry of syntrophin binding to the dystrophin protein complex. Curr Biol 2000; 10(20):1295-1298.
67. Sadoulet-Puccio HM, Rajala M, Kunkel LM. Dystrobrevin and dystrophin: An interaction through coiled-coil motifs. Proc Natl Acad Sci USA 1997; 94(23):12413-12418.
68. Straub V, Duclos F, Venzke DP et al. Molecular pathogenesis of muscle degeneration in the delta-sarcoglycan-deficient hamster. Am J Pathol 1998; 153(5):1623-1630.
69. Ibraghimov-Beskrovnaya O, Ervasti JM, Leveille CJ et al. Primary structure of dystrophin-associated glycoproteins linking dystrophin to the extracellular matrix. Nature 1992; 355(6362):696-702.
70. Smalheiser NR, Schwartz NB. Cranin: A laminin-binding protein of cell membranes. Proc Natl Acad Sci USA 1987; 84(18):6457-6461.
71. Thompson TG, Chan YM, Hack AA et al. Filamin 2 (FLN2): A muscle-specific sarcoglycan interacting protein. J Cell Biol 2000; 148(1):115-126.
72. Bonnemann CG, Thompson TG, van der Ven PF et al. Filamin C accumulation is a strong but nonspecific immunohistochemical marker of core formation in muscle. J Neurol Sci 2003; 206(1):71-78.
73. Speer MC, Vance JM, Grubber JM et al. Identification of a new autosomal dominant limb-girdle muscular dystrophy locus on chromosome 7. Am J Hum Genet 1999; 64(2):556-562.
74. Palenzuela L, Andreu AL, Gamez J et al. A novel autosomal dominant limb-girdle muscular dystrophy (LGMD 1F) maps to 7q32.1-32.2. Neurology 2003; 61(3):404-406.
75. Straub V, Rafael JA, Chamberlain JS et al. Animal models for muscular dystrophy show different patterns of sarcolemmal disruption. J Cell Biol 1997; 139(2):375-385.
76. Matsuda R, Nishikawa A, Tanaka H. Visualization of dystrophic muscle fibers in mdx mouse by vital staining with Evans blue: Evidence of apoptosis in dystrophin-deficient muscle. J Biochem (Tokyo) 1995; 118(5):959-964.
77. Petrof BJ, Shrager JB, Stedman HH et al. Dystrophin protects the sarcolemma from stresses developed during muscle contraction. Proc Natl Acad Sci USA 1993; 90(8):3710-3714.
78. Lynch GS, Rafael JA, Chamberlain JS et al. Contraction-induced injury to single permeabilized muscle fibers from mdx, transgenic mdx, and control mice. Am J Physiol Cell Physiol 2000; 279(4):C1290-1294.
79. Hack AA, Cordier L, Shoturma DI et al. Muscle degeneration without mechanical injury in sarcoglycan deficiency. Proc Natl Acad Sci USA 1999; 96(19):10723-10728.
80. Ervasti JM. Costameres: The Achilles' heel of Herculean muscle. J Biol Chem 2003; 278(16):13591-13594.
81. Bloch RJ, Capetanaki Y, O'Neill A et al. Costameres: Repeating structures at the sarcolemma of skeletal muscle. Clin Orthop 2002; (403 Suppl):S203-210.
82. Hoshijima M, Pashmforoush M, Knoll R et al. The MLP family of cytoskeletal Z disc proteins and dilated cardiomyopathy: A stress pathway model for heart failure progression. Cold Spring Harb Symp Quant Biol 2002; 67:399-408.
83. Knoll R, Hoshijima M, Hoffman HM et al. The cardiac mechanical stretch sensor machinery involves a Z disc complex that is defective in a subset of human dilated cardiomyopathy. Cell 2002; 111(7):943-955.

CHAPTER 11

Sarcomeric Proteins in LGMD

Olli Carpén

Introduction

L imb girdle muscular dystrophies are caused by mutations in several genes that encode for proteins with divergent functions. The most common LGMD forms result from sarcolemmal adhesion complex defects (sarcoglycans, Chapter 10), but also other sarcolemmal molecules (caveolin-3 and dysferlin; Chapters 8 and 9), nuclear membrane (lamin A/C; Chapter 12) components and other types of molecules underlie LGMD. To date, three sarcomeric proteins have been shown to be directly involved in this form of muscular dystrophy (Fig. 1, Table 1). Mutations in myotilin cause dominant LGMD (1A), whereas mutations in telethonin and titin result in recessive LGMD (2G and 2J, respectively). A fourth molecule, the protease calpain-3, may also be linked to sarcomeric abnormalities, as many of its substrates are cytoskeletal proteins. Sarcomeric protein mutations may also result in other clinical phenotypes, such as nemaline myopathies. For titin, two different disease forms exist; homozygous mutations cause LGMD, whereas heterozygous mutations of titin gene result in distal myopathy. Mutations in the known disease-associated sarcomeric proteins are a rare cause of LGMD and their pathogenetic mechanisms remain to be elucidated.

Myotilin – LGMD1A

LGMD1A is a rare autosomal dominant muscular dystrophy caused by mutations in *myotilin* (*TTID, MYOT*) gene. So far, two large families with different missense mutations, both of which result in a single residue substitution (S55F, T57I), have been identified.[1,2]

Clinical Picture

The affected individuals in both families exhibit proximal leg and arm weakness as the presenting symptom. The disease progresses later to include distal weakness.[1,3] The mean age of onset is from 27 (first family) to around 50 years (second family). In both families, approximately half of the affected individuals exhibit a distinctive nasal, dysarthric pattern of speech. Tightened heel chords and reduced knee and elbow deep tendon reflexes are frequently seen, but nerve conduction studies are normal. Creatine kinase levels are 1.6-15-fold elevated. EMG changes are indicative of primary myopathy.

Morphological features in LGMD1A include nonspecific myopathic changes, such as degeneration of myofibers, variations in fiber size, fiber splitting and large numbers of centrally located myonulei. At later stages, interstitial fibrosis and extensive fatty infiltration is seen. There is no evidence for selective myopathic involvement of specific fiber types or for any alteration in the relative numbers of fiber types. Large numbers of rimmed, possibly autophagic vacuoles are seen. An interesting feature is the presence of striking Z-disc irregularity, which includes Z-disc streaming and the presence of filament aggregates, which apparently contain sarcomeric thin filaments and morphologically resemble nemaline rods.[1] In patient muscle, both the normal and the mutated allele are transcribed and therefore immunohistochemical or western blot analysis is not helpful as a diagnostic procedure.

Molecular Mechanisms of Muscular Dystrophies, edited by Steve J. Winder. ©2006 Eurekah.com.

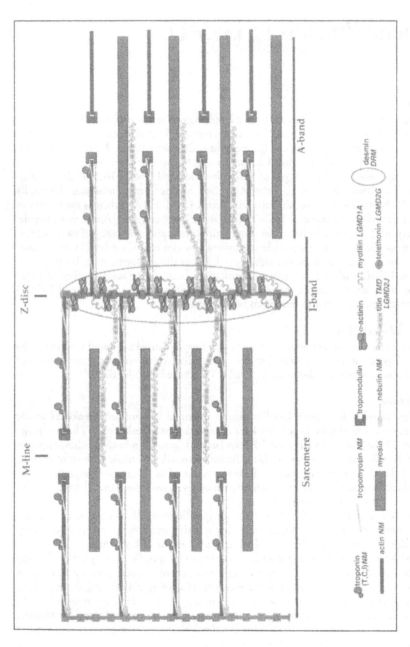

Figure 1. Schematic structure of the sarcomere and localization of sarcomeric proteins associated with LGMD and other muscular disorders. Thin filaments, which consist of filamentous actin and actin-associated proteins, are aligned and cross-linked by Z-disc proteins. The sarcomeric architecture is stabilized by titin, which associates with both thin and thick filaments. Thin filaments are stabilized by nebulin. Z-discs are surrounded by desmin and intermediate filament-associated proteins. Z-disc is linked to sarcolemmal attachment complex and extracellular matrix. Only some of the Z-disc components are presented. The italic letters depict muscle disorders associated with specific proteins. *NM* = nemaline myopathy, *LGMD* = Limb-Girdle Muscular Dystrophy, *TMD* = tibial muscular dystrophy *DRM* = desmin related myopathy.

Table 1. Characteristics of LGMD-associated sarcomeric proteins

	Myotilin	Telethonin	Titin
Disease	LGMD1A	LGMD2G	LGMD2J/TMD
Inheritance	dominant	recessive	recessive/dominant
Age of onset	3rd-6th decade	1st-2nd decade	1st-3rd/4th-5th decade
Typical mutation protein	missense	truncating	missense/del/ins at last exon
Location	Z-disc	Z-disc	spans the sarcomere
Function (major)	actin cross-linking	unclear	spring and ruler

Gene and Protein

Myotilin gene is located in chromosome 5q31.[4] The size of the entire gene is <20 000 bp without the promoter region. The gene is composed of 10 exons. There is no evidence for alternative splicing. The gene encodes for a 498 amino acid protein with molecular size of 57 kDa. Myotilin is composed of a unique, serine-rich amino-terminus and a carboxy-terminus that contains two Ig-like domains, most homologous to Ig-domains of palladin, myopalladin and Z-disc Ig-domains 7 and 8 of titin.[5-8] Expression pattern is restricted, with highest expression in skeletal muscle, moderate expression in heart and little or no expression in other tissues. However, during development, a much wider expression pattern is seen.[9]

Interactions and Functions

So far, three interaction partners, all of which are Z-disc components, have been described for myotilin. Myotilin binds to α-actinin, an actin cross-linking protein and a backbone of the Z-disc structure.[4] α-actinin has also several other binding partners in the Z-disc, including titin. A second binding partner is filamin C, a muscle specific filamin isoform.[10] Filamins serve as actin cross-linking proteins and anchor actin cytoskeleton to the cell membrane. Myotilin also directly binds actin, and very efficiently bundles and stabilizes actin filaments.[11] The bundling effect is enhanced, when myotilin acts together with α-actinin. Myotilin forms dimers, which require its carboxy-terminal half, and dimerization is apparently necessary for the actin-bundling activity. The disease mutations do not affect any of the known interactions. Interestingly, the mutations alter residues, which are putative phosphorylation sites. Due to its strong actin-bundling activity, myotilin is thought to serve as a stabilizer of the Z-disc.

Pathogenetic Mechanism

The disease association and functional characteristics indicate an indispensable role for myotilin in stabilization and anchorage of thin filaments, which may be a prerequisite for correct Z-disc organization. The disease results from the presence of dysfunctional myotilin instead of loss of myotilin protein, but the pathogenetic mechanism is not yet understood.

Teletonin (T-Cap) – LGMD2G

LGMD2G is a rare form of autosomal recessive LGMD, caused by homozygous mutations in *telethonin* (*TCAP*) gene. Patients from four unrelated families with null mutations in the telethonin gene and deficiency of this protein in muscle have been identified.

Clinical Picture

There is a wide spectrum of inter- and intrafamilial clinical variability. All patients have shown proximal involvement and marked weakness but half of the individuals also have atrophy in the distal muscles of the legs, with a phenotype resembling Mioshi myopathy, while others have calf hypertrophy.[12,13] The age at onset ranges from 9 to 15 years. Some patients have lost the ability to walk within the third or fourth decade. Serum creatine kinase is increased 3 to 30-fold. Heart involvement has been observed in half of the patients. The course may be somewhat milder in affected females than males.

The affected muscles show dystrophic changes and presence of rimmed vacuoles.[12,13] Immunohistochemical and western blot analysis indicates loss of telethonin protein expression.[13] However, the ultrastructure of sarcomeres is preserved and Z-disc components such as myotilin, α-actinin and titin appear intact.[14] Muscular dystrophy associated proteins dystrophin, sarcoglycans, dysferlin and calpain-3 are normally expressed in telethonin deficient muscle. Telethonin expression is preserved in most neuromuscular dystrophies, although it is secondarily downregulated in neurogenic atrophy.[15] This has to be taken into account, when evaluating immunohistochemical or western blot findings.

Gene and Protein

The human *telethonin* gene resides in 17q12. The gene consists of 2 exons and encodes for a 167 amino acid protein with molecular size of 19 kDa. There is no homology to other proteins and no structural domains have been identified in the sequence. The expression is strictly restricted to skeletal and cardiac muscle. The protein is located in sarcomeric Z discs.[16] Telethonin mRNA is one of the most abundant transcripts in skeletal muscle,[16] however the transcript and protein expression is downregulated in neurogenic atrophy.[15,17]

Interactions and Functions

Telethonin binds to the Z-disc associated Ig-domains Z1Z2 of titin.[18,19] The interaction may allow telethonin to cross-link titin molecules at their N-termini and thereby provide stability to Z-disc.

Myostatin is a growth factor, which negatively regulates myoblast proliferation. Telethonin interacts with myostatin and blocks myostatin secretion in myoblasts.[20] In this way, telethonin may promote muscle cell growth and regeneration.

Calsarcin 1, FATZ (myozenin, calsarcin 2), and calsarcin 3 are three homologous proteins, all of which directly interact with telethonin.[21-23] In addition, all three calcarcin family members interact with structural Z-disc proteins α-actinin, filamin C (γ-filamin), and ZASP/Cypher/Oracle and with calcineurin, a signaling molecule involved in cardiac hypertrophy and skeletal muscle differentiation.[22]

Telethonin serves also as a link between sarcomeric Z-discs and the T-tubule system, which couples excitation and contraction by regulating the intracellular ion flow. The link is formed by an association between telethonin and MinK, a component of K^+ channel, which regulates repolarization of myocytes.[24] The interaction between telethonin and MinK is regulated by phosphorylation of telethonin at S157 (the substrate of titin-kinase), which results in loss of binding.[24]

In summary, telethonin appears to be an important Z-disc adapter protein, which links together and regulates Z-disc structural components, ion channels and signaling pathways involved in myocyte growth and differentiation.

Pathogenetic Mechanism

The disease results from loss of telethonin expression. The pathogenic mechanism is still under investigation. However, since the diseased muscles appear to have preserved their sarcomere integrity, it is possible that telethonin's association with ion channel functions and/or signaling pathways, rather than its structural functions, is the underlying cause.

Titin – LGMD2J

LGMD2J is an autosomal recessive disorders resulting from mutation of both alleles of *titin* (*TTN*) gene. So far, one family with several affected members has been described. Heterozygous *titin* mutations are more common and result in autosomal dominant distal myopathy, termed tibial muscular dystrophy (TMD).[25] All known mutations reside in the last exon of the gene (Mex6).[26-27] Two of them are missense mutations, which result in a single residue substitution and one is a deletion/insertion mutation, which results in change of four amino acid residues.

Clinical Picture

LGMD2J manifests during first to third decade. There is weakness in all proximal muscles. Distal muscles are less severely affected, and facial muscles remain unaffected. Loss of ambulation occurs between 3rd and 4th decade.[25] Histopathological changes include severe nonspecific dystrophic alterations in proximal muscles, with less alterations in distal muscle compartments.[25] The integrity of the sarcomere appears preserved.[26]

TMD is a late onset disorder, with age of onset typically between 35-45 years, but sometimes presenting much later. Muscle weakness and atrophy are usually confined to the anterior compartment of the lower leg, in particular, the tibialis anterior muscle.[28] Cardiac manifestations have not been described in LGMD2J or TMD patients, although some other titin mutations are associated with dilated cardiomyopathy.[29,30] Histopathological alterations vary from mild myopathic changes to severe muscular dystrophy. Rimmed vacuoles may be present. At late stage, anterior tibial muscles are wasted and replaced by adipose and connective tissue.

Immunostaining of LGMD2J muscle has shown that some M-band specific epitopes in titin are lost, whereas staining with other titin antibodies is normal. This finding may be useful in diagnosis. TMD samples do not show alterations in titin staining.[26] A secondary calpain 3 loss is seen in TMD/LGMD2J muscle.[31]

Gene and Protein

Titin gene resides in 2q31. The gene has several sites for alternative splicing, causing isoforms of different lengths to appear in different muscles.[32] The entire coding region of titin (connectin) consists of 363 exons, which encodes the largest known polypeptide. The size of the isoforms varies from 6,000 to 42,000 amino acid residues (700-4,200 kDa). The structure consists of repetitive domains of Ig-type (166 copies), fibronectin 3 repeats (FN3, 132 copies) and nonrepetitive segments including a titin-kinase domain.[8] The expression is restricted to striated muscle. The protein stretches over the length of one-half of the sarcomere, from the Z disc to the M line.[8]

Interactions and Functions

Titin interacts via its different domains with several structural and signaling proteins. The known M-line interaction partners include structural proteins myomesin and M-protein. Known regulatory and signaling ligands of titin in the M-line are p94/CAPN3 (calpain-3)[33] and MURF-1[34,35] and MURF2.[36] MURFs are transcriptional regulators implicated in myofibrillogenesis. The interaction site of MURF-1 is within A168-170 of titin, adjacent to the kinase domain and somewhat amino-terminal of the known titin mutations. Epitopes in A169/170 and myomesin binding are retained in affected muscles. Defective p94/CAPN3 interaction with titin remains a possible effector of the pathogenesis LGMD2J/TMD.[26]

Titin serves several crucial functions in the sarcomere.[37] First, the titin I-band domains have elastic properties, which allow titin to serve as a molecular spring and ruler that governs some aspects of myofibrillar stiffness. Second, titin plays an important role in early assembly during myofibrillogenesis, and it may act as a primary sarcomeric template and stabilizer. Third, the titin C-terminal region contains a Ser/Thr kinase domain whose function remains elusive, but whose presence suggests that titin also is involved in signaling pathways.

Pathogenetic Mechanism

Both mutant and wild-type titin RNAs are transcribed in muscle from heterozygous patients with TMD, and in LGMD2J muscle, most of the protein is intact.[26] Thus, the disease appears to results from dysfunction of the very carboxy-terminal M-line associated titin protein sequence. Possibly, the pathogenetic mechanism involves perturbation of titin-based myofibrillar signaling and/or defective calpain-3 regulation.

Conclusions

Recent years have shed light on the role of sarcomeric proteins in LGMD and other muscle disorders. Many open questions, however, still exist. For instance, it is not known whether defects in myotilin, telethonin and titin result in a common pathogenetic mechanism and what are the crucial steps in the molecular pathogenesis of each disease form. The limited amount of identified patients and mutations does not yet allow us to form a complete picture of the phenotypic variability of sarcomeric LGMDs. Finally, further understanding of the molecular complexity of the Z-disc, and clarification of the signaling pathways involved in Z-disc assembly, turnover and maintenance are needed, before the molecular pathways that result in LGMD can be defined.

Note Added in Proof

A recent study demonstrated that myotilin is mutated also in another form of muscle disease termed myofibrillar myopathy (MFM) or desmin-related myopathy.[38] MFM is a heterogenous group of disorders characterized by a pathological pattern of myofibrillar dissolution associated with accumulation of myofibrillar degradation products. Previously, mutations in desmin and alphaB-crystallin genes have been shown to be a cause of a minority of MFM cases. Selgen and Engel identified four different missense mutations in six of 57 MFM patients. All mutations resulted in a change of N-terminal Ser or Thr (S55F, S60C/F or S951) residue and thus the type of mutations resemble those detected in LGMD1A. Clinical characteristics of MFM patients with myotilin mutations included: (1) age of onset between 50 and 80 years, (2) progressive muscle weakness, in most cases starting from proximal muscles, (3) decreased reflexes, (4) evidence of peripheral neuropathy in all patients, and (5) cardiac symptoms in the majority of patients. The mechanism, by which myotilin mutations result in two different clinical phenotypes remains to be clarified.

References

1. Hauser MA, Horrigan SK, Salmikangas P et al. Myotilin is mutated in limb girdle muscular dystrophy 1A. Hum Mol Genet 2000; 9:2141-2147.
2. Hauser MA, Conde CB, Kowaljow V et al. Myotilin mutation found in second pedigree with LGMD1A. Am J Hum Genet 2002; 71:1428-1432.
3. Gilchrist JM, Pericak-Vance MA, Silverman L et al. Clinical and genetic investigation in autosomal dominant limb-girdle muscular dystrophy. Neurology 1988; 38:5-8.
4. Salmikangas P, Mykkänen OM, Grönholm M et al. Myotilin, a novel sarcomeric protein with two Ig-like domains, is encoded by a candidate gene for limb-girdle muscular dystrophy. Hum Mol Genet 1999; 8:1329-1336.
5. Parast MM, Otey CA. Characterization of palladin, a novel protein localized to stress fibers and cell adhesions. J Cell Biol 2000; 150:643-656.
6. Mykkänen OM, Grönholm M, Rönty M et al. Characterization of human palladin, a microfilament-associated protein. Mol Biol Cell 2001; 12:3060-3073.
7. Bang ML, Mudry RE, McElhinny AS et al. Myopalladin, a novel 145-kilodalton sarcomeric protein with multiple roles in Z-disc and I-band protein assemblies. J Cell Biol 2001; 153:413-427.
8. Labeit S, Kolmerer B. Titins: Giant proteins in charge of muscle ultrastructure and elasticity. Science 1995; 13(270):293-296.
9. Mologni L, Salmikangas P, Fougerousse F et al. Developmental expression of myotilin, a gene mutated in limb-girdle muscular dystrophy type 1A. Mech Dev 2001; 103:121-125.
10. van der Ven PF, Wiesner S, Salmikangas P et al. Indications for a novel muscular dystrophy pathway. Gamma-filamin, the muscle-specific filamin isoform, interacts with myotilin. J Cell Biol 2000; 151:235-248.
11. Salmikangas P, van der Ven PF, Lalowski M et al. Myotilin, the limb-girdle muscular dystrophy 1A (LGMD1A) protein, cross-links actin filaments and controls sarcomere assembly. Hum Mol Genet 2003; 12:189-203.
12. Moreira ES, Vainzof M, Marie SK et al. The seventh form of autosomal recessive limb-girdle muscular dystrophy is mapped to 17q11-12. Am J Hum Genet 1997; 61:151-159.

13. Moreira ES, Wiltshire TJ, Faulkner G et al. Limb-girdle muscular dystrophy type 2G is caused by mutations in the gene encoding the sarcomeric protein telethonin. Nature Genet 2000; 24:163-166.
14. Vainzof M, Moreira ES, Suzuki OT et al. Telethonin protein expression in neuromuscular disorders. Biochim Biophys Acta 2002; 1588:33-40.
15. Schröder R, Reimann J, Iakovenko A et al. Early and selective disappearance of telethonin protein from the sarcomere in neurogenic atrophy. J Muscle Res Cell Motil 2001; 22:259-264.
16. Valle G, Faulkner G, De Antoni A et al. Telethonin, a novel sarcomeric protein of heart and skeletal muscle. FEBS Lett 1997; 415:163-168.
17. Mason P, Bayol S, Loughna PT. The novel sarcomeric protein telethonin exhibits developmental and functional regulation. Biochem Biophys Res Commun 1999; 257:699-703.
18. Mues A, van der Ven PF, Young P et al. Two immunoglobulin-like domains of the Z-disc portion of titin interact in a conformation-dependent way with telethonin. FEBS Lett 1998; 428:111-114.
19. Zou P, Gautel M, Geerlof A et al. Solution scattering suggests cross-linking function of telethonin in the complex with titin. J Biol Chem 2003; 278:2636-2644.
20. Nicholas G, Thomas M, Langley B et al. Titin-cap associates with, and regulates secretion of, Myostatin. J Cell Physiol 2002; 193:120-131.
21. Faulkner G, Pallavicini A, Comelli A et al. FATZ, a filamin-, actinin-, and telethonin-binding protein of the Z-disc of skeletal muscle. J Biol Chem 2000; 275:41234-41242.
22. Frey N, Richardson JA, Olson EN. Calsarcins, a novel family of sarcomeric calcineurin-binding proteins. Proc Nat Acad Sci 2000; 97:14632-14637.
23. Takada F, Vander Woude DL, Tong H-Q et al. Myozenin: An alpha-actinin- and gamma-filamin-binding protein of skeletal muscle Z lines. Proc Nat Acad Sci 2001; 98:1595-1600.
24. Furukawa T, Ono Y, Tsuchiya H et al. Specific interaction of the potassium channel beta-subunit minK with the sarcomeric protein T-cap suggests a T-tubule-myofibril linking system. J Mol Biol 2001; 313:775-784.
25. Udd B, Käärianen H, Somer H. Muscular dystrophy with separate clinical phenotypes in a large family. Muscle Nerve 1991; 14:1050-1058.
26. Hackman P, Vihola A, Haravuori H et al. Tibial muscular dystrophy is a titinopathy caused by mutations in TTN, the gene encoding the giant skeletal-muscle protein titin. Am J Hum Genet 2002; 71:492-500.
27. van den Bergh PY, Bouquiaux O, Verellen C et al. Tibial muscular dystrophy in a Belgian family. Ann Neurol 2003; 54:248-251.
28. Udd B. Limb-girdle type muscular dystrophy in a large family with distal myopathy: Homozygous manifestation of a dominant gene? J Med Genet 1992; 29:383-389.
29. Gerull B, Gramlich M, Atherton J et al. Mutations of TTN, encoding the giant muscle filament titin, cause familial dilated cardiomyopathy. Nature Genet 2002; 30:201-204.
30. Itoh-Satoh M, Hayashi T, Nishi H et al. Titin mutations as the molecular basis for dilated cardiomyopathy. Biochem Biophys Res Commun 2002; 291:385-393.
31. Haravuori H, Vihola A, Straub V et al. Secondary calpain3 deficiency in 2q-linked muscular dystrophy: Titin is the candidate gene. Neurology 2001; 56:869-877.
32. Bang ML, Centner T, Fornoff F et al. The complete gene sequence of titin, expression of an unusual approximately 700-kDa titin isoform, and its interaction with obscurin identify a novel Z-line to I-band linking system. Circ Res 2001; 89:1065-1072.
33. Sorimachi H, Kinbara K, Kimura S et al. Muscle-specific calpain, p94, responsible for limb girdle muscular dystrophy type 2A, associates with connectin through IS2, a p94-specific sequence. J Biol Chem 1995; 270:31158-31162.
34. Centner T, Yano J, Kimura E et al. Identification of muscle specific ring finger proteins as potential regulators of the titin kinase domain. J Mol Biol 2001; 306:717-726.
35. McElhinny AS, Kakinuma K, Sorimachi H et al. Muscle-specific RING finger-1 interacts with titin to regulate sarcomeric M-line and thick filament structure and may have nuclear functions via its interaction with glucocorticoid modulatory element binding protein-1. J Cell Biol 2002; 157:125-136.
36. Pizon V, Iakovenko A, van Der Ven PF et al. Transient association of titin and myosin with microtubules in nascent myofibrils directed by the MURF2 RING-finger protein. J Cell Sci 2002; 115:4469-4482.
37. Granzier H, Labeit D, Wu Y et al. Titin as a modular spring: Emerging mechanisms for elasticity control by titin in cardiac physiology and pathophysiology. J Muscle Res Cell Motil 2002; 23:457-471.
38. Selcen D, Engel AG. Mutations in ZASP define a novel form of muscular dystrophy in humans. Ann Neurol 2005; 57(2):269-76.

CHAPTER 12

Lamins and Emerin in Muscular Dystrophy:
The Nuclear Envelope Connection

Josef Gotzmann and Roland Foisner

Abstract

Lamins are nuclear intermediate filaments which form a network like structure underneath the nuclear membrane, the nuclear lamina, as well as complexes in the nuclear interior. Lamins associate with numerous proteins in the inner nuclear membrane, including emerin, and lamina-associated polypeptides. Beyond their structural roles lamins likely serve essential functions in many nuclear activities, such as DNA replication and transcription. Mutations in A-type lamins and emerin have been linked to at least eight rare human diseases (laminopathies), affecting skeletal and cardiac muscle, as well as fat, bone and neuronal tissues, or causing premature ageing. There is no clear genotype-phenotype relation of mutations, and it is still unknown, how mutations in the ubiquitously expressed lamin A and emerin cause tissue-restricted phenotypes. Focusing on neuromuscular laminopathies, we discuss potential disease mechanisms, including defects in lamin structure and assembly, misregulation of gene expression, perturbation of cell cycle control during tissue regeneration, and deregulation of metabolic and signaling functions of the endoplasmic reticulum.

Organization of the Nucleus

In eukaryotic cells the nucleus is the largest organelle that contains the chromosomes and harbors essential cellular processes, such as DNA replication, transcription, and RNA splicing. These functions largely depend on a dynamic structural organization, which is to a great part established by lamins and lamin-binding proteins.[1-5] Lamins are nuclear-specific (type V) intermediate filaments[6,7] and form the nuclear lamina, a meshwork-like structure underlying the inner nuclear membrane that provides mechanical stability for the nuclear envelope (NE). The nuclear lamina, and two concentric lipid bilayers, inner (IM) and outer (OM) nuclear membrane, with nuclear pore complexes (NPC) constitute the NE in higher eukaryotic cells[5] (Fig. 1). Although inner and outer membrane are connected at NPCs, they are functionally and structurally distinct.[8] While the OM can be considered as a part of the endoplasmic reticulum (ER), the IM harbors specific integral membrane proteins,[9] which bind lamins and mediate tight association of the membrane to the underlying lamina scaffold.

Apart from the peripheral lamina, lamins have also been identified in the nuclear interior,[3] where they are located on nucleoplasmic filaments[10] and form stable higher order structures.[11,12]

Components of Lamin Structures

Several lamin proteins evolved during evolution.[13,14] The human genome contains three lamin genes (LMNA, LMNB1, LMNB2) giving rise to seven proteins by alternative splicing. Two types of lamins can be distinguished based on their sequence, expression pattern, post-translational modification, and biochemical properties. B-type lamins, are expressed ubiquitously throughout development. They are stably farnesylated and carboxy-methylated at a

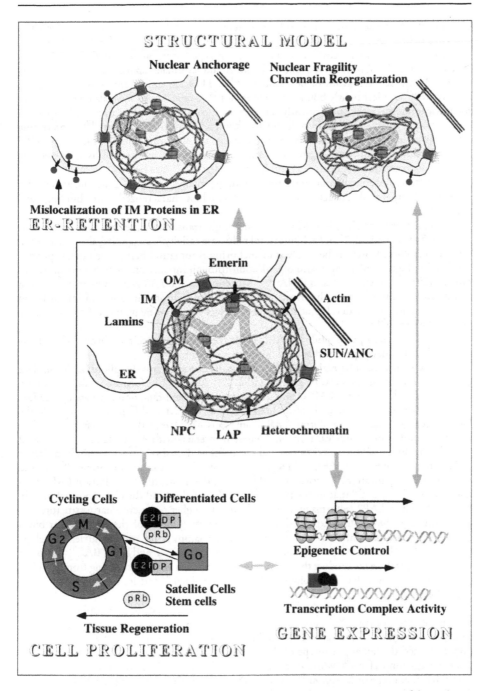

Figure 1. Disease models. The cartoon in the center depicts the schematic organization of the nucleus in normal eukaryotic cells, focusing on proteins and nuclear structures relevant for laminopathies. The images outside represent the major phenotypes in laminopathic cells, according to the different disease models. For details, see text. ER: endoplasmic reticulum; IM: inner nuclear membrane; LAP: Lamina-associated polypeptide; NPC: nuclear pore complex; OM: outer nuclear membrane

C-terminal CAAX motif and stay tightly associated with membranes.[15,16] Genetic analyses in *Caenorhabdidtis elegans*[17] and *Drosophila melanogaster*,[18] and RNA-interference (RNAi) experiments in mammalian cells[19] have shown that B-type lamins are essential for embryonic development and viability. A-type lamins, lamin A and its smaller splice variant lamin C, are expressed only at later stages of development.[20-22] They are only transiently or not at all farnesylated[23] and less tightly bound to membranes.[24] A-type lamins have also been found in the nuclear interior either permanently throughout the cell cycle [10,11] or transiently upon incorporation of injected lamins into the nuclear lamina,[25,26] during G1 phase,[12,27] and during lamin processing.[28,29] A-type lamins are not essential for embryonic development, but fulfill important functions in tissue organization and homeostasis in adult organisms as revealed by *LMNA* knockout mouse models [30,31] and human diseases[32-41] (see below).

Lamin structures contain numerous lamin-binding proteins, which cofractionate with lamins even after detergent-salt extraction of nuclei.[42-44] The best characterized lamin binding partners include:

Lamin B receptor (LBR) is a IM protein with eight transmembrane domains[45] that interacts with B-type lamins[46,47] and exhibits sterol reductase activity,[48] suggesting also functions in sterol metabolism. LBR may be involved in anchoring chromatin fibers at the nuclear periphery[49] and in higher order chromatin organization through interactions with heterochromatin protein HP1,[50,51] histones,[52] and DNA.[47,53] Mutations in the *LBR* gene in humans[54] and in ichthyosis *ic/ic* mice[55] cause abnormal nuclear shape and chromatin organization in blood granulocytes and severe developmental defects (Pelger Huet anomaly and HEM/Greenberg Skeletal Dysplasia[54,56]).

Lamina-associated-polypeptide 1 (LAP1) comprise three IM protein isoforms differentially expressed during development that bind both types of lamins.[43,57,58]

Lamina-associated polypeptide 2 (LAP2) is a family of 6 alternatively spliced lamin-binding proteins in mammals, LAP2α, β, γ, δ, ϵ, and ζ.[59-62] All LAP2 isoforms contain a common N-terminus and, except for LAP2α and ζ, a single transmembrane domain passing the IM. While LAP2β, ϵ, δ, γ and ζ are structurally related, LAP2α is different sharing only the N-terminus with the other isoforms. LAP2β interacts with lamin B at the NE,[43,63-65] while LAP2α is located in the nuclear interior,[44] where it interacts with A-type lamins.[12] All LAP2 proteins contain a ~ 40 amino acid structural motif (LEM domain),[66,67] which binds to Barrier to Autointegration Factor (BAF),[68,69] a highly conserved DNA-cross linking protein[70] involved in chromatin organization,[71] nuclear assembly,[72] and development.[73] In addition LAP2 contains a LEM-like motif[66] that interacts with DNA. LAP2 also bind the chromosomal protein HA95.[74,75] Thus, LAP2 isoforms may be key factors in higher order chromatin structure.

Emerin is a ubiquitously expressed integral membrane protein of the IM[76,77] that binds preferentially A-type lamins in vitro[78-81] and requires lamin A for proper nuclear envelope targeting.[31,82-86] Emerin also contains an N-terminal LEM domain and binds BAF.[81,87] Emerin was identified as the gene product that is missing or mutated in patients suffering from X-linked Emery-Dreifuss Muscular Dystrophy (X-EDMD, see below),[88] but its cellular functions remain elusive.

MAN1 is a lamina-associated protein[89] in the IM with a LEM domain and two putative transmembrane domains.[90,91] Association of MAN1 to BAF and lamins has been demonstrated in *C. elegans*.[92] Simultaneous knockdown of MAN1 and emerin in the worm dramatically enhanced the lethal phenotype of the single MAN1 knockdown due to severe chromosome segregation defects,[92] while in contrast to humans, emerin knockdown alone has no detectable phenotype in *C. elegans*.[86] Thus, MAN1 and emerin may have partially overlapping functions in chromatin organization. In *Xenopus* MAN1-type proteins have also been implicated in the Bone Morphogenetic Protein (BMP) signaling through interaction with Smads.[93,94]

A novel class of NE proteins, containing a highly conserved 176 residues long SUN (Sad1-UNC-84) domain were recently identified in *S. pombe* (Sad1p), *Drosophila*, *C. elegans* (UNC 84) and mammals (SUN1 and SUN2).[95,96] In yeast and *C. elegans*, these proteins are involved in nuclear migration and anchorage during development,[97] likely by connecting the nucleus to the actin or microtubule cytoskeleton. Since UNC-84 localization at the NE depends on lamin,[98] these proteins may directly or indirectly connect the lamina with the cytoskeleton.

C. elegans ANC-1, a member of a novel spectrin-repeat-containing protein family[99] at the NE also links nuclei to the actin cytoskeleton in a lamin-dependent manner.[100] ANC-1-like proteins in mammals have been called Syne-1 and -2,[101] Nesprin-1 and -2,[102] Myne-1-and -2,[103] and Nuance.[104] Although nesprin-1alpha has been shown to bind lamin A and emerin in vitro,[105] other studies have localized these proteins in the OM and the Golgi.[106] The emerging scheme is the existence of several isoforms of spectrin repeat-domain proteins, which localize to both IM and OM, and directly or indirectly link lamins to actin, alone or together with SUN-domain proteins.

Proteomic approaches[107,108] and visual screening[109] have identified numerous novel potential NE proteins, but their interaction with lamins has to be tested in the future.

Laminopathies—Nuclear Lamina Proteins and Disease

At least eight rare human hereditary diseases that arise from mutation in genes encoding nuclear lamina proteins (collectively termed "laminopathies") have been described. These include in isolation or combination: muscular dystrophy, cardiomyopathy, peripheral neuropathy, lipodystrophy, insulin-resistant diabetes, liver steatosis, bone dysplasia, osteolysis, and premature ageing.

Historically, X-linked recessive Emery-Dreifuss-Muscular Dystrophy (X-EDMD) was the first heritable disorder linked to a mutation in a gene (*EMD*, formerly *STA*) encoding a NE protein,[110] later named emerin[111,112] after its discoverer Alan Emery.[113] The estimated incidence of X-EDMD is 1:50000,[114] and thus far lower than those of the two other most common muscular dystrophies, Duchenne/Becker and myotonic. Emerin is ubiquitously expressed in human and mouse tissues, but its molecular functions are still largely unknown. More than 70 different mutations leading to X-EDMD have been described[115] and mapped throughout the 2.1 kb long gene. The majority of these comprise nonsense-mutations, resulting in emerin deficient cells.[116] Only in a few cases, missense mutations yield decreased levels of emerin. In all reported cases, mutated emerin is mis-localized to the ER and has increased solubility in detergent-containing buffers compared to wild-type emerin.[79,117,118] A similar mis-localization of emerin has been described in cells of lamin A knockout mice and in cells expressing some laminopathy-linked mutated lamin A.[31,82-85] This may explain why some mutations in lamin A cause identical or similar pathological phenotypes as X-EDMD (see below). Loss or significant reduction of emerin expression in X-EDMD patients affect only skeletal and heart muscle, an unexpected observation in view of emerin's ubiquitous expression. This may be explained by the potential interaction of emerin with unknown muscle-specific factors, or by a unique lamina composition in muscle cells, rendering them more susceptible to NE defects upon loss of emerin.[36]

Beside the X-linked, autosomal dominant (AD-EDMD) and rare cases of autosomal-recessive (AR-EDMD)[119] cases of EDMD have been reported. Positional cloning linked AD-EDMD to the *LMNA* gene on chromosome 1q21-q22 encoding A-type lamins.[120] The clinical phenotype of AD-EDMD patients is identical to that of X-EDMD patients.

The spectrum of laminopathies linked to lamin mutations is steadily increasing, including other neuromuscular diseases, such as Dilated Cardiomyopathy with conduction defect Type 1A (CMD1A),[121] Limb Girdle Muscular Dystrophy 1B (LGMD1B)[122] and Charcot-Marie-Tooth disorder type 2B1 (CMT2B1),[123] a neuropathy associated with muscle weakness and wasting. In addition, nonmuscle laminopathies have been described, such as familial partial lipodystrophy of the Dunnigan-type (FPLD)[124-126] and Mandibuloacral dysplasia (MAD),[127] and very recently lamin A mutations were linked to premature ageing in Hutchinson-Gilford progeria syndrome (HGPS)[128,129] and Atypical Werner Syndrome[130] (see below).

Is There a B-Type Laminopathy?

So far, disease-linked lamin mutations were exclusively found in A-type lamins. As B-type lamins are important for development, mutations in B-type lamins may be embryonic lethal. Embryonic lethality upon functional disruption of lamin B has been shown in *Drosophila*[18] and *C. elegans*.[17] Accordingly, lamin B function has not been tested in knockout mice, but recently nuclear lamina defects and early lethality have been described in mice deficient for Rce1 or Icmt,[131,132] two enzymes involved in post-transcriptional processing of B-type lamins.[15] At least parts of the pathogenic defects in these mice may be attributed to loss of lamin B function due to improper processing.

Also mutations in the lamin B-binding protein LBR were linked to the human diseases, Pelger-Huet anomaly (PHA OMIM 169400)[54] and Greenberg Skeletal Dysplasia.[56] It is however unclear, if the pathogenic phenotype arises through interference with lamin B function or through effects on lamin B-independent functions of LBR in chromatin organization[54] or sterol metabolism.[56]

Genetics of Laminopathies

The Universal Mutation Database (UMD)[133] contains a comprehensive collection of identified mutations in exon and intron sequences of *LMNA* and *EMD*.[134] The current records on X-EDMD include data from 84 patients, representing a total of 31, mainly null mutations in *EMD*. So far, statistical analyses failed to detect mutational hot spot regions.

For mutations in lamin A, 537 patients covering six different laminopathic diseases (not including premature ageing) with a total of 124 different mutations have been registered. Eighty percent of these represent missense mutations, the rest are in frame deletions/insertions, nonsense mutations, frameshift and splice site mutations. So far, two hot spot mutation regions could be identified in *LMNA*:

- An R453W exchange was detected in 16% of the reported AD-EDMD cases (equivalent to 11% of all laminopathy cases).
- In 84% of FPLD cases, an R482W/Q/L mutation in *LMNA* was identified, matching to about 13% of all laminopathies. Moreover, FPLD-causing mutations were restricted to the C-terminal tail of lamin A encoded by exons 8-11.

Genetic alterations in *EMD* and *LMNA* account for only 40% of clinically confirmed EDMD cases. The remaining 60% of EDMD cases are likely linked to other NE components functionally and structurally overlapping with nuclear lamin A. Interestingly, a recent proteomic screen, identified 67 novel NE proteins, 23 of which mapped to chromosome regions linked to a spectrum of dystrophies.[108] Thus, the number of NE-proteins with a direct linkage to disease is likely to grow in coming years.

18 out of 20 classical cases of Hutchinson Gilford premature ageing syndrome harbored an identical de novo silent single-base substitution, G608G (GGC > GGT) within exon 11 of *LMNA*, which result in the generation of a cryptic splice site, and the production of a protein with a 50 amino acids deletion near the C-terminus.[129] Since the deleted region contained the posttranslational cleavage site in lamin A, the mutated protein is not correctly processed. However, other rare cases of HGPS have also been reported, carrying single base mutations in exon 11 or in other exons.[41]

The Cellular Phenotype of Neuromuscular Laminopathies

The most prominent alterations found in EDMD muscle biopsies include variations in muscle fiber size, decreased diameters of type I fibers, and an increased number of internal nuclei.[135-138] Fibrosis-like depositions and augmentation of necrotic tissue are not as prominent in EDMD as in other muscular dystrophies. EDMD cells typically display nuclear abnormalities, such as a thickened lamina, lobulations and herniations of the NE, and distortions of the nuclear membrane with extruding chromatin.[139,140] Electron microscopy studies show abnormal distributions of heterochromatic regions, frequently aggregated in the nuclear interior and displaced from the nuclear periphery.[137,140]

Clinical Phenotypes of Laminopathies

(Links to the OMIM database at http://www.ncbi.nlm.nih.gov/Omim/)

Neuromuscular Diseasess

Emery-Dreifuss Muscular Dystrophy (EDMD, OMIM #310300)

EDMD is characterized by three diagnostic principles: (1) early contractures of the Achilles tendons, the elbows, spine, and postcervical muscles; (2) variable slowly progressive muscle wasting and weakness starting in the upper arms and lower legs; (3) cardiomyopathy at progressed stages of the disease associated with conduction abnormalities. Implantation of a pacemaker is a currently accepted therapeutic intervention, although the risk of sudden death due to heart failure remains relatively high.[39,134] The clinical symptoms for the X-linked, and the autosomal-dominant and -recessive forms of EDMD are identical.

Dilated Cardiomyopathy with Conduction Defect Type 1A (CMD1A, OMIM #115200)

CMD1A is a condition with autosomal dominant dilated cardiomyopathy and conduction system defects including sinus bradycardia, atrioventricular conduction block, or atrial arrhythmias. Joint contractures or skeletal myopathy are not part of the phenotype.

Limb Girdle Muscular Dystrophy 1B (LGMD1B, OMIM #159001)

LGMD1B is an autosomal dominant condition characterized by a limb-girdle phenotype with absent or late mild contractures, and heart involvement including atrial paralysis with dilated cardiomyopathy and sudden death.

Neuropathies

Charcot-Marie-Tooth Disorder Type 2B1 (CMT2B1, OMIM #605588)

Autosomal recessive CMT2B1 is an axonal neuropathy with the age of onset in the second decade, with distal wasting and weakness more prominent in the lower than in the upper limbs, pes cavus, and areflexia. Motor nerve conduction velocity is normal or slightly reduced and sensory nerve action potential is decreased or absent clearly reflecting an axonal degenerative process.

Lipodystrophies

Dunnigan-Type Familial Partial Lipodystrophy (FPLD, OMIM #151660)

FPLD is an autosomal dominant condition characterized by partial lipodystrophy and insulin-resistance. Patients have a normal fat distribution in early childhood, but with puberty almost all subcutaneous adipose tissue from the upper and lower extremities and gluteal and truncal areas gradually disappears. Simultaneously, adipose tissue accumulates on the face and neck, and intraabdominal region. Affected patients are insulin-resistant and may develop glucose intolerance and diabetes mellitus, hypertriglyceridemia, and low levels of high density lipoprotein (HDL) cholesterol. Women have a higher prevalence of diabetes and atherosclerotic vascular disease and have higher serum triglycerides and lower HDL cholesterol concentrations.

Mandibuloacral Dysplasia (MAD, OMIM #248370)

MAD is a very rare condition, inherited as an autosomal-recessive trait, characterized by midchildhood onset of skeletal malformations and lipodystrophic phenotypes, including mandibular and clavicular dysplasia, acroosteolysis, club-shaped terminal phalanges, stiff joints, alopecia, hand and feet skin atrophy, lipodystrophy, and insulin-resistant diabetes.

Premature Ageing

Hutchinson-Gilford Progeria (HGPS, OMIM #176670)

HGP is a dominant condition characterized by precocious senility, coronary artery disease, absence of subcutaneous fat, congenital contractures, and persistence of fontanellae with pseudohydrocephaly. Affected children age up to ten-times faster, than normal individuals show retarded growth, develop wrinkled skin and lose their hairs. Most patients die early (aged 10 to 13) from heart attacks or stroke.

(Atypical) Werner-Syndrome (OMIM #277700)

Werner's Syndrome is an autosomal, recessively inherited, segmental progeroid syndrome, involving multiple symptoms of ageing, such as scleroderma-like skin, cataract formation, shortened stature, subcutaneous calcification, premature arteriosclerosis, diabetes mellitus, and a prematurely aged faces. The disease was originally named after a clinically similar pathogenic phenotype caused by mutations in the Werner-DNA-helicase.[141]

Animal Models of Laminopathies

Lamin-A Knockout-Mice

The first attempt to dissect the molecular functions of A-type lamins in vivo was the generation of *LMNA*-knockout mice by targeting exons 8 to 11.[31] In contrast to laminopathy patients, heterozygous mice developed normally and were undistinguishable from their wild-type littermates. Homozygous *LMNA* null mice began to display signs of muscular dystrophy by 4 weeks after birth. In addition, irregular nuclear shapes and NE abnormalities similar to those observed in FPLD and MAD cells[127,142] were detectable in primary fibroblast derived from these mice. Moreover, sciatic nerve histology of mice tissue resembled that of CMT2B-affected individuals.[123] Thus *LMNA* knockout animals may represent a mosaic phenotype of laminopathies.

Progeria-Type Lamin A Knock-In Mice

In an attempt to create a model for AD-EDMD, a mutated *LMNA* gene encoding lamin A with the EDMD-linked L530P mutation,[120] was introduced into mice.[30] Heterozygous mice did not show any signs of muscular dystrophy, while homozygous mice showed a phenotype markedly reminiscent of symptoms observed in progeria patients. These mice showed severe growth retardation by 4 to 6 days after birth and died within 4 to 5 weeks. Other progeria-like features were a slight waddling gait, micrognathia and unusual dentition, loss of subcutaneous fat, reduced numbers of eccrine and sebaceous glands, increased collagen deposition in skin, and decreased hair follicle density. Together, the phenotype of transgenic *LMNA* L530P/L530P mice significantly overlapped with clinical symptoms of HGPS patients.

Zmpste 24 Knockout Mouse

Targeted inactivation of the metalloprotease Zmpste 24, which is most likely involved in posttranslational cleavage of the farnesylated lamin A C-terminus, generates mice with muscular dystrophy and dilated cardiomyopathy.[143] Similar to lamin A knockout mice, fibroblasts of Zmpste 24 knockout mice have nuclei with irregular shapes, herniations, and membranous blebs. Although loss of Zmpste 24 lead to accumulation of unprocessed prelamin A in nuclei,[143] Zmpste 24 may not directly cleave prelamin A, but activate an unknown endoproteinase. The link between lamin A processing and laminopathies is underscored by the recent discovery of Zmpste 24 mutations in MAD patients.[144] Thus unprocessed lamin A may have the same functional deficiencies as mutated lamin A.

Molecular Models for Laminopathies

In view of the near ubiquitous expression of emerin and lamins A/C, it is unclear how ablation or malfunction of these proteins can lead to tissue-restricted diseases. Several models have been put forward, but none of these can fully explain the complex disease phenotypes. Based on the mosaic phenotype of the lamin-A knockout mouse it is likely that different functions of lamins may be deregulated to different extents by the mutations in the different diseases. In addition, secondary molecular events that may affect specific interaction partners of A-type lamins, may account for the diverse clinical outcomes. Thus, even within families of EDMD-affected patients, the clinical pictures may vary substantially.[145]

To fully understand the molecular mechanisms of laminopathic diseases, it is essential to know the molecular functions of lamins and lamin complexes at the NE and in the nuclear interior. However, functions of lamins that go beyond their purely structural role are just beginning to emerge.[6] We will briefly summarize the potential functions of lamin complexes, before we discuss potential mechanisms of the disease.

Functions of Lamin Complexes

Similar to cytoplasmic intermediate filaments nuclear lamins have structural functions by forming a nucleoskeletal scaffold that provides physical stability, defines nuclear shape,[2,3,14] and positions nuclear pore complexes.[146] Consistent with this function, nuclei assembled in vitro in the absence of lamins were rather fragile,[147,148] and nuclei of lamin A knockout mice showed an irregular shape.[31] In higher eukaryotes, which disassemble nuclear lamin structures during mitosis, the lamins and lamina associated proteins have been found to be essential for proper reassembly of nuclear structure after sister chromatid separation.[5]

The complex interactions of lamins and lamin-binding proteins with DNA[149] and with chromatin-associated proteins, such as histones, HP1, HA95, and BAF[5,150] implicate these proteins in higher order chromatin organization. Lamin complexes may provide chromatin docking sites at the NE and/or structurally organize chromatin fibers. Since higher order chromatin organization is ultimately linked to control of gene expression, lamina proteins might also be involved in epigenetic control mechanisms.[151,152] Although there is no strict correlation of gene expression and gene localization, transcriptionally poor, heterochromatin regions are often localized at the nuclear periphery, while transcriptionally active chromatin is more centrally located.[153,154]

Components of the NE may directly influence the activity of transcription complexes. Disorganization of the lamina upon injection of headless lamin mutants into BHK 21 cells inhibited RNA polymerase II-, but not RNA polymerase I- and III-dependent transcription.[155] Thus, lamins may form a nuclear scaffold required for transcription, but one may also speculate that lamin complexes influence transcription complex activity directly. The idea is supported by the finding that lamin A – LAP2α complexes bind directly to retinoblastoma protein (Rb)[156,157] and mediate nuclear anchorage of Rb, which is required for pRb's repressive activity towards E2F-DP-dependent transcription in arrested cells.[158-160] Furthermore, the NE protein LAP2β interacted with the E2F/DP-interacting[161] transcriptional repressor, mouse germ cell less (mGCL)[162] and downregulated E2F-dependent reporter activity.[162] Thus, transcription complexes may be inhibited by being tethered to the nuclear lamina.[14]

Lamina proteins may also be involved in DNA replication. Nuclei assembled in the absence of lamins failed to replicate their DNA,[148,163] and expression of lamin mutants causing reorganization of endogenous lamins inhibited DNA replication.[164-167] Since these lamin mutants caused reorganization of factors involved in the elongation phase, but not in the initiation phase of replication,[165] it is possible that either lamins directly interact with and regulate replication elongation complexes or that a lamin-network-mediated nuclear architecture is essential for assembly of these complexes. Also, ectopic expression of lamin-binding LAP2β fragments in mammalian cells inhibited progression into S-phase,[65] and addition of LAP2β fragments to *Xenopus* in vitro nuclear assembly reactions affected DNA replication.[168] Although the potential regulation of

DNA replication by LAP2 is still unclear, recent reports provide evidence that the interaction of LAP2β with the chromosomal protein HA95 is important for DNA replication.[74,75] LAP2β - HA95 complexes may stabilize Cdc6, a component of the prereplication complex.[74]

Lamina proteins might also be involved in apoptosis. Lamins, LAP2α and LAP2β are early targets of apoptosis.[169-171] Expression of uncleavable lamin mutants[170] or depletion of caspase 6, which is required for apoptotic cleavage of lamin A,[172] significantly delayed or inhibited apoptosis. Also prevention of nuclear lamina formation,[173] or RNAi mediated downregulation of lamin B[19] initiated apoptosis. Thus, mislocalization or lack of expression of B-type lamins are ultimately linked to apoptosis, and lamin degradation is required for proper apoptosis.

Aside from the role of lamins in intranuclear activities, the recently discovered direct or indirect interaction of lamins with SUN-domain and the spectrin-repeat containing proteins[95] also point to a function of lamins in linking the nucleus to the cytoskeleton. This function may be important for correct positioning of nuclei in a syncytium, such as skeletal muscle,[174] or for nuclear migration during development.

Altogether, a growing number of essential nuclear activities have been directly or indirectly linked to lamina structures in the past years. It is thus conceivable that disturbance of any of these functions can contribute to the disease phenotype to different degrees. Mutations in emerin and lamin A for instance do not solely affect these two proteins but may have significant impact on other nuclear complexes tightly linked to lamin A structures. Furthermore, as lamin A complexes are located at the nuclear periphery (e.g., lamin A - emerin) as well as in the nuclear interior (e.g., lamin A – LAP2α), different disease-linked mutations, that preferentially affect either peripheral or intranuclear interactions and functions of lamins, may have different consequences and thus, contribute to the high diversity of disease phenotypes observed.

The following, not mutually exclusive models have been suggested to explain the molecular basis of the disease (Fig. 1).

The Structural Model

This model suggests that disease-specific mutations destroy the atomic structure of emerin and lamin or interfere with lamin assembly. While emerin structure has not been solved on the molecular level (with the exception of the LEM domain), the tri-partite structural organization of lamins in N-terminal head, coiled coil rod, and C-terminal tail is well known. Yet only a region in lamin's C-terminal domain has been structurally resolved at the atomic level. Amino acids 430 to 545 in lamin A tail adopt an Ig-like fold.[175,176] Interestingly arginine at position 453, which is often mutated to tryptophan (R453W) in EDMD is localized within the core of the Ig-fold and the mutation is predicted to destabilize and disrupt the molecular structure of the lamin tail. Also many mutations in the rod domain are predicted to disrupt the molecular structure, but this has not yet been tested experimentally.

In contrast, R482W/Q mutations, a hot spot in FPLD, are localized at the surface of the Ig-fold and are predicted to change the interaction of the Ig-fold structure due to a change of surface charge. In line with this model an adipocyte-specific transcription factor, sterol response element binding protein 1 (SREBP-1) has been found to interact with lamin A – C-terminus, and FPLD-specific mutations apparently decrease the interaction.[177] Other known lamin interaction partners binding to the C-terminal tail are emerin,[83] the nucleoskeletal protein LAP2α[12] and DNA.[149] These interactions could therefore also be affected by disease specific mutations in the lamin tail. In line with this model, emerin mislocalized to the ER in some AD-EMD cells,[82] and DNA was found to bind with lower affinity to the R483W/Q lamin mutant.[149] LAP2α's interaction with mutated lamins has not yet been tested.

Structure-perturbing mutations in lamin can interfere with the assembly of lamin complexes. Accordingly, lamins containing some disease-linked mutations aggregate in the nucleoplasm when expressed in cells,[84,85,178] while others assembled properly but affected emerin localization.[82] However, not only structure perturbing mutations may cause defects in lamin assembly. Mutations affecting the interaction of lamins with proteins that are likely involved in lamin assembly, such as nuclear membrane or chromatin proteins, may also interfere with proper assembly.[38] A lot more in vitro and in vivo assembly assays with mutated lamins will have to be performed in the future to resolve this issue.

In any case, assembly defects due to lamin mutations can destabilize lamina structure to different degrees. Also, the loss of emerin expression in X-EDMD and the mislocalization of emerin in AD-EDMD may cause destabilization of the peripheral lamina. Consequently this may lead to more fragile nuclei, causing deformation or breakage of nuclei and abnormal chromatin organization.[139,179] This model is particularly attractive for explaining pathological phenotypes in muscle tissue, which are exposed to external physical forces during contraction, making muscle nuclei with a weakened lamina prone to mechanical destruction.

Another possible consequence of a weakened lamina is the destabilization of the lamina – actin cytoskeleton connection, leading to a disturbance of nuclear anchorage or even to a reorganization of cytoplasmic actin. Most of the more common muscular dystrophies have been linked to mutations in actin associated protein complexes, such as the dystrophin-associated protein complexes at the plasma membrane. Mutations perturb the integrity of these complexes and render muscle fiber membranes more fragile.[180] A similar scenario can be envisaged at the actin filament anchorage sites in the NE, where spectrin-repeat domain proteins and SUN domain proteins provide a link between the lamina and the actin cytoskeleton.[95] EDMD-linked lamin A mutations may destabilize this link and disturb nuclear anchorage and migration.[97] Consistent with such a model, muscle fibers in EDMD patients contain a large number of incorrectly positioned internal nuclei as compared to a mostly peripheral localization in wild type tissue.[135-138] However, as regenerating muscle fibers also contain a higher number of internal nuclei, it is unclear, whether this is a specific effect due to a structurally perturbed lamina-cytoskeleton link, or whether it reflects increased muscle fiber regeneration in EDMD patients (see below).

Gene Expression Model

This model suggests that mutations in lamins and emerin can change gene expression by several mechanisms. First, on the epigenetic level the disturbance of lamin structure or of lamin interactions with chromatin proteins may interfere with heterochromatin formation. This is supported by the observed dislocation of heterochromatin in EDMD nuclei.[179] As silencing of genes is intimately linked to heterochromatin formation and positioning at the nuclear periphery,[181-183] gene expression could be grossly affected in EDMD cells. In support of this hypothesis, cDNA microarray analysis of X-EDMD versus control fibroblasts revealed 60, out of 2400 genes, whose expression changed, and expression of 28 of the affected genes was rescued upon emerin expression.[184]

Besides epigenetic control mechanisms, emerging data implicate lamins in direct transcriptional control, by interaction with transcriptional complexes, such as Rb,[157,185] Germ-cell-less (gcl),[87] and SREBP-1.[177] While loss of emerin in X-EDMD could affect gcl repressor activity and thus cause gross perturbations of gene expression, changes in lamin – Rb interactions may specifically affect E2F-dependent gene expression, which in turn regulates cell cycle progression (see below). As adipocyte-specific transcription factor SREBP-1 bound FPLD lamin A mutants less efficiently than wild type lamin A,[177] it is also tempting to speculate that SREBP-1 activity is affected in FPLD patients leading to defects in adipocyte differentiation and in fatty acid biosynthesis.

Proliferation – Cell Cycle – Ageing Model

The disease model suggests that mutations in emerin or lamin A impair the control of cell proliferation. Since embryonic development is not grossly affected in laminopathy patients, the basic cell cycle regulation machinery is most likely not affected. However, regulatory mechanisms controlling the exit from and reentry into the cell cycle of specific cells in the adult organism during tissue homeostasis and regeneration may be affected. In this model, the pathological phenotype is caused by a defect in regeneration of damaged muscle, rather than muscle degeneration. Impaired regenerative potential of satellite cells in skeletal muscle,[186] and limited regeneration of cardiomyocytes from adult stem cells[187] may be affected in laminopathy patients. The molecular players involved in the control of regeneration are still unknown, but in view of the reported interaction of lamins with Rb,[157] which regulates exit from the cell cycle, it is tempting to speculate that lamin complexes are involved in Rb-mediated mechanisms.

As ageing and replicative senescence are intimately linked to cell cycle control, one may speculate that this disease model may also explain some phenotypes of premature ageing. As human cells can undergo a limited number of cell cycles before they enter senescence and stop dividing, the regeneration capacity of muscle satellite cells in EDMD may be exhausted due to the larger number of cell cycles these cells had to go through in order to regenerate destroyed muscle fibers. The defect in regeneration would only be visible upon exhaustion of the cells' capacity to cycle, which would be consistent with the late onset of the disease during childhood. However, this intriguing model has not been tested experimentally yet.

ER-Retention Theory

In this model, pathogenic features of laminopathies are caused by the failure of mutated lamin A to retain nuclear membrane proteins in the NE. As the ER forms a continuous membrane system with the OM, proteins that are usually tightly linked to the lamina and thus retained in the IM, may upon loss of interaction with the lamina diffuse along the OM to the ER. Incorrectly localized proteins in the ER may interact with ER resident proteins and affect their functions (e.g., biochemical pathways of the fatty acid metabolism in adipocytes or Ca^{2+} release during contraction cycles in muscle). This model is supported by the mislocalization of emerin to the ER in EDMD cells,[79,117,188] however defects in ER resident proteins have not been shown yet.[189]

Taken together, several models have been proposed to explain the diverse molecular phenotypes of laminopathies, based on clinical and cell biological studies in laminopathic tissues and cells. However, we still cannot link a particular model to a specific subset of laminopathies. Thus, the actual disease causing mechanisms may be more complex and represent a combination of several models, which may contribute to different degrees to the pathological phenotypes of different laminopathies. We will have to learn a lot more about the cellular functions of lamins and their structure-function relationship, until we can realistically aim at unraveling the molecular mechanisms behind these diseases in detail and eventually develop novel diagnostic tools and successful therapeutic intervention strategies.

Acknowledgements

Work in the author's laboratory was supported by grants from the "Hochschuljubiläumsstiftung der Stadt Wien" to JG and by grants from the Austrian Science Research Fund (FWF, P15312 to RF), the Association for International Cancer Research (AICR, UK), the Jubiläumsfonds of the Austrian National Bank, and from the "Österreichische Muskelforschung" to RF.

Further Reading—Internet Links

Comprehensive information about muscular dystrophies and mutations of emerin and lamin A/C, and specifically-linked inherited diseases can be found at following sites:

    http://www.ncbi.nlm.nih.gov/htbin-post/Omim/dispmim?300384
    http://www.ncbi.nlm.nih.gov/htbin-post/Omim/dispmim?150330
    http://www.dmd.nl/lmna_home.html
    http://www.enmc.org/nmd/diagnostic.cfm
    http://www.neuro.wustl.edu/neuromuscular/musdist/lg.html#1b
    http://archive.uwcm.ac.uk/uwcm/mg/search/119108.html
    http://www.myocluster.org/f_euro.html
    http://www.neuro.wustl.edu/neuromuscular/maltbrain.html
    http://telethon.bio.unipd.it/

References

1. Gotzmann J, Foisner R. Lamins and lamin-binding proteins in functional chromatin organization. Crit Rev Eukaryot Gene Expr 1999; 9(3-4):257-265.
2. Hutchison CJ. Lamins: Building blocks or regulators of gene expression? Nat Rev Mol Cell Biol 2002; 3(11):848-858.
3. Shumaker DK, Kuczmarski ER, Goldman RD. The nucleoskeleton: Lamins and actin are major players in essential nuclear functions. Curr Opin Cell Biol 2003; 15(3):358-366.
4. Goldman RD, Gruenbaum Y, Moir RD et al. Nuclear lamins: Building blocks of nuclear architecture. Genes Dev 2002; 16(5):533-547.
5. Foisner R. Cell cycle dynamics of the nuclear envelope. Scientific World Journal 2003; 3(3):1-20.
6. Herrmann H, Foisner R. Intermediate filaments: Novel assembly models and exciting new functions for nuclear lamins. Cell Mol Life Sci Aug 2003; 60(8):1607-1612.
7. Stuurman N, Heins S, Aebi U. Nuclear lamins: Their structure, assembly, and interactions. J Struct Biol 1998; 122(1-2):42-66.
8. Gerace L, Foisner R. Integral membrane proteins and dynamic organization of the nuclear envelope. Trends Cell Biol 1994; 4:127-131.
9. Foisner R. Inner nuclear membrane proteins and the nuclear lamina. J Cell Sci 2001; 114:3791-3792.
10. Hozak P, Sasseville AM, Raymond Y et al. Lamin proteins form an internal nucleoskeleton as well as a peripheral lamina in human cells. J Cell Sci 1995; 108(Pt 2):635-644.
11. Moir RD, Yoon M, Khuon S et al. Nuclear lamins A and B1: Different pathways of assembly during nuclear envelope formation in living cells. J Cell Biol 2000; 151(6):1155-1168.
12. Dechat T, Korbei B, Vaughan OA et al. Lamina-associated polypeptide 2alpha binds intranuclear A-type lamins. J Cell Sci 2000; 113(12):3473-3484.
13. Cohen M, Lee KK, Wilson KL et al. Transcriptional repression, apoptosis, human disease and the functional evolution of the nuclear lamina. Trends Biochem Sci 2001; 26(1):41-47.
14. Gruenbaum Y, Goldman RD, Meyuhas R et al. The nuclear lamina and its functions in the nucleus. Int Rev Cytol 2003; 226:1-62.
15. Maske CP, Hollinshead MS, Higbee NC et al. A carboxyl-terminal interaction of lamin B1 is dependent on the CAAX endoprotease Rce1 and carboxymethylation. J Cell Biol 2003; 162(7):1223-1232.
16. Kitten GT, Nigg EA. The CaaX motif is required for isoprenylation, carboxyl methylation, and nuclear membrane association of lamin B2. J Cell Biol 1991; 113(1):13-23.
17. Liu J, Ben-Shahar TR, Riemer D et al. Essential roles for Caenorhabditis elegans lamin gene in nuclear organization, cell cycle progression, and spatial organization of nuclear pore complexes. Mol Biol Cell 2000; 11(11):3937-3947.
18. Lenz-Bohme B, Wismar J, Fuchs S et al. Insertional mutation of the Drosophila nuclear lamin Dm0 gene results in defective nuclear envelopes, clustering of nuclear pore complexes, and accumulation of annulate lamellae. J Cell Biol 1997; 137(5):1001-1016.
19. Harborth J, Elbashir SM, Bechert K et al. Identification of essential genes in cultured mammalian cells using small interfering RNAs. J Cell Sci 2001; 114(Pt 24):4557-4565.
20. Stewart C, Burke B. Teratocarcinoma stem cells and early mouse embryos contain only a single major lamin polypeptide closely resembling lamin B. Cell 1987; 51(3):383-392.
21. Rober RA, Weber K, Osborn M. Differential timing of nuclear lamin A/C expression in the various organs of the mouse embryo and the young animal: A developmental study. Development 1989; 105(2):365-378.
22. Lehner CF, Stick R, Eppenberger HM et al. Differential expression of nuclear lamin proteins during chicken development. J Cell Biol 1987; 105(1):577-587.
23. Moir RD, Spann TP, Goldman RD. The dynamic properties and possible functions of nuclear lamins. Int Rev Cytol 1995:141-182.
24. Izumi M, Vaughan OA, Hutchison CJ et al. Head and/or CaaX domain deletions of lamin proteins disrupt preformed lamin A and C but not lamin B structure in mammalian cells. Mol Biol Cell 2000; 11(12):4323-4337.
25. Goldman AE, Moir RD, Montag-Lowy M et al. Pathway of incorporation of microinjected lamin A into the nuclear envelope. J Cell Biol 1992; 119:725-735.
26. Pugh GE, Coates PJ, Lane EB et al. Distinct nuclear assembly pathways for lamins A and C lead to their increase during quiescence in Swiss 3T3 cells. J Cell Sci 1997; 110(Pt 19):2483-2493.
27. Bridger JM, Kill IR, O'Farrell M et al. Internal lamin structures within G1 nuclei of human dermal fibroblasts. J Cell Sci 1993; 104(Pt 2):297-306.
28. Sasseville AM, Raymond Y. Lamin A precursor is localized to intranuclear foci. J Cell Sci 1995; 108(Pt 1):273-285.

29. Lutz RJ, Trujillo MA, Denham KS et al. Nucleoplasmic localization of prelamin A: Implications for prenylation- dependent lamin A assembly into the nuclear lamina [published erratum appears in Proc Natl Acad Sci USA 1992; 89(12):5699]. Proc Natl Acad Sci USA 1992; 89(7):3000-3004.
30. Mounkes LC, Kozlov S, Hernandez L et al. A progeroid syndrome in mice is caused by defects in A-type lamins. Nature 2003; 423(6937):298-301.
31. Sullivan T, Escalante-Alcalde D, Bhatt H et al. Loss of A-type lamin expression compromises nuclear envelope integrity leading to muscular dystrophy. J Cell Biol 1999; 147(5):913-920.
32. Bonne G, Mercuri E, Muchir A et al. Clinical and molecular genetic spectrum of autosomal dominant Emery-Dreifuss muscular dystrophy due to mutations of the lamin A/C gene. Ann Neurol 2000; 48(2):170-180.
33. Burke B, Stewart CL. Life at the edge: The nuclear envelope and human disease. Nat Rev Mol Cell Biol 2002; 3(8):575-585.
34. Hegele RA. Molecular basis of partial lipodystrophy and prospects for therapy. Trends Mol Med 2001; 7(3):121-126.
35. Worman HJ, Courvalin JC. The nuclear lamina and inherited disease. Trends Cell Biol 2002; 12(12):591-598.
36. Morris GE. The role of the nuclear envelope in Emery-Dreifuss muscular dystrophy. Trends Mol Med 2001; 7(12):572-577.
37. Wilson KL, Zastrow MS, Lee KK. Lamins and disease: Insights into nuclear infrastructure. Cell 2001; 104(5):647-650.
38. Hutchison CJ, Alvarez-Reyes M, Vaughan OA. Lamins in disease: Why do ubiquitously expressed nuclear envelope proteins give rise to tissue-specific disease phenotypes? J Cell Sci 2001; 114(Pt 1):9-19.
39. Emery AE. Emery-Dreifuss muscular dystrophy - a 40 year retrospective. Neuromuscul Disord 2000; 10(4-5):228-232.
40. Mounkes L, Kozlov S, Burke B et al. The laminopathies: Nuclear structure meets disease. Curr Opin Genet Dev 2003; 13(3):223-230.
41. Novelli G, D'Apice MR. The strange case of the "lumper" lamin A/C gene and human premature ageing. Trends Mol Med °2003; 9(9):370-375.
42. Senior A, Gerace L. Integral membrane proteins specific to the inner nuclear membrane and associated with the nuclear lamina. J Cell Biol 1988; 107:2029-2036.
43. Foisner R, Gerace L. Integral membrane proteins of the nuclear envelope interact with lamins and chromosomes, and binding is modulated by mitotic phosphorylation. Cell 1993; 73(7):1267-1279.
44. Dechat T, Gotzmann J, Stockinger A et al. Detergent-salt resistance of LAP2alpha in interphase nuclei and phosphorylation-dependent association with chromosomes early in nuclear assembly implies functions in nuclear structure dynamics. EMBO J 1998; 17(16):4887-4902.
45. Worman HJ, Evans CD, Blobel G. The lamin B receptor of the nuclear envelope inner membrane: A polytopic protein with eight potential transmembrane domains. J Cell Biol 1990; 111(4):1535-1542.
46. Simos G, Georgatos SD. The inner nuclear membrane protein p58 associates in vivo with a p58 kinase and the nuclear lamins. EMBO J 1992; 11(11):4027-4036.
47. Ye Q, Worman HJ. Primary structure analysis and lamin B and DNA binding of human LBR, an integral protein of the nuclear envelope inner membrane. J Biol Chem 1994; 269(15):11306-11311.
48. Silve S, Dupuy PH, Ferrara P et al. Human lamin B receptor exhibits sterol C14-reductase activity in Saccharomyces cerevisiae. Biochim Biophys Acta 1998; 1392(2-3):233-244.
49. Pyrpasopoulou A, Meier J, Maison C et al. The lamin B receptor (LBR) provides essential chromatin docking sites at the nuclear envelope. EMBO J 1996; 15(24):7108-7119.
50. Ye Q, Callebaut I, Pezhman A et al. Domain-specific interactions of human HP1-type chromodomain proteins and inner nuclear membrane protein LBR. J Biol Chem 1997; 272(23):14983-14989.
51. Ye Q, Worman HJ. Interaction between an integral protein of the nuclear envelope inner membrane and human chromodomain proteins homologous to Drosophila HP1. J Biol Chem 1996; 271(25):14653-14656.
52. Polioudaki H, Kourmouli N, Drosou V et al. Histones H3/H4 form a tight complex with the inner nuclear membrane protein LBR and heterochromatin protein 1. EMBO Rep 2001; 2(10):920-925.
53. Duband-Goulet I, Courvalin JC. Inner nuclear membrane protein LBR preferentially interacts with DNA secondary structures and nucleosomal linker. Biochemistry 2000; 39(21):6483-6488.
54. Hoffmann K, Dreger CK, Olins AL et al. Mutations in the gene encoding the lamin B receptor produce an altered nuclear morphology in granulocytes (Pelger-Huet anomaly). Nat Genet 2002; 31(4):410-414.

55. Shultz LD, Lyons BL, Burzenski LM et al. Mutations at the mouse ichthyosis locus are within the lamin B receptor gene: A single gene model for human Pelger-Huet anomaly. Hum Mol Genet 1 2003; 12(1):61-69.
56. Waterham HR, Koster J, Mooyer P et al. Autosomal recessive HEM/Greenberg skeletal dysplasia is caused by 3 beta-hydroxysterol delta 14-reductase deficiency due to mutations in the lamin B receptor gene. Am J Hum Genet 2003; 72(4):1013-1017.
57. Martin L, Crimaudo C, Gerace L. cDNA cloning and characterization of lamina-associated polypeptide 1C (LAP1C), an integral protein of the inner nuclear membrane. J Biol Chem 1995; 270(15):8822-8828.
58. Powell L, Burke B. Internuclear exchange of an inner nuclear membrane protein (p55) in heterokaryons: In vivo evidence for the interaction of p55 with the nuclear lamina. J Cell Biol 1990; 111(6Pt 1):2225-2234.
59. Berger R, Theodor L, Shoham J et al. The characterization and localization of the mouse thymopoietin/lamina- associated polypeptide 2 gene and its alternatively spliced products. Genome Res 1996; 6(5):361-370.
60. Furukawa K, Pante N, Aebi U et al. Cloning of a cDNA for lamina-associated polypeptide 2 (LAP2) and identification of regions that specify targeting to the nuclear envelope. EMBO J 1995; 14(8):1626-1636.
61. Harris CA, Andryuk PJ, Cline S et al. Three distinct human thymopoietins are derived from alternatively spliced mRNAs. Proc Natl Acad Sci USA 1994; 91(14):6283-6287.
62. Dechat T, Vlcek S, Foisner R. Review: Lamina-Associated Polypeptide 2 Isoforms and Related Proteins in Cell Cycle-Dependent Nuclear Structure Dynamics. J Struct Biol 2000; 129(2/3):335-345.
63. Furukawa K, Fritze CE, Gerace L. The major nuclear envelope targeting domain of LAP2 coincides with its lamin binding region but is distinct from its chromatin interaction domain. J Biol Chem 1998; 273(7):4213-4219.
64. Furukawa K, Kondo T. Identification of the lamina-associated-polypeptide-2-binding domain of B-type lamin. Eur J Biochem 1998; 251(3):729-733.
65. Yang L, Guan T, Gerace L. Lamin-binding fragment of LAP2 inhibits increase in nuclear volume during the cell cycle and progression into S phase. J Cell Biol 1997; 139(5):1077-1087.
66. Cai M, Huang Y, Ghirlando R et al. Solution structure of the constant region of nuclear envelope protein LAP2 reveals two LEM-domain structures: One binds BAF and the other binds DNA. EMBO J 2001; 20(16):4399-4407.
67. Laguri C, Gilquin B, Wolff N et al. Structural characterization of the lem motif common to three human inner nuclear membrane proteins. Structure 2001; 9(6):503-511.
68. Shumaker DK, Lee KK, Tanhehco YC et al. LAP2 binds to BAF-DNA complexes: Requirement for the LEM domain and modulation by variable regions. EMBO J 2001; 20(7):1754-1764.
69. Furukawa K. LAP2 binding protein 1 (L2BP1/BAF) is a candidate mediator of LAP2- chromatin interaction. J Cell Sci 1999; 112(Pt):2485-2492.
70. Zheng R, Ghirlando R, Lee MS et al. Barrier-to-autointegration factor (BAF) bridges DNA in a discrete, higher-order nucleoprotein complex. Proc Natl Acad Sci USA 2000; 97(16):8997-9002.
71. Segura-Totten M, Kowalski AK, Craigie R et al. Barrier-to-autointegration factor: Major roles in chromatin decondensation and nuclear assembly. J Cell Biol 2002; 158:475-485.
72. Haraguchi T, Koujin T, Segura-Totten M et al. BAF is required for emerin assembly into the reforming nuclear envelope. J Cell Sci 2001; 114(Pt 24):4575-4585.
73. Furukawa K, Sugiyama S, Osouda S et al. Barrier-to-autointegration factor plays crucial roles in cell cycle progression and nuclear organization in Drosophila. J Cell Sci 2003; 116(Pt 18):3811-3823.
74. Martins S, Eikvar S, Furukawa K et al. HA95 and LAP2 beta mediate a novel chromatin-nuclear envelope interaction implicated in initiation of DNA replication. J Cell Biol 2003; 160(2):177-188.
75. Martins SB, Eide T, Steen RL et al. HA95 is a protein of the chromatin and nuclear matrix regulating nuclear envelope dynamics. J Cell Sci 2000; 113:3703-3713.
76. Manilal S, Nguyen TM, Sewry CA et al. The Emery-Dreifuss muscular dystrophy protein, emerin, is a nuclear membrane protein. Hum Mol Genet 1996; 5(6):801-808.
77. Ostlund C, Ellenberg J, Hallberg E et al. Intracellular trafficking of emerin, the Emery-Dreifuss muscular dystrophy protein. J Cell Sci 1999; 112(Pt 11):1709-1719.
78. Sakaki M, Koike H, Takahashi N et al. Interaction between emerin and nuclear lamins. J Biochem (Tokyo) 2001; 129(2):321-327.
79. Fairley EA, Kendrick-Jones J, Ellis JA. The Emery-Dreifuss muscular dystrophy phenotype arises from aberrant targeting and binding of emerin at the inner nuclear membrane. J Cell Sci 1999; 112(Pt 15):2571-2582.

80. Clements L, Manilal S, Love DR et al. Direct interaction between emerin and lamin A. Biochem Biophys Res Commun 2000; 267(3):709-714.
81. Lee KK, Haraguchi T, Lee RS et al. Distinct functional domains in emerin bind lamin A and DNA-bridging protein BAF. J Cell Sci 2001; 114(Pt 24):4567-4573.
82. Holt I, Ostlund C, Stewart CL et al. Effect of pathogenic mis-sense mutations in lamin A on its interaction with emerin in vivo. J Cell Sci 2003; 116(Pt 14):3027-3035.
83. Vaughan OA, MAlvarez-Reyes M, Bridger JM et al. Both emerin and lamin C depend on lamin A for localization at the nuclear envelope. J Cell Sci 2001; 114:2577-2590.
84. Raharjo WH, Enarson P, Sullivan T et al. Nuclear envelope defects associated with LMNA mutations cause dilated cardiomyopathy and Emery-Dreifuss muscular dystrophy. J Cell Sci 2001; 114(Pt 24):4447-4457.
85. Ostlund C, Bonne G, Schwartz K et al. Properties of lamin A mutants found in Emery-Dreifuss muscular dystrophy, cardiomyopathy and Dunnigan-type partial lipodystrophy. J Cell Sci 2001; 114(Pt 24):4435-4445.
86. Gruenbaum Y, Lee KK, Liu J et al. The expression, lamin-dependent localization and RNAi depletion phenotype for emerin in C. elegans. J Cell Sci 2002; 115(Pt 5):923-929.
87. Holaska JM, Lee KK, Kowalski AK et al. Transcriptional Repressor Germ Cell-less (GCL) and Barrier to Autointegration Factor (BAF) compete for binding to emerin in vitro. J Biol Chem 2003; 278(9):6969-6975.
88. Nagano A, Koga R, Ogawa M et al. Emerin deficiency at the nuclear membrane in patients with Emery-Dreifuss muscular dystrophy. Nat Genet 1996; 12(3):254-259.
89. Paulin-Levasseur M, Blake DL, Julien M et al. The MAN antigens are nonlamin constituents of the nuclear lamina in vertebrate cells. Chromosoma 1996; 104:367-379.
90. Lin F, Blake DL, Callebaut I et al. MAN1, an inner nuclear membrane protein that shares the LEM domain with lamina-associated polypeptide 2 and emerin. J Biol Chem 2000; 275(7):4840-4847.
91. Wu W, Lin F, Worman HJ. Intracellular trafficking of MAN1, an integral protein of the nuclear envelope inner membrane. J Cell Sci 2002; 115(Pt 7):1361-1371.
92. Liu J, Lee KK, Segura-Totten M et al. MAN1 and emerin have overlapping function(s) essential for chromosome segregation and cell division in Caenorhabditis elegans. Proc Natl Acad Sci USA 2003; 100(8):4598-4603.
93. Raju GP, Dimova N, Klein PS et al. SANE, a novel LEM domain protein, regulates bone morphogenetic protein signaling through interaction with Smad1. J Biol Chem 2003; 278(1):428-437.
94. Osada S, Ohmori SY, Taira M. XMAN1, an inner nuclear membrane protein, antagonizes BMP signaling by interacting with Smad1 in Xenopus embryos. Development 2003; 130(9):1783-1794.
95. Starr DA, Han M. ANChors away: An actin based mechanism of nuclear positioning. J Cell Sci 2003; 116(Pt 2):211-216.
96. Raff JW. The missing (L) UNC? Curr Biol 1999; 9(18):R708-710.
97. Malone CJ, Fixsen WD, Horvitz HR et al. UNC-84 localizes to the nuclear envelope and is required for nuclear migration and anchoring during C. elegans development. Development 1999; 126(14):3171-3181.
98. Lee KK, Starr D, Cohen M et al. Lamin-dependent localization of UNC-84, a protein required for nuclear migration in Caenorhabditis elegans. Mol Biol Cell 2002; 13(3):892-901.
99. Roper K, Gregory SL, Brown NH. The 'spectraplakins': Cytoskeletal giants with characteristics of both spectrin and plakin families. J Cell Sci 2002; 115(Pt 22):4215-4225.
100. Starr DA, Han M. Role of ANC-1 in tethering nuclei to the actin cytoskeleton. Science 2002; 298(5592):406-409.
101. Apel ED, Lewis RM, Grady RM et al. Syne-1, a dystrophin- and Klarsicht-related protein associated with synaptic nuclei at the neuromuscular junction. J Biol Chem 2000; 275(41):31986-31995.
102. Zhang Q, Skepper JN, Yang F et al. Nesprins: A novel family of spectrin-repeat-containing proteins that localize to the nuclear membrane in multiple tissues. J Cell Sci 2001; 114(Pt 24):4485-4498.
103. Mislow JM, Kim MS, Davis DB et al. Myne-1, a spectrin repeat transmembrane protein of the myocyte inner nuclear membrane, interacts with lamin A/C. J Cell Sci 2002; 115(Pt 1):61-70.
104. Zhen YY, Libotte T, Munck M et al. NUANCE, a giant protein connecting the nucleus and actin cytoskeleton. J Cell Sci 2002; 115(Pt 15):3207-3222.
105. Mislow JM, Holaska JM, Kim MS et al. Nesprin-1alpha self-associates and binds directly to emerin and lamin A in vitro. FEBS Lett 2002; 525(1-3):135-140.
106. Gough LL, Fan J, Chu S et al. Golgi localization of Syne-1. Mol Biol Cell 2003; 14(6):2410-2424.
107. Dreger M, Bengtsson L, Schoneberg T et al. Nuclear envelope proteomics: Novel integral membrane proteins of the inner nuclear membrane. Proc Natl Acad Sci USA 2001; 98(21):11943-11948.
108. Schirmer EC, Florens L, Guan T et al. Nuclear membrane proteins with potential disease links found by subtractive proteomics. Science 2003; 301(5638):1380-1382.

109. Rolls MM, Stein PA, Taylor SS et al. A visual screen of a GFP-fusion library identifies a new type of nuclear envelope membrane protein. J Cell Biol 1999; 146(1):29-44.
110. Bione S, Maestrini E, Rivella S et al. Identification of a novel X-linked gene responsible for Emery-Dreifuss muscular dystrophy. Nature Genet 1994; 8:323-327.
111. Manilal S, thi Man N, Sewry CA et al. The Emery-Dreifuss muscular dystrophy protein, emerin, is a nuclear membrane protein. Hum Mol Genet 1996; 5:801-808.
112. Nagano A, Koga R, Ogawa M et al. Emerin deficiency at the nuclear membrane in patients with Emery-Dreyfuss muscular dystrophy. Nature Genet 1996; 12:254-259.
113. Emery AE, Dreifuss FE. Unusual type of benign x-linked muscular dystrophy. J Neurol Neurosurg Psychiatry 1966; 29(4):338-342.
114. Maidment SL, Ellis JA. Muscular dystrophies, dilated cardiomyopathy, lipodystrophy and neuropathy: The nuclear connection. Exp Rev Mol Med 2002; http://www.expertreviews.org/02004842h.htm:1-21.
115. Yates JR. 43rd ENMC International Workshop on Emery-Dreifuss Muscular Dystrophy, 22 June 1996, Naarden, The Netherlands. Neuromuscul Disord 1997; 7(1):67-69.
116. Yates JR, Bagshaw J, Aksmanovic VM et al. Genotype-phenotype analysis in X-linked Emery-Dreifuss muscular dystrophy and identification of a missense mutation associated with a milder phenotype. Neuromuscul Disord 1999; 9(3):159-165.
117. Di Blasi C, Morandi L, Raffaele di Barletta M et al. Unusual expression of emerin in a patient with X-linked Emery-Dreifuss muscular dystrophy. Neuromuscul Disord 2000; 10(8):567-571.
118. Ellis JA, Brown CA, Tilley LD et al. Two distal mutations in the gene encoding emerin have profoundly different effects on emerin protein expression. Neuromuscul Disord 2000; 10(1):24-30.
119. Raffaele Di Barletta M, Ricci E, Galluzzi G et al. Different mutations in the LMNA gene cause autosomal dominant and autosomal recessive Emery-Dreifuss muscular dystrophy. Am J Hum Genet 2000; 66(4):1407-1412.
120. Bonne G, Di Barletta MR, Varnous S et al. Mutations in the gene encoding lamin A/C cause autosomal dominant Emery- Dreifuss muscular dystrophy. Nat Genet 1999; 21(3):285-288.
121. Fatkin D, MacRae C, Sasaki T et al. Missense mutations in the rod domain of the lamin A/C gene as causes of dilated cardiomyopathy and conduction-system disease. N Engl J Med 1999; 341(23):1715-1724.
122. Muchir A, Bonne G, van der Kooi AJ et al. Identification of mutations in the gene encoding lamins A/C in autosomal dominant limb girdle muscular dystrophy with atrioventricular conduction disturbances (LGMD1B). Hum Mol Genet 2000; 9(9):1453-1459.
123. De SandreGiovannoli A, Chaouch M, Kozlov S et al. Homozygous defects in LMNA, encoding lamin A/C nuclear-envelope proteins, cause autosomal recessive axonal neuropathy in human (Charcot-Marie-Tooth disorder type 2) and mouse. Am J Hum Genet 2002; 70(3):726-736.
124. Speckman RA, Garg A, Du F et al. Mutational and haplotype analyses of families with familial partial lipodystrophy (Dunnigan variety) reveal recurrent missense mutations in the globular C-terminal domain of lamin A/C. Am J Hum Genet 2000; 66(4):1192-1198.
125. Shackleton S, Lloyd DJ, Jackson SN et al. LMNA, encoding lamin A/C, is mutated in partial lipodystrophy. Nat Genet 2000; 24(2):153-156.
126. Cao H, Hegele RA. Nuclear lamin A/C R482Q mutation in canadian kindreds with dunnigan-type familial partial lipodystrophy. Hum Mol Genet 2000; 9(1):109-112.
127. Novelli G, Muchir A, Sangiuolo F et al. Mandibuloacral Dysplasia is caused by a mutation in LMNA-encoding lamin A/C. Am J Hum Genet 2002; 71(2):426-431.
128. De SandreGiovannoli A, Bernard R, Cau P et al. Lamin A Truncation in Hutchinson-Gilford Progeria. Science 2003.
129. Eriksson M, Brown WT, Gordon LB et al. Recurrent de novo point mutations in lamin A cause Hutchinson-Gilford progeria syndrome. Nature 2003.
130. Chen L, Lee L, Kudlow BA et al. LMNA mutations in atypical Werner's syndrome. Lancet 2003; 362(9382):440-445.
131. Bergo MO, Leung GK, Ambroziak P et al. Isoprenylcysteine carboxyl methyltransferase deficiency in mice. J Biol Chem 2001; 276(8):5841-5845.
132. Kim E, Ambroziak P, Otto JC et al. Disruption of the mouse Rce1 gene results in defective Ras processing and mislocalization of Ras within cells. J Biol Chem 1999; 274(13):8383-8390.
133. Beroud C, Collod-Beroud G, Boileau C et al. UMD (Universal mutation database): A generic software to build and analyze locus-specific databases. Hum Mutat 2000; 15(1):86-94.
134. Bonne G, Yaou RB, Beroud C et al. 108th ENMC International Workshop, 3rd Workshop of the MYO-CLUSTER project: EUROMEN, 7th international emery-dreifuss muscular dystrophy (EDMD) workshop, 13-15 September 2002, Naarden, The Netherlands. Neuromuscul Disord 2003; 13(6):508-515.

135. Sewry CA. The role of immunocytochemistry in congenital myopathies. Neuromuscul Disord 1998; 8(6):394-400.
136. Sewry CA. Immunocytochemical analysis of human muscular dystrophy. Microsc Res Tech 2000; 48(3-4):142-154.
137. Sewry CA, Brown SC, Mercuri E et al. Skeletal muscle pathology in autosomal dominant Emery-Dreifuss muscular dystrophy with lamin A/C mutations. Neuropathol Appl Neurobiol 2001; 27(4):281-290.
138. Bonne G, Mercuri E, Muchir A et al. Clinical and molecular genetic spectrum of autosomal dominant Emery-Dreifuss muscular dystrophy due to mutations of the lamin A/C gene [In Process Citation]. Ann Neurol 2000; 48(2):170-180.
139. Fidzianska A, Toniolo D, Hausmanowa-Petrusewicz I. Ultrastructural abnormality of sarcolemmal nuclei in Emery-Dreifuss muscular dystrophy (EDMD). J Neurol Sci Jul 15 1998; 159(1):88-93.
140. Ognibene A, Sabatelli P, Petrini S et al. Nuclear changes in a case of X-linked Emery-Dreifuss muscular dystrophy. Muscle Nerve 1999; 22(7):864-869.
141. Chen L, Oshima J. Werner Syndrome. J Biomed Biotechnol 2002; 2(2):46-54.
142. Vigouroux C, Auclair M, Dubosclard E et al. Nuclear envelope disorganization in fibroblasts from lipodystrophic patients with heterozygous R482Q/W mutations in the lamin A/C gene. J Cell Sci 2001; 114(Pt 24):4459-4468.
143. Pendas AM, Zhou Z, Cadinanos J et al. Defective prelamin A processing and muscular and adipocyte alterations in Zmpste24 metalloproteinase-deficient mice. Nat Genet 2002; 31(1):94-99.
144. Agarwal AK, Fryns JP, Auchus RJ et al. Zinc metalloproteinase, ZMPSTE24, is mutated in mandibuloacral dysplasia. Hum Mol Genet 2003; 12(16):1995-2001.
145. Brodsky GL, Muntoni F, Miocic S et al. Lamin A/C gene mutation associated with dilated cardiomyopathy with variable skeletal muscle involvement. Circulation 2000; 101(5):473-476.
146. Daigle N, Beaudouin J, Hartnell L et al. Nuclear pore complexes form immobile networks and have a very low turnover in live mammalian cells. J Cell Biol 2001; 154(1):71-84.
147. Meier J, Campbell KH, Ford CC et al. The role of lamin LIII in nuclear assembly and DNA replication, in cell- free extracts of Xenopus eggs. J Cell Sci 1991; 98(Pt 3):271-279.
148. Newport JW, Wilson KL, Dunphy WG. A lamin-independent pathway for nuclear envelope assembly. J Cell Biol 1990; 111(6Pt 1):2247-2259.
149. Stierle V, Couprie J, Ostlund C et al. The carboxyl-terminal region common to lamins A and C contains a DNA binding domain. Biochemistry 2003; 42(17):4819-4828.
150. Vlcek S, Dechat T, Foisner R. Nuclear envelope and nuclear matrix: Interactions and dynamics. Cell Mol Life Sci 2001; 58(12-13):1758-1765.
151. Lachner M, O'Carroll D, Rea S et al. Methylation of histone H3 lysine 9 creates a binding site for HP1 proteins. Nature 2001; 410(6824):116-120.
152. Lachner M, O'Sullivan RJ, Jenuwein T. An epigenetic road map for histone lysine methylation. J Cell Sci 2003; 116(Pt 11):2117-2124.
153. Croft JA, Bridger JM, Boyle S et al. Differences in the localization and morphology of chromosomes in the human nucleus. J Cell Biol 1999; 145(6):1119-1131.
154. Boyle S, Gilchrist S, Bridger JM et al. The spatial organization of human chromosomes within the nuclei of normal and emerin-mutant cells. Hum Mol Genet 2001; 10(3):211-219.
155. Spann TP, Goldman AE, Wang C et al. Alteration of nuclear lamin organization inhibits RNA polymerase II-dependent transcription. J Cell Biol 2002; 156(4):603-608.
156. Mancini MA, Shan B, Nickerson JA et al. The retinoblastoma gene product is a cell cycle-dependent, nuclear matrix-associated protein. Proc Natl Acad Sci USA 1994; 91(1):418-422.
157. Markiewicz E, Dechat T, Foisner R et al. The lamin A/C binding protein LAP2alpha is required for the nuclear anchorage of the retinoblastoma protein. Mol Biol Cell 2002; 13:4401-4413.
158. Lavia P, Jansen-Durr P. E2F target genes and cell-cycle checkpoint control. BioEssays 1999; 21(3):221-230.
159. Kaelin Jr WG. Functions of the retinoblastoma protein. BioEssays 1999; 21(11):950-958.
160. Dyson N. The regulation of E2F by pRB-family proteins. Genes Dev 1998; 12(15):2245-2262.
161. de la Luna S, Allen KE, Mason SL et al. Integration of a growth-suppressing BTB/POZ domain protein with the DP component of the E2F transcription factor. EMBO J 1999; 18(1):212-228.
162. Nili E, Cojocaru GS, Kalma Y et al. Nuclear membrane protein LAP2beta mediates transcriptional repression alone and together with its binding partner GCL (germ-cell-less). J Cell Sci 2001; 114(Pt 18):3297-3307.
163. Goldberg M, Jenkins H, Allen T et al. Xenopus lamin B3 has a direct role in the assembly of a replication competent nucleus: Evidence from cell-free egg extracts. J Cell Sci 1995; 108(Pt 11):3451-3461.

164. Spann TP, Moir RD, Goldman AE et al. Disruption of nuclear lamin organization alters the distribution of replication factors and inhibits DNA synthesis. J Cell Biol 1997; 136(6):1201-1212.
165. Moir RD, Spann TP, Herrmann H et al. Disruption of nuclear lamin organization blocks the elongation phase of DNA replication. J Cell Biol 2000; 149(6):1179-1192.
166. Ellis DJ, Jenkins H, Whitfield WG et al. GST-lamin fusion proteins act as dominant negative mutants in Xenopus egg extract and reveal the function of the lamina in DNA replication. J Cell Sci 1997; 110(Pt 20):2507-2518.
167. Schirmer EC, Guan T, Gerace L. Involvement of the lamin rod domain in heterotypic lamin interactions important for nuclear organization. J Cell Biol 2001; 153(3):479-489.
168. Gant TM, Harris CA, Wilson KL. Roles of LAP2 proteins in nuclear assembly and DNA replication: Truncated LAP2beta proteins alter lamina assembly, envelope formation, nuclear size, and DNA replication efficiency in Xenopus laevis extracts. J Cell Biol 1999; 144(6):1083-1096.
169. Buendia B, Santa-Maria A, Courvalin JC. Caspase-dependent proteolysis of integral and peripheral proteins of nuclear membranes and nuclear pore complex proteins during apoptosis. J Cell Sci 1999; 112(Pt 11):1743-1753.
170. Rao L, Perez D, White E. Lamin proteolysis facilitates nuclear events during apoptosis. J Cell Biol 1996; 135(6Pt 1):1441-1455.
171. Gotzmann J, Vlcek S, Foisner R. Caspase-mediated cleavage of the chromosome-binding domain of lamina-associated polypeptide 2 alpha. J Cell Sci 2000; 113(Pt 21):3769-3780.
172. Ruchaud S, Korfali N, Villa P et al. Caspase-6 gene disruption reveals a requirement for lamin A cleavage in apoptotic chromatin condensation. EMBO J 2002; 21(8):1967-1977.
173. Steen RL, Collas P. Mistargeting of B-type lamins at the end of mitosis: Implications on cell survival and regulation of lamins A/C expression. J Cell Biol 2001; 153(3):621-626.
174. Hughes SM, Schiaffino S. Control of muscle fibre size: A crucial factor in ageing. Acta Physiol Scand 1999; 167(4):307-312.
175. Krimm I, Ostlund C, Gilquin B et al. The Ig-like structure of the C-terminal domain of lamin a/c, mutated in muscular dystrophies, cardiomyopathy, and partial lipodystrophy. Structure 2002; 10(6):811-823.
176. Dhe-Paganon S, Werner ED, Chi YI et al. Structure of the globular tail of nuclear lamin. J Biol Chem 2002; 277(20):17381-17384.
177. Lloyd DJ, Trembath RC, Shackleton S. A novel interaction between lamin A and SREBP1: Implications for partial lipodystrophy and other laminopathies. Hum Mol Genet 2002; 11(7):769-777.
178. Favreau C, Dubosclard E, Ostlund C et al. Expression of lamin A mutated in the carboxyl-terminal tail generates an aberrant nuclear phenotype similar to that observed in cells from patients with Dunnigan-type partial lipodystrophy and Emery-Dreifuss muscular dystrophy. Exp Cell Res 2003; 282(1):14-23.
179. Fidzianska A, Hausmanowa-Petrusewicz I. Architectural abnormalities in muscle nuclei. Ultrastructural differences between X-linked and autosomal dominant forms of EDMD. J Neurol Sci 2003; 210(1-2):47-51.
180. Dalkilic I, Kunkel LM. Muscular dystrophies: Genes to pathogenesis. Curr Opin Genet Dev 2003; 13(3):231-238.
181. Andrulis ED, Neiman AM, Zappulla DC et al. Perinuclear localization of chromatin facilitates transcriptional silencing. Nature 1998; 394(6693):592-595.
182. Cockell M, Gasser SM. Nuclear compartments and gene regulation. Curr Opin Genet Dev 1999; 9(2):199-205.
183. Teixeira MT, Dujon B, Fabre E. Genome-wide nuclear morphology screen identifies novel genes involved in nuclear architecture and gene-silencing in Saccharomyces cerevisiae. J Mol Biol 2002; 321(4):551-561.
184. Tsukahara T, Tsujino S, Arahata K. CDNA microarray analysis of gene expression in fibroblasts of patients with X-linked Emery-Dreifuss muscular dystrophy. Muscle Nerve 2002; 25(6):898-901.
185. Ozaki T, Saijo M, Murakami K et al. Complex formation between lamin A and the retinoblastoma gene product: Identification of the domain on lamin A required for its interaction. Oncogene 1994; 9(9):2649-2653.
186. Asakura A. Stem cells in adult skeletal muscle. Trends Cardiovasc Med Apr 2003; 13(3):123-128.
187. Beltrami AP, Barlucchi L, Torella D et al. Adult cardiac stem cells are multipotent and support myocardial regeneration. Cell 2003; 114(6):763-776.
188. Ellis JA, Craxton M, Yates JR et al. Aberrant intracellular targeting and cell cycle-dependent phosphorylation of emerin contribute to the Emery-Dreifuss muscular dystrophy phenotype. J Cell Sci 1998; 111(Pt 6):781-792.
189. Mounkes LC, Burke B, Stewart CL. The A-type lamins: Nuclear structural proteins as a focus for muscular dystrophy and cardiovascular diseases. Trends Cardiovasc Med 2001; 11(7):280-285.

CHAPTER 13

Distinct Mechanisms Downstream of the Repeat Expansion Are Implicated in the Molecular Basis of Myotonic Dystrophy Type 1

Keith Johnson and Rami Jarjour

Abstract

Myotonic dystrophy type 1 (DM1) is the most common inherited muscular dystrophy affecting adults. The underlying mutation is the same in all patients with DM1, namely a trinucleotide (CTG) repeat expansion. There is now conclusive evidence that there are several distinct molecular mechanisms that probably occur concomitantly in the tissues of individuals affected by DM1 leading to the array of symptoms they exhibit. These include chromatin mediated and RNA-processing defects that alter the normal expression patterns of numerous genes, some of which explain specific symptoms such as myotonia, insulin resistance and cataract that are part of the DM1 phenotype. A range of animal models have been generated that replicate many of the key molecular and phenotypic features of DM1.

Clinical Signs and Symptoms in Myotonic Dystrophy Type 1 (DM1)

With an estimated incidence of 1 in 8,000 myotonic dystrophy (DM1, OMIM 160900) is the commonest form of adult onset muscular dystrophy. It is a multisystemic disease, characterised clinically by myotonia, muscle wasting and weakness, cataracts, insulin resistance, testicular atrophy, heart conduction defects and mental disturbances.[1] The heart symptoms are common and often pacemakers need to be fitted to prevent fatal dysrhythmias. The Scottish Muscle Centre, in collaboration with the UK Myotonic Dystrophy Support Group and the Muscular Dystrophy Campaign have produced a fact sheet to inform the patients, their families and the clinicians treating them, to make them aware of these symptoms and their potential clinical complications (www.gla.ac.uk/muscle/dm.htm).

Although DM1 is a multisystemic disease, it most often presents clinically as a progressive neuromuscular disorder with a characteristic pattern of muscle wasting and weakness. Weakness and wasting of facial, jaw and sternomastoid muscles contribute to one of the classical features of DM1, the hatchet facial appearance. Involvement of the extraocular muscles produces ptosis and weakness of eyelid closure. Distal muscle weakness is prominent in the early stages of the disease but proximal muscle involvement occurs as the disease progresses. Other muscles affected are the small muscles of the hand (e.g., the exterior muscles of the wrist) and feet (e.g., the dorsiflexors). Many of the DM1 muscle biopsy changes are distinctive but not specific. In 70% of biopsies there is atrophy of type I muscle fibres and hypertrophy of predominantly histochemical type II muscle fibres.

Molecular Mechanisms of Muscular Dystrophies, edited by Steve J. Winder. ©2006 Eurekah.com.

Mapping of the DM1 Locus, Identification of Underlying Mutation and Molecular Explanation of Anticipation

Myotonic dystrophy is inherited as an autosomal dominant trait. Historically DM1 was one of the first human inherited traits to be genetically linked with markers by the work of Mohr and colleagues using classical blood groups. Mohr demonstrated linkage between DM and the Lutheran and Secretor blood group markers.[2] This linkage group was extended and mapped to chromosome 19 with the inclusion of the complement component 3 (C3) gene.[3] Further mapping refined the location of the DM gene to chromosome 19q 13.3 and the underlying mutation was identified as a triplet repeat expansion.[4-9]

This discovery led to an immediate molecular explanation for anticipation, a hallmark of families with DM1. Anticipation is when the severity of a condition increases and the age of onset decreases from generation to generation. To date trinucleotide repeat expansion is the only known cause of anticipation and close scrutiny of pedigrees for this phenomenon may therefore give a significant clue in conditions where the mutation has not yet been identified. Anticipation is particularly striking in DM1 families where the age of onset decreases on average 20 years per generation.[10] As a consequence the typical presentation of a family with DM is of a grandfather with a cataract as his only symptom, his daughter would be classically affected with myotonia and some weakness and wasting, particularly of the muscles of the face and neck, and her child congenitally affected, with severe respiratory failure and delayed muscle maturation at birth.[1] It is still common for all three generations to be diagnosed after the birth of a congenitally affected infant, the symptoms in the older generations having been sufficiently mild to be undiagnosed.

It is important to note that the symptoms of congenital DM1 are not the same as those of adult patients. There is usually no myotonia until their teens and the incidence of congenital cataracts does not appear to be higher than that in the general population.[1] Muscle from patients with congenital DM1 shows pathological changes not present in adult patients. There are also differences in the histopathological findings in muscle biopsies from congenital and classical onset DM1. In congenital cases the muscles are hypoplastic and the fibres have an immature appearance, with centrally placed nuclei and abundant satellite cells, suggesting arrested development rather than degeneration that is seen in adult biopsies.[11]

Another consequence of the discovery of the mutation was the realization that there was more than one form of dominantly inherited myotonic dystrophy, with the much rarer DM2 form mapping to a distinct locus, now known to be on chromosome 3. DM2 is due to the expansion of a tetranucleotide repeat.[12] For the purposes of this review we will focus on DM1.

Repeat Expansion and the Dynamic Nature of the Mutation in DM1

The underlying mutation in all individuals affected by DM1 is the expansion of a CTG repeat at chromosome 19q13.3. In the general population this locus contains alleles with repeat arrays ranging in size from 5 to 37 copies, with the 5 repeat allele being the most common.[13] In individuals affected by DM1 the repeat array contains from 50 to several thousands of CTGs, with the number of repeats correlating reasonably well with age at onset and severity of symptoms. Normal CTG repeat numbers[5-37] are inherited in a stable manner but expansions containing over 40 repeats tend to expand further in successive generations. The germline instability of the CTG repeat has been shown to be dependant on the sex of the parent and the size of the CTG repeat in the parent.[14] The congenital form of the disease is almost always transmitted maternally. When the size of the CTG repeats is more than 166 CTG repeats, length increases during transmission are more likely through female meioses. However, smaller-size repeats (<100 CTG repeats), tend to be transmitted through the male germline.[15] Larger repeat lengths are not transmitted by males, possibly due to nonviability of sperm carrying large repeats. A molecular basis for this is offered by the identification of the *radial spokehead-like* gene (*RSHL1*) distal of *DMWD* gene (Fig. 1 and ref. 16). *RSHL-1* was identified due to its homology to genes encoding proteins that form part of the ciliary motility flagella in

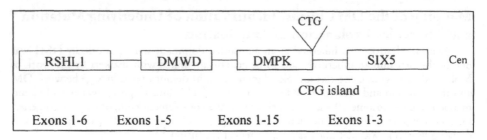

Figure 1. A schematic diagram of the DM1 region showing the relative positions of the *RSHL-1*, *DMWD*, *DMPK* and *SIX5* genes and the location of the unstable CTG repeat and the CpG island covering the 3'-end of DMPK and the 5" end of SIX5 and including the repeat.

sea urchins and protozoa, including key functional roles in sperm motility in the former. There is genetic evidence that a gene for familial primary ciliary dyskinesia (FPCD) maps to the *RSHL-1* region of 19q13.3.[16]

Somatic size heterogeneity is one of the features of the DM1 mutation. It occurs within and between different tissues and continues throughout the lifetime of an individual, almost certainly underlying the progressive nature of the condition.[17] Expansions in muscle are greater than in lymphocytes from the same DM1 patient[18] although there is no such increase in normal muscle.[19] In general, most tissues show a greater degree of repeat expansion than seen in blood. It has been shown that the CTG repeat expansion is less in sperm than in blood of DM1 patients.[20] However, cerebellum and some parts of the brain (i.e., frontal cortex and thalamus) show a smaller expansion number compared to blood.[20,21]

Small pool PCR (SP-PCR) analysis has shown that somatic mutation length variability occurs within tissues.[22] It shows that a smear of CTG repeat expansion in the blood of DM1 patients can be dissected into a series of separate signals, each representing the mutation length carried by individual chromosomes. Conventional techniques (e.g., Southern blotting and PCR) analyse the expansion size in at least 10^4 cells simultaneously (close to 10^6 cells using Southern blotting), whereas SP-PCR can analyse very small amounts of input DNA (i.e., down to ~1-2 genomic equivalents).

Genes at the Locus and Their Functions

The DM1 repeat expansion occurs in a very gene-rich region of the human genome and alters the expression of a number of genes due to effects at the level of chromatin structure. Three genes at the locus have been studied in some depth, *DMPK*, *SIX5* and *DMWD*, although effects on other genes further from the repeat cannot be ruled out, including *RSHL-1* as discussed in the previous section. A map summarizing the genes and their positions relative to the repeat expansion is shown in Figure 1.

DMPK

The *DMPK* gene encodes a member of the subfamily of Rho binding kinases. The *DMPK* gene is about 13 kb in length, comprising 15 exons which undergo extensive alternative splicing (see Fig. 1 and ref. 23). Members of the Rho kinase family have a role in cell shape determination and in the regulation of actin-myosin contractility. Expression of *DMPK* has been studied at the RNA and protein level. RNA studies showed that *DMPK* is expressed in a wide range of tissues but is mainly found in tissues involved in the clinical manifestations of DM1 (heart, brain, muscle and eye).

However, there is considerable confusion in the literature about the size of DMPK protein and its distribution and localisation within tissues. This confusion has almost certainly arisen because DMPK displays very significant homology to members of the Rho-binding class of protein kinases and cross-reaction of antisera is a common problem, especially with polyclonals targeted against the kinase domain.[24] A panel of anti-DMPK monoclonal antibodies with appropriate controls, including tissues from *Dmpk* knockout mice showed DMPK to be almost exclusively expressed in skeletal and cardiac muscle.[25] DMPK has an almost identical kinase domain to that of the human Cdc42 binding protein beta (CDC42BPB) or MRCKβ.[26,27] Human CDC42BPB was isolated using primers designed from an alignment of a 33 amino acid stretch spanning kinase domains VIII and IX from DMPK and two members of the Rho kinase family, PK428 and Rhop160[ROCK] as these members of this family share a unique 3 amino acid insert in this domain region. Lam et al[25] showed that MRCKα and MRCKβ were two key cross-reacting species with one of their monoclonals with an epitope in the catalytic (kinase) domain. It is almost certain that the confusion in the literature regarding the localization and distribution of DMPK is due to cross-reaction of some anti-DMPK polyclonals with MRCKα and MRCKβ which are far more widely expressed than DMPK.

DMWD (59)

The human dystrophia myotonica-containing WD repeat motif (*DMWD*) gene (previously known as 59) is located 500bp upstream of the *DMPK* gene. WD repeats are highly conserved domains in a family of proteins engaged in signal transduction and cell regulatory functions. Members of this family usually contain between 4 and 8 WD repeats. DMWD has four WD repeats, each repeat forms a propeller blade in other WD family members. *DMWD* is homologous to the murine gene *DMR-N9*.[23] The *DMWD* gene is about 10 kb in length and is expressed in tissues affected in DM1 such as the heart, brain, testis, liver, kidney, and spleen.[28] However, there is little known about the function of DMWD or its involvement in DM1 pathogenesis. Analysis of DMWD expression in the mouse brain has shown that from postnatal day 7 to postnatal day 21 the levels of mRNA expression remain constant but the levels of DMWD protein gradually increase. DMWD protein is expressed throughout the brain, at low levels in glia, more prominently in neurons and significantly in neuropil with high density of synaptic connections. A potential role for DMWD in transcriptional regulation of synaptic growth and maturation has been proposed based on functional analysis of homologues of DMWD in *Aspergillus nidulans* and *Drosophila*.[29]

SIX5 (DMAHP)

Human and mouse *SIX5* (previously known as DM locus-associated homeodomain protein (*DMAHP*)) genes were identified by sequencing the extensive CpG island at 3' end of the *DMPK* gene (Fig. 1) and by the recognition of highly conserved regions across the two species, including homology with a homeodomain encoding gene, *sine oculis* (*so*), from *Drosophila*.[30] The CTG repeat lies within a 3.5 kb CpG island in the human gene, but the CpG island is less extensive in the mouse gene (1.8 kb).[31] The human and mouse *SIX5* exonic sequences are very similar and encode proteins that are 85% identical at the amino acid level. Sequence data show that *SIX5* encodes a homeodomain containing protein that shows significant similarity to the *Drosophila* eye development gene *sine oculis* (*so*), the founding member of the *SIX* subfamily of homeodomain genes. All members of the *SIX* subfamily share a region of a high sequence homology that encodes a 60 amino acid homeodomain and another domain located immediately upstream of the homeodomain called the SIX domain which comprises 116 amino acids. RT-PCR and western blotting studies have shown that *SIX5* is expressed in skeletal muscle, heart, brain, fibroblasts and lymphocytes and at several additional locations in foetal tissues.[30,31] Hence, *SIX5* and *DMPK* expression patterns overlap significantly. Correlations of these effects on SIX5 target genes with DM1 symptoms are listed in Table 1 and discussed further in the section describing target genes of SIX5.

Effects of the Repeat Expansion on Gene Expression

The repeat expansion is transcribed within the 3'-UTR of the *DMPK* gene and has effects on both the transcription of *DMPK* and on the processing of a number of other transcripts (effects in *trans*). The repeat expansion has also been shown to have an effect on the chromatin

Table 1. *DM symptoms and potential links with targets of SIX5*

Gene/Motif	Accession Number	DM Symptoms That It May Contribute to
Dopamine D5 receptor	U21164	- mental disturbances - loss of memory
Ca²⁺-activated K channel	NT_004524	Myotonia
Myogenin	AF050501	Muscular abnormality
MEF-3 aldolase	-	Muscular abnormality
MEF-3 myogenin	-	Muscular abnormality
DOCK180	NT_008818	Muscular abnormality
Myogenic repressor	AF060154	Muscular abnormality
ZNF9	U19765	DM2
Insulin receptor	M23100	Insulin resistance
Tissue plasminogen activator	K03021	Increased levels of tPA

structure, which in turn affects expression of several of the genes at the locus, including *DMPK*, *DMWD* and *SIX5*. There is a great deal of evidence that decreases in expression of *SIX5* are due to chromatin structure changes caused by the repeat expansion. Regulatory elements of *SIX5* have been identified upstream of the gene, within the introns towards the 3'-end of the *DMPK* gene. These elements become disconnected from *SIX5* by expanding repeats with associated decreases in transcription of *SIX5* leading to functional haploinsufficiency. The consequences of this will be discussed further when we look at the target genes regulated by SIX5.

A DNAase I hypersensitive site is positioned downstream of the CTG repeat (but upstream of *SIX5*) in the wildtype DM1 locus but the large CTG repeat expansions eliminate the hypersensitive site by altering the chromatin structure in that area. This hypersensitive site contains an enhancer element that regulates transcription of the *SIX5* gene.[32] Analysis of *SIX5* expression in fibroblast and skeletal muscle cells of DM1 patients revealed a 2- to 4-fold reduction in *SIX5* transcript levels compared to wildtype controls. Moreover, it has been reported that levels of *SIX5* mRNA are reduced in primary myoblast cultures, post-mortem muscle, heart, and brain tissues taken from DM1 patients using allele-specific RT-PCR which exploits a coding polymorphism in exon C of *SIX5*.[33] These results have also been confirmed by showing a reduction of *SIX5* expression levels in muscle tissues obtained at autopsy from DM1 patients.[34] Reduced levels of *SIX5* transcripts have been reported in different brain areas (frontal, occipital, temporal and cerebellum lobes) of DM1 patients.[35,36] However, a number of studies have shown different results from what has been previously reported. One study observed no changes of *SIX5* expression levels from DM1 muscle biopsies.[37] Another study showed that the gross levels of transcripts from *SIX5* were not altered in cell lines from DM1 patients.[38] These differences may be due to differences in the methods used. Reanalysing the DM1 samples used in the latter study utilising allele-specific PCR showed that the *SIX5* transcripts are reduced in both nuclear and cytoplasmic compartments compared to the control samples,[39] indicating an effect on SIX5 expression at the transcriptional level. In normal myoblasts, the levels of mRNA synthesised from each *SIX5* allele are similar. In contrast, myoblasts from DM1 patients show reduced expression of the DM1-linked *SIX5* allele.[33] Therefore, the more sensitive allele-specific analysis detects changes in SIX5 mRNA levels that are not detectable in gross level analyses.

Two CTCF zinc finger binding sites have been identified in the 3'-UTR of *DMPK* flanking the CTG repeat.[40] A combination of the CTG repeats and the CTCF sites forms an insulator element between *DMPK* and *SIX5* and mediates the inhibition of the promoter-enhancer interactions. Methylation of these sites prevents binding of CTCF. CpGs in the region of the CTG repeat are methylated in congenital DM1 but not in the adult form of the disease.[41] Loss of binding of CTCF at the DM1 locus may contribute to the distinct phenotype of congenital

DM1 through loss of normal regulation of *SIX5*, which from model organism studies of the orthologous gene, *D-Six4* in *Drosophila*, is a critical determinant of normal muscle development.

There is also evidence for affects on the expression of *DMPK* and *DMWD* but, in contrast to the direct effect on transcription for *SIX5*, for these two genes the effect is achieved by a post-transcriptional mechanism mediated at the level of mRNA processing. Conflicting results have been reported on the levels of *DMPK* transcripts in congenital DM1 patients. Increased *DMPK* mRNA levels have been reported.[42] Increased expression levels of *DMPK* have been reported in brains of patients with congenital DM1.[43] In contrast to the previous data, reduced levels of *DMPK* have been shown in congenital DM1 myoblasts[44] and in muscle samples obtained from congenital DM1 foetuses.[45] Different molecular techniques were used in these studies. RT-PCR was used in the first and third studies, immunohistochemistry was used in the second study and western analysis was used in the last study. Moreover, samples were taken from foetuses and infants and from different tissues (muscle, brain, and myoblasts) adding more difficulties in comparing these data.

Effects of the Repeat Expansion at the mRNA Level

There is a growing body of evidence that a significant part of the molecular pathogenesis of DM1 involves a dominant gain of function at the RNA level, acting in trans.[46] A gain of function model involves the formation of extended hairpin loops within the CUG tract of the 3'-UTR of *DMPK* transcripts which impair nuclear cytoplasmic transport, resulting in nuclear retention of both the repeat and associated RNA-binding proteins, resulting in altered RNA processing of potentially many transcripts (for recent review of this area see 47 and references cited therein).

In vitro studies have shown that CUG repeats form hairpin loops with extended regions of double-stranded RNA.[48,49] However, the structure of the expanded CUG repeat expansion in vivo is unknown. Foci of CTG transcripts have been detected in the nuclei of fibroblast and skeletal muscle biopsies from DM1 patients but not from normal controls.[50] It has been speculated that the CUG repeat expansion in the *DMPK* mRNA acts dominantly at the RNA level by sequestering CUG binding proteins from other transcripts that require them for processing. CUG binding protein 1 (CUG-BP1) was the first protein identified with an ability to bind nonexpanded CUG repeats in vitro.[51,52] CUG-BP1 expression is induced 3- to 4-fold in DM1 tissues and in cells that express expanded CUG repeats.[53-55] CUG-BP1 and related RNA binding proteins are involved in the regulation of splicing and translation of RNA. However, recent data question the role of CUG-BP1 in the molecular pathology of DM1 as a sequestered protein. Visualisation of CUG-BP1 by electron microscopy has shown that CUG-BP1 is a single stranded RNA-binding protein that has a binding preference for CUG-rich RNA elements but not duplex CUG hairpins.[56] Moreover, CUG-BP1 localised to the base of the hairpin loop and not along the repeat tract stem. It was also shown that CUG-BP does not bind extensively to the CUG repeat expansion in vitro[56] and that it preferentially binds to UG dinucleotide repeats in a yeast three hybrid system.[57] More recently, CUG-BP1 failed to be detected in the inclusions of DM1 cells.[58,59] Recent studies have shown that the family of human muscleblind (MBNL), muscleblind-like (MBLL) and muscleblind-like-X-linked (MBXL) proteins interact with expanded CUG repeats in vitro[58] and colocalise in vivo with the nuclear foci of expanded CUG repeats in DM1 and DM2 cells.[59-61] These proteins are homologous to the *Drosophila* muscleblind (mbl) proteins that have roles in visual development and embryonic muscle differentiation.[62]

Abnormal Regulation of Alternative Splicing

CUG-BP1 has been shown to bind to pre-mRNAs that contain CUG repeats (such as the cardiac troponin T (cTnT) pre-mRNA) and regulate their alternative splicing.[63] Two cTnT alternative isoforms have been reported: the foetal form containing the alternatively spliced exon 5, and the adult form, which does not contain exon 5. An increase in the foetal form has been reported in the striated muscle from adult DM1 patients and cell lines derived from DM1 patient muscles. However, effects of cTnT alternative splicing have not been implicated in the molecular pathology of cardiac abnormalities. CUG-BP1 has also been reported to bind to the insulin receptor (*IR*) gene.[53] IR is a heterotetramer composed of two α- and two β- subunits. Alternative splicing of exon 11 of the α-subunit results in the expression of two isoforms: IR-A lacking exon 11, and IR-B including exon 11. IR-B is the predominant isoform in normal skeletal muscle and is expressed in the insulin responsive tissues (skeletal muscle, adipose tissue, and liver). IR-A is the predominant isoform in DM1 skeletal muscle[53] and in congenital DM1 skeletal muscle[45] and does not signal effectively. CUG-BP1 mediates this switch through an intronic element located upstream of exon 11, and specifically binds to this element in vitro. Therefore, mis-splicing of *IR* pre-mRNA and the increased abundance of the IR-A isoform may explain the unusual form of insulin resistance in DM1 patients. In addition, the CUG repeat expansion has been reported to bind to the muscle-specific chloride channel (*ClC-1*) and induce aberrant splicing of its pre-mRNA, resulting in loss of ClC-1 protein in the surface membrane.[64,65] Hence, it has been postulated that aberrant splicing of the *ClC-1* pre-mRNA leads to chloride channelopathy and membrane hyper-excitability in DM1. ClC-1 protein is reduced in skeletal muscle tissues from DM1 patients[64] and in transgenic mice expressing an expanded CUG repeat[65] and therefore its altered splicing could be responsible for the myotonia seen in DM1.

Target Genes of SIX5 and Their Potential Contributions to the DM1 Phenotype

Mammalian SIX *Genes*

Mammals possess 6 *SIX* genes which can be subdivided into 3 different sub-classes based on the similarities of their SIX and homeodomains.[66] These subclasses are: *SIX1/SIX2*, *SIX3/SIX6* and *SIX4/SIX5*. These 6 mammalian *SIX* genes and *Drosophila so* demonstrate high levels of sequence similarity in the SIX domain and the homeodomain (Fig. 2). However, outside of these two regions the structures of the proteins are completely different. Like the *Drosophila so* gene, the mammalian homeodomains of the SIX family have altered two highly conserved amino acids, the arginine at position 5 and the glutamine at position 12 in helix I. There is a valine at position 5 in SIX5, SIX4, and D-Six4, a serine in SIX1, SIX2 and so, and a threonine in SIX3 and SIX6. All SIX proteins have a serine at position 12 except SIX3 and SIX6 where there is a threonine. These critical amino acid positions in the homeodomain are thought to be very important for determining the binding specificity of the SIX proteins.

Murine Six2, Six4, Six5 and human SIX5 have been shown to bind to the Na$^+$ K$^+$ ATPase α1 subunit gene regulatory element ARE.[67,68] It has been reported that both the SIX domain and the homeodomain are required for binding activity of Six4 and SIX5. Six1 and Six4 have been shown to bind to a myogenic enhancing factor 3 (MEF-3) site and transactivate a reporter gene containing MEF-3 sites.[69] MEF-3 motifs are found in the promoter region of *Myogenin*, one of *MyoD* family of genes. MEF-3 motifs are also found in other skeletal muscle-specific regulatory regions and have shown to be involved in the transcriptional regulation of the *cardiac troponin C* gene and the *aldolase* A muscle specific promoter both in vitro and in vivo. There are several other binding sites for transcription factors in the *Myogenin* promoter region such as MEF-2 and E-box motifs that are bound by MEF-2 and MyoD proteins, respectively. Myogenin is a basic helix-loop-helix transcription factor and is required for myoblast fusion in vivo.

A whole genome screen identified 14 human DNA fragments which formed specific complexes with an expressed fragment of SIX5 (GST-SIX+HD).[68] One of these DNA fragments (TspE) has been mapped to chromosome 1 and is from the 5'-UTR of the dopamine D5 receptor transcribed pseudogene 2 (*φDRD5-2*). TspE also shows 92% and 97% identity to the

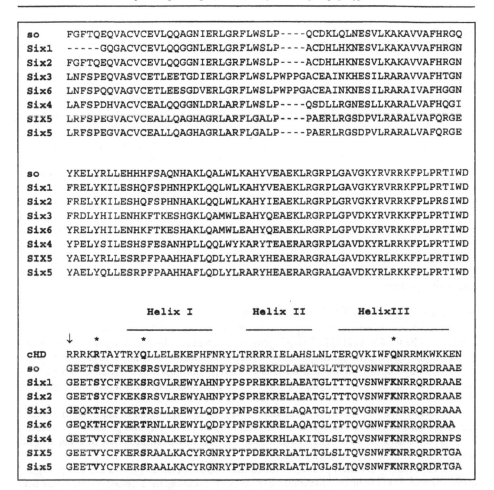

```
so     FGFTQEQVACVCEVLQQAGNIERLGRFLWSLP----QCDKLQLNESVLKAKAVVAFHRGQ
Six1   -----GQGACVCEVLQQGGNLERLGRFLWSLP----ACDHLHKNESVLKAKAVVAFHRGN
Six2   FGFTQEQVACVCEVLQQGGNIERLGRFLWSLP----ACDHLHKNESVLKAKAVVAFHRGN
Six3   LNFSPEQVASVCETLEETGDIERLGRFLWSLPWPPGACEAINKHESILRARAVVAFHTGN
Six6   LNFSPQQVAGVCETLEESGDVERLGRFLWSLPWPPGACEAINKNESILRARAIVAFHGGN
Six4   LAFSPDHVACVCEALQQGGNLDRLARFLWSLP----QSDLLRGNESLLKARALVAFHQGI
SIX5   LRFSPEGVACVCEALLQAGHAGRLARFLGALP----PAERLRGSDPVLRARALVAFQRGE
Six5   LRFSPEGVACVCEALLQAGHAGRLARFLGALP----PAERLRGSDPVLRARALVAFQRGE


so     YKELYRLLEHHHFSAQNHAKLQALWLKAHYVEAEKLRGRPLGAVGKYRVRRKFPLPRTIWD
Six1   FRELYKILESHQFSPHNHPKLQQLWLKAHYVEAEKLRGRPLGAVGKYRVRRKFPLPRTIWD
Six2   FRELYKILESHQFSPHNHAKLQQLWLKAHYIEAEKLRGRPLGRVGKYRVRRKFPLPRSIWD
Six3   FRDLYHILENHKFTKESHGKLQAMWLEAHYQEAEKLRGRPLGPVDKYRVRKKFPLPRTIWD
Six6   YRELYHILENHKFTKESHAKLQAMWLEAHYQEAEKLRGRPLGPVDKYRVRKKFPLPRTIWD
Six4   YPELYSILESHSFESANHPLLQQLWYKARYTEAERARGRPLGAVDKYRLRRKFPLPRTIWD
SIX5   YAELYRLLESRPFPAAHHAFLQDLYLRARYHEAERARGRALGAVDKYRLRKKFPLPRTIWD
Six5   YAELYQLLESRPFPAAHHAFLQDLYLRARYHEAERARGRALGAVDKYRLRKKFPLPRTIWD


           Helix I          Helix II         HelixIII
          _____        _____        _____

        ↓    *        *                                    *
cHD     RRRKRTAYTRYQLLELEKEFHFNRYLTRRRRIELAHSLNLTERQVKIWFQNRRMKWKKEN
so      GEETSYCFKEKSRSVLRDWYSHNPYPSPREKRDLAEATGLTTTQVSNWFKNRRQRDRAAE
Six1    GEETSYCFKEKSRGVLREWYAHNPYPSPREKRELAEATGLTTTQVSNWFKNRRQRDRAAE
Six2    GEETSYCFKEKSRSVLREWYAHNPYPSPREKRELAEATGLTTTQVSNWFKNRQRDRAAE
Six3    GEQKTHCFKERTRSLLREWYLQDPYPNPSKKRELAQATGLTPTQVGNWFKNRRQRDRAAA
Six6    GEQKTHCFKERTRNLLREWYLQDPYPNPSKKRELAQATGLTPTQVGNWFKNRRQRDRAA
Six4    GEETVYCFKEKSRNALKELYKQNRYPSPAEKRHLAKITGLSLTQVSNWFKNRRQRDRNPS
SIX5    GEETVYCFKERSRAALKACYRGNRYPTPDEKRRLATLTGLSLTQVSNWFKNRRQRDRTGA
Six5    GEETVYCFKERSRAALKACYRGNRYPTPDEKRRLATLTGLSLTQVSNWFKNRRQRDRTGA
```

Figure 2. Comparison of the predicted amino acids sequences of the SIX subfamily proteins. The homologous regions of the predicted mouse SIX subfamily proteins (Six1, Six2, Six3, Six4, Six5, Six6) and human SIX5 are aligned. cHD is a consensus homeodomain sequence compiled from 346 homeodomain sequences.[98] The arrowhead indicates the start of the homeodomain and the asterisks indicate amino acids 5, 12, and 50 of the homeodomain (shown in bold). The three helices of the homeodomain are shown. The SIX domain is the region upstream of the homeodomain.

5'-UTR of the human dopamine D5 receptor gene (*DRD5*) and the 5'-UTR of the DRD5 transcribed pseudogene 1 (*φDRD5-1*). Other fragments from this screen have been identified in the human genome.[70,71]

Insulin like growth factor 2 (*Igf2*) and insulin like growth factor binding protein 5 (*Igfbp5*) have been reported as targets of Six5.[72] The *Igfbp5* promoter can also be activated by Six1. Igfbp5 and Igf2 are components of insulin like growth factor (IGF). Expression levels of *Igfbp5* and *Igf2* were reduced in fibroblasts from homozygous *Six5* knockout mice and expression of human IGFBP5 was altered in cells of DM1 patients. IGFBP5 and IGF2 may contribute to the insulin resistance seen in DM1 patients, in addition to the RNA effects. Moreover, *Six2* and *Six4* are also potential transcriptional targets of Six5.[72] Expression levels of Six2 and Six4 are elevated in the skeletal muscle of homozygous *Six5* knockout mice.[73]

SIX5 Protein Binding Activities

Several reports have shown that the SIX proteins interact with eyes absent (EYA) proteins. The *Drosophila* eyes absent (*eya*) gene is the founding member of the *EYA* gene family. *eya* is essential for formation of compound eyes in *Drosophila*.[74] *Drosophila* sine oculis (so) and eya interact with one another and form a transcription factor complex that regulates multiple steps in *Drosophila* eye development.[75] Yeast two-hybrid assay and in vitro GST-fusion protein binding assays have demonstrated that eya and so interact directly through conserved eya and six domains respectively.

In mammals, 4 homologues (*EYA1*, *EYA2*, *EYA3*, *EYA4*) of the *Drosophila eya* gene have been identified.[76,77] It has been shown that Six1, Six2, Six4 and Six5 but not Six3 cooperate with Eya1, Eya2 and Eya3 to activate the transcription of their target genes (e.g., the *myogenin* promoter).[78] Six proteins are located in the nucleus, whereas Eya proteins are located in the cytoplasm when they are expressed separately in COS7 cells. However, coexpression of Six proteins induces a nuclear translocation of Eya indicating a direct interaction between Six and Eya. Six3 does not interact with any known member of the Eya family. However, it has been reported that Six3 interacts with Grg4 and Grg5, which are the mouse homologues of the *Drosophila* transcriptional corepressor Groucho.[79] Hence, it is possible that some Six proteins play roles as both transcriptional repressors and activators,[80] dependant on the binding of interacting proteins.

Identification of SIX5 Target Genes

The SIX5 consensus binding sequence has been determined by random oligonucleotide selection to be CCGGTGTCT.[70] Target genes were identified by searching the databases for genes containing the SIX5 binding site and by sequence analysis of functional candidate genes based on knowledge of the clinical features of DM1. Although some potential targets were unequivocally identified, many BLAST hits are still to be linked to known genes or regulatory elements.[70]

The *Drosophila Six4* (*D-Six4*) model described later in this chapter could be used to study the target genes since *D-Six4* has no other homologs in the *Drosophila* genome.[81] However, functional similarities between *SIX5* and *D-Six4* should be determined first in order to gain insights into the relevance of such studies.

The GST-SIX+HD fusion peptide binds to the 5'-UTR of the dopamine D5 receptor gene (*DRD-5*) and 5'-UTR of the dopamine D5 receptor pseudogene 2 (*φDRD5-2*).[68] Furthermore, the dopamine D5 receptor pseudogene 1 (*φDRD5-1*) contains a SIX5 binding site in its 5'-UTR. Although the human genome is estimated to contain up to 20,000 pseudogenes,[82] the potential functional role of pseudogenes is not well understood. In a recent study, it has been shown that a transcribed noncoding pseudogene (*Makorin-p1*) regulates the mRNA stability of its homologous coding gene (*Makorin1*).[83] *Makorin-p1* and *Makorin1* overlap in the 5' region. *Makorin-p1* expression protects *Makorin1* from degradation in trans, and a cis-acting element that facilitates degradation of mRNA is present in the 5' region of *Makorin1*.[83] However, the mechanism by which a pseudogene protects the mRNA decay of its homologous coding gene is still unknown. Interestingly, the two pseudogenes (*φDRD5-2*) and (*φDRD5-1*) of the dopamine D5 receptor gene (*DRD5*) are transcribed. The 5'-UTR of *DRD5* shares 95% identity with (*φDRD5-2*) and (*φDRD5-1*).[84] It is therefore possible that one or both pseudogenes could play a role in stabilising *DRD5* mRNA in a manner similar to that by which *Makorin-p1* regulates *Makorin1* mRNA. If this is the case, SIX5 might be a key regulator of expression of these transcribed (*φDRD5-2*) and (*φDRD5-1*) pseudogenes in addition to regulating *DRD5* expression.

Further intrigue may be added to the situation if SIX5, like other homeodomain proteins turns out to be an RNA binding protein as well as a transcriptional regulator. RNA-binding proteins appear to bind to AU-rich elements in the 3'-UTR of mRNAs and influence mRNA stability. More recently, the RNA-binding protein HuR has been shown to play an important role in skeletal myogenesis by binding to the 3'-UTR of *Myogenin* and *MyoD* transcripts.[85] Interestingly, GST-SIX+HD also binds to the 3'-UTR of *Myogenin*. Therefore, if SIX5 binds to the 3'-UTR of *Myogenin* transcript in vivo, it may influence *Myogenin* mRNA stability and thus play a role in skeletal myogenesis through that pathway in addition to acting through transcriptional regulation of myogenic factors.

Animal Models of DM1

To further investigate DM1 pathological mechanisms, *Dmpk* and *Six5* knockout mice and expanded CTG repeat models have been created.

Dmpk *Knockout Mice*

Dmpk knockout mice show only mild myopathic changes and abnormal cardiac conduction.[86,87] These changes are only seen in homozygous knockout mice indicating these traits display a recessive mode of inheritance rather than the dominant mode evident in DM1 pedigrees. Furthermore, none of the major features of DM1 were seen in these models. It was therefore concluded that the loss of *Dmpk* may be necessary, but not sufficient for development of DM1.

Six5 *Knockout Mice*

Two *Six5* knockout mouse models have been generated and characterised. The first was created by replacing *Six5* with the *Neo* gene.[88] Both heterozygous and homozygous knockout mice develop cataracts and show increased steady state levels of the $Na^+ K^+$ ATPase α1 (*Atp1a1*) subunit gene transcript and decreased *Dmpk* mRNA levels at an early stage of life. This suggests that altered ion homeostasis within the lens may contribute to the cataract formation. These mice are sterile and later develop gonadal, muscle and skeletal pathologies.[89] The second model was created by replacing the first exon of *Six5* with a beta-galactosidase reporter gene.[73] Homozygous mice develop cataracts but do not have muscle or reproductive defects. Hence, these two knockout models support the hypothesis that DM1 is a contiguous gene syndrome associated with the partial loss of *DMPK* and *SIX5* and strongly support a role of SIX5 haploinsufficiency in the development of cataracts in DM1 patients, in accordance with a detailed study of expression patterns of *DMPK* and *SIX5* in the human eye.[90]

Expanded CTG Repeat Models

Several independent studies have reported the behaviour of the DM1 CTG repeat in transgenic mice. Transgenic mice were generated with a fragment of the 3'-UTR of the human *DMPK* gene containing 162 CTG repeats.[91] Further mouse models were generated by using a 45 kb genomic fragment, containing *DMWD*, *DMPK* with 55 CTG repeats and *SIX5*, as a transgene.[92] Both studies have reported intergenerational and somatic instability of the CTG repeats in the transgenic mice but no demonstrable phenotype. In a more recent study, transgenic mice, that express human skeletal actin (*HSA1*) with either a nonexpanded (CTG)5 repeat or an expanded (CTG)250 repeat in the final exon of the *HSA1* gene, were created.[93] Mice that express the long (CTG)250 repeat transgene develop myotonia and histologically defined myopathy whereas those expressing short repeats do not.

Muscleblind KO Model

Disruption of the *Mbnl1* gene in mice leads to muscle, eye and mRNA splicing abnormalities.[94] Muscleblind proteins have been shown to colocalize with mutant DM transcripts in vivo and with CUG repeats in vitro.[58,59,61] The *Mbnl1* knockout mice show myotonia by EMG analysis and reduction of ClC-1 protein on the surface of muscle fibres from the vastus muscle by immunocytochemistry. This is consistent with altered splicing patterns detected by RT-PCR that lead to retention of cryptic exons predicted to lead to the expression of truncated nonfunctional ClC-1.[94] These data further implicate a role for muscleblind in the manifestation of myotonia but the finding of cataract in this model may not relate the cataract in DM. Muscleblind mutations in *Drosophila* lead to significant eye and muscle phenotypes. The cataract in DM1 patients would be predicted by the RNA-mediated model to arise via interactions between muscleblind and DMPK transcripts containing expanded CUG repeats. This is not consistent with the expression pattern of DMPK in the adult human eye. In a series of adult eyes DMPK expression was not detected in the lens at either the mRNA or protein levels using immunocytochemistry, western blotting, RT-PCR and northern analysis[90] The muscleblind KO mice add significant weight to the hypothesis that muscleblind is involved in the aetiology of myotonia in both DM1 and DM2, but the interpretation with regard to the cause of the cataract in DM1 requires further investigation.

Drosophila Models

Drosophila Six4 (*D-Six4*) is the *Drosophila* homologue of *SIX5*.[95,81] D-Six4 shows 67% and 65% identity to SIX4 and SIX5, respectively. A mutagenesis screen identified 4 *Drosophila* lines mutant in *D-Six4*.[81] One of mutations alters an arginine residue in the SIX domain which is conserved in all SIX proteins. The phenotype affects muscle and gonad development and is lethal. Males show severe testicular reduction. Moreover, myoblasts fail to fuse with developing myotubes. The production of these phenotypes in *Drosophila* mutants makes them a valuable resource for studying the function of *D-Six4* in development in the absence of a compensating gene. In contrast to the situation which prevails in mice carrying *Six4* or *Six5* knockouts which are viable, almost certainly due to the fact that these genes are significantly homologous and can functionally compensate for each other during development.

Remaining Questions and Issues

Our understanding of many of the puzzling features of DM1 has increased significantly with the identification of the underlying mutation. The unstable repeat is obviously key to the development and progression of the symptoms, with the rate of progression varying between tissues and individuals. A plausible route to therapy for many of those affected by DM1 would be to identify the factors that lead to this continuing expansion of the repeat in somatic cells and to prevent further expansion, effectively halting further progression of the disease. If this becomes possible then early diagnostic screening of newborns and a regimen of preventative therapy would have dramatic effects on the outcomes for those children, given that many document "educational difficulties" before the onset of any physical symptoms. Repeat instability has been modeled in mice and in cell lines established from those models.[96] There is an emerging body of evidence that cis-acting factors contribute to repeat instability during replication.[97] There is a good chance that such studies will lead to the identification of compounds that are able to slow or halt the progression of the repeat expansions.

The consequences of the repeat expansion at the level of gene expression and function are becoming clearer although DM1 remains a complex disease to study despite its "simple" Mendelian inheritance. Several effects on gene expression have been clearly implicated as contributing to the phenotype. At the transcriptional level haploinsufficiency for SIX5 leads to downstream effects that are most strongly implicated in cataract formation but which may also play important roles in muscle development and hence contribute to the congenital form, as well as to other symptoms, such as the cognitive deficits and insulin resistance. At the level of RNA processing mediated by *DMPK* mRNA CUG repeats and associated perturbations of key RNA processing proteins, most probably muscleblind, are effects on insulin receptors and chloride channels that may contribute to the whole-body insulin resistance and myotonia. In many tissues both of these effects may be at play.

Hence in the decade or so since the identification of the repeat expansion underlying DM1 we have accumulated substantial evidence that implicates different pathways in the molecular pathology of this condition. We have an array of animal and cellular models in hand that replicate different clinical features of the disease and a detailed understanding of the disease process is emerging. In the next decade hopefully this understanding will increase in its granularity to the point where new treatments based on a firm understanding of the disease mechanism start to be deployed in the clinic as many of the symptoms of this condition remain clinically unresponsive at this time.

Acknowledgements

The authors wish to acknowledge the many members of the Johnson lab past and present, colleagues in the Dynamic Mutation Group in Anderson College, clinical colleagues in the Scottish Muscle Centre and many collaborators and funding agencies, including the UK Myotonic Dystrophy Support Group, Muscular Dystrophy Campaign, Association Francais contre les Myopathies, Wellcome Trust, Muscular Dystrophy Association of America and BBSRC. Finally, and most importantly, we wish to acknowledge the many patients and their families who have supported our research.

References

1. Harper PS. Myotonic dystrophy. Philadelphia: Saunders 1989.
2. Mohr J. A study of linkage in man. Munksgaard: Copenhagen 1954.
3. Whitehead AS, Solomon E, Chambers S et al. Assignment of the structural gene for the third component of human complement to chromosome 19. Proc Natl Acad Sci USA 1982; 79:5021-5026.
4. Aslanidis C, Jansen G, Amemiya C et al. Cloning of the essential myotonic dystrophy region and mapping of the putative defect. Nature 1992; 355:548-551.
5. Brook JD, McCurrach ME, Harley HG et al. Molecular basis of myotonic dystrophy: Expansion of a trinucleotide (CTG) repeat at the 3' end of a transcript encoding a protein kinase family member. Cell 1992; 68:799-808.
6. Buxton J, Shelbourne P, Davies J et al. Detection of an unstable fragment of DNA specific to individuals with myotonic dystrophy. Nature 1992; 355:547-548.
7. Fu YH, Friedman DL, Richards S et al. An unstable triplet repeat in a gene related to myotonic dystrophy. Science 1992; 255:1256-1258.
8. Harley HG, Brook JD, Rundle SA et al. Expansion of an unstable DNA region and phenotypic variation in myotonic dystrophy. Nature 1992; 355:545-546.
9. Mahadevan M, Tsilfidis C, Sabourin L et al. Myotonic dystrophy mutation: An unstable CTG repeat in the 3' untranslated region of the gene. Science 1992; 255:1253-1255.
10. Howeler CJ, Busch HFM, Geraedts JPM et al. Anticipation in myotonic dystrophy: Fact or Fiction? Brain 1989; 112:779-797.
11. Sarnat HB, Silbert SW. Maturational arrest of fetal muscle in neonatal myotonic dystrophy. A pathologic study of four cases. Arch Neurol 1976; 33:466-474.
12. Liquori CL, Ricker K, Moseley ML et al. Myotonic dystrophy type 2 caused by a CCTG expansion in intron 1 of ZNF9. Science 2001; 293:864-867.
13. Davies J, Yamagata H, Shelbourne P et al. Comparison of the myotonic dystrophy associated CTG repeats in European and Japanese populations. J Med Genet 1992; 29:766-769.
14. Lavedan C, Hofmann-Radvanyi H, Shelbourne P et al. Myotonic dystrophy: Size-and sex-dependant dynamics of CTG meiotic instability, and somatic mosaicism. Am J Hum Genet 1993; 52:875-883.
15. Brunner HG, Bruggenwirth HT, Nillesen W et al. Influence of sex of the transmitting parent as well as of parental allele size on the CTG expansion in myotonic dystrophy (DM). Am J Hum Genet 1993; 53:1016-1023.
16. Eriksson M, Ansved T, Anvret M et al. A Mammalian Radial Spokehead-Like gene, RSHL1, at the myotonic dystrophy-1 locus. Biochem and Biophys Res Comm 2001; 281:835-841.
17. Martorell L, Monckton DG, Games J et al. Progression of somatic CTG repeat length heterogeneity in the blood cells of myotonic dystrophy patients. Hum Mol Genet 1998; 7:307-312.
18. Anvret M, Ahlberg G, Grandell U et al. Larger expansions of the CTG repeat in muscle compared to lymphocytes from patients with myotonic dystrophy. Hum Mol Genet 1993; 2:1397-1400.
19. Ansved T, Ahlberg G, Grandell U. Variation of CTG-repeat number of the DMPK gene in muscle tissue. Neuromusc Disord 1997; 7:152-155.
20. Jansen G, Willems P, Coerwinkel M. Gonosomal mosaicism in myotonic dystrophy patients: Involvement of mitotic events in (CTG)n repeat variation and selection against extreme expansion in sperm. Am J Hum Genet 1994; 54:575-585.
21. Ishii S, Nishio T, Sunohara N. Small increase in triplet repeat length of cerebellum from patients with myotonic dystrophy. Hum Genet 1996; 98:138-140.

22. Monckton DG, Wong LJ, Ashizawa T et al. Somatic mosaicism, germline expansions, germline reversions and intergenerational reductions in myotonic dystrophy males: Small pool PCR analyses. Hum Mol Genet 1995; 4:1-8.

23. Jansen G, Mahadevan M, Amemiya C et al. Characterisation of the myotonic dystrophy region predicts multiple protein isoform-encoding mRNAs. Nat Genet 1992; 1:261-266.

24. Pham YC, Man N, Lam LT et al. Localization of myotonic dystrophy protein kinase in human and rabbit tissues using a new panel of monoclonal antibodies. Hum Mol Genet 1998; 7:1957-1965.

25. Lam LT, Pham YC, Man N et al. Characterisation of a monoclonal antibody panel shows that the myotonic dystrophy protein kinase, DMPK, is expressed almost exclusively in muscle and heart. Hum Mol Genet 2000; 9:2167-2173.

26. Moncrieff CL. Cloning and characterization of a novel DMPK-related gene: CDC42BPB. PhD Thesis: University of Glasgow 1999.

27. Moncrieff CL. Bailey MES. Morrison N et al. Cloning and chromosomal localization of human Cdc42-binding protein kinase β. Genomics 1999; 57:297-300.

28. Shaw DJ, McCurrach M, Rundle SA et al. Genomic organization and transcriptional units at the myotonic dystrophy locus. Genomics 1993; 18:673-679.

29. Jolanda HAM, Westerlaken CEEM, Van der Zee WP et al. The DMWD protein from the myotonic dystrophy (DM1) gene region is developmentally regulated and is present most prominently in synapse-dense brain areas. Brain Res 2003; 971:116-127.

30. Boucher CA, King SK, Carey N et al. A novel homeodomain-encoding gene is associated with a large CpG island interrupted by the myotonic dystrophy unstable (CTG)n repeat. Hum Mol Genet 1995; 4:1919-1925.

31. Heath SK, Carne S, Hoyle C et al. Characterisation of expression of mDMAHP, a homeodomain -encoding gene at the murine DM locus. Hum Mol Genet 1997; 6:651-657.

32. Klesert TR, Otten AD, Bird TD et al. Trinucleotide repeat expansion at the myotonic dystrophy locus reduces expression of DMAHP. Nat Genet 1997; 16:402-406.

33. Thornton CA, Wymer JP, Simmons Z et al. Expansion of the myotonic dystrophy CTG repeat reduces expression of the flanking DMAHP gene. Nat Genet 1997; 16:407-409.

34. Korade-Mirnics Z, Tarleton J, Servidei S et al. Myotonic dystrophy: Tissue-specific effect of somatic CTG expansions on allele-specific DMAHP/SIX5 expression. Hum Mol Genet 1999; 8:1017-1023.

35. Gennarelli M, Pavoni M, Amicucci P et al. Reduction of the DM-associated homeo domain protein (DMAHP) mRNA in different brain areas of myotonic dystrophy patients. Neuromuscul Disord 1999; 9:215-219.

36. Tachi N, Ohya K, Chiba S. Expression of the myotonic dystrophy locus-associated homeodomain protein in congenital myotonic dystrophy. J Child Neurology 1999; 14:471-473.

37. Eriksson M, Ansved T, Edstrom L et al. Simultaneous analysis of expression of the three myotonic dystrophy locus genes in adult skeletal muscle samples: The CTG expansion correlates inversely with DMPK and 59 expression levels, but not DMAHP levels. Hum Mol Genet 1999; 8:1053-1060.

38. Hamshere MG, Newman EE, Alwazzan M et al. Transcriptional abnormality in myotonic dystrophy affects DMPK but not neighboring genes. Proc Natl Acad Sci USA 1997; 94:7394-7399.

39. Alwazzan M, Newman E, Hamshere MG et al. Myotonic dystrophy is associated with a reduced level of RNA from the DMWD allele adjacent to the expanded repeat. Hum Mol Genet 1999; 8:1491-1497.

40. Filippova GN, Thienes CP, Penn BH et al. CTCF-binding sites flank CTG/CAG repeats and form a methylation-sensitive insulator at the DM1 locus. Nat Genet 2001; 28:335-343.

41. Steinbach P, Glaser D, Walther V et al. The DMPK gene of severely affected myotonic dystrophy patients is hypermethylated proximal to the largely expanded CTG repeat. Am J Hum Genet 1998; 62:278-285.

42. Sabourin LA, Mahadevan MS, Narang M et al. Effect of the myotonic dystrophy (DM) mutation on mRNA levels of the DM gene. Nat Genet 1993; 4:233-238.

43. Endo A, Motonaga K, Arahata K et al. Developmental expression of myotonic dystrophy protein kinase in brain and its relevance to clinical phenotype. Acta Neuropathol (Berlin) 2000; 100:513-520.

44. Frisch R, Singleton KR, Moses PA et al. Effect of triplet repeat expansion on chromatin structure and expression of dmpk and neighboring genes, six5 and dmwd, in myotonic dystrophy. Mol Genet Metab 2001; 74:281-291.

45. Furling D, Lam LT, Agbulut O et al. Changes in myotonic dystrophy protein kinase levels and muscle development in congenital myotonic dystrophy. Am J Path 2003; 162:1001-1009.

46. Wang J, Pegoraro E, Menegazzo E et al. Myotonic dystrophy: Evidence for a possible dominant-negative RNA mutation. Hum Mol Genet 1995; 4:599-606.

47. Faustino NA, Cooper TA. PremRNA splicing and human disease. Genes and Development 2003; 17:419-437.
48. Napierala M, Krzyosiak WJ. CUG repeats present in myotonin kinase RNA form metastable "slippery" hairpins. J Biol Chem 1997; 272:31079-31085.
49. Tian B, White RJ, Xia T et al. Expanded CUG repeat RNAs form hairpins that activate the double-stranded RNA-dependent protein kinase PKR. RNA 2000; 6:79-87.
50. Taneja KL, McCurrach M, Schalling M et al. Foci of trinucleotide repeat transcripts in nuclei of myotonic dystrophy cells and tissues. J Cell Biol 1995; 128:995-1002.
51. Timchenko LT, Miller JW, Timchenko NA et al. Identification of a (CUG)n triplet repeat RNA-binding protein and its expression in myotonic dystrophy. Nucl Acids Res 1996; 24:4407-4414.
52. Timchenko LT, Timchenko NA, Caskey CT et al. Novel proteins with binding specificity for DNA CTG repeats and RNA CUG repeats: Implications for myotonic dystrophy. Hum Mol Genet 1996; 5:115-121.
53. Savkur RS, Philips AV, Cooper TA. Aberrant regulation of insulin receptor alternative splicing is associated with insulin resistance in myotonic dystrophy. Nat Genet 2001; 29:40-47.
54. Timchenko LT. Human Genetics '99: Trinucleotide repeats. Myotonic Dystrophy: The role of RNA CUG triplet repeats. Am J Hum Genet 1999; 64:360-364.
55. Timchenko NA, Cai ZJ, Welm AL et al. RNA CUG repeats sequester CUGBP1 and alter protein levels and activity of CUGBP1. J Biol Chem 2001; 276:7820-7826.
56. Michalowski S, Miller JW, Urbinati CR et al. Visualization of double-stranded RNAs from the myotonic dystrophy protein kinase gene and interactions with CUG-binding protein. Nucl Acids Res 1999; 27:3534-3542.
57. Takahashi N, Sasagawa N, Suzuki K et al. The CUG-binding protein binds specifically to UG dinucleotide repeats in a yeast three hybrid mapping. Biochem Biophys Res Comm 2000; 277:518-523.
58. Miller JW, Urbinati CR, Teng-Umnuay P et al. Recruitment of human muscleblind proteins to (CUG)(n) expansions associated with myotonic dystrophy. Embo J 2000; 19:4439-4448.
59. Fardaei M, Larkin K, Brook JD et al. In vivo localisation of MBNL protein with DMPK expanded-repeat transcripts. Nucl Acids Res 2001; 29:2766-2771.
60. Mankodi A, Urbinati CR, Yuan QP et al. Muscleblind localizes to nuclear foci of aberrant RNA in myotonic dystrophy types 1 and 2. Hum Mol Genet 2001; 10:2165-2170.
61. Fardaei M, Rogers MT, Thorpe HM et al. Three proteins, MBNL, MBLL and MBXL, colocalize in vivo with nuclear foci of expanded-repeat transcripts in DM1 and DM2 cells. Hum Mol Genet 2002; 11:805-814.
62. Begemann G, Paricio N, Artero R et al. Muscleblind, a gene required for photoreceptor differentiation in Drosophila, encodes novel nuclear Cys3His-type zinc-finger-containing proteins. Development 1997; 124:4321-4331.
63. Philips AV, Timchenko LT, Cooper TA. Disruption of splicing regulated by a CUG-binding protein in myotonic dystrophy. Science 1998; 280:737-741.
64. Charlet BN, Savkur RS, Singh G et al. Loss of the muscle-specific chloride channel in type 1 myotonic dystrophy due to misregulated alternative splicing. Mol Cell 2002; 10:45-53.
65. Mankodi A, Takahashi MP, Jiang H et al. Expanded CUG repeats trigger aberrant splicing of ClC-1 chloride channel pre-mRNA and hyperexcitability of skeletal muscle in myotonic dystrophy. Mol Cell 2002; 10:35-44.
66. Kawakami K, Sato S, Ozaki H et al. SIX family genes-structure and function as transcription factors and their role in development. Bioessays 2000; 22:616-626.
67. Kawakami K, Ohto H, Takizawa T et al. Identification and expression of six family genes in mouse retina. FEBS Lett 1996; 393:259-263.
68. Harris SE, Winchester CL, Johnson KJ. Functional analysis of the homeodomain protein SIX5. Nucl Acids Res 2000; 28:1871-1878.
69. Spitz F, Demignon J, Porteu A et al. Expression of myogenin during embryogenesis is controlled by Six/sine oculis homeoproteins through a conserved MEF3 binding site. Proc Natl Acad Sci USA 1998; 95:14220-14225.
70. Jarjour R. Identification of SIX5 binding site, target genes, and functional links with myotonic dystrophy (DM1) symptoms. PhD thesis: University of Glasgow 2003.
71. Jarjour R, Pall G, Hamilton GM et al. Identification of the SIX5 DNA binding site and target genes: Further evidence for role of SIX5 haploinsufficiency in DM1. Manuscript submitted 2004.
72. Sato S, Nakamura M, Cho DH et al. Identification of transcriptional targets for Six5: Implication for the pathogenesis of myotonic dystrophy type 1. Hum Mol Genet 2002; 11:1045-58.
73. Klesert TR, Cho DH, Clark JI et al. Mice deficient in Six5 develop cataracts: Implications for myotonic dystrophy. Nat Genet 2000; 25:105-109.

74. Bonini NM, Leiserson WM, Benzer S. The eyes absent gene: Genetic control of cell survival and differentiation in the developing Drosophila eye. Cell 1993; 72:379-395.
75. Pignoni F, Hu B, Zavitz KH et al. The eye-specification proteins So and Eya form a complex and regulate multiple steps in Drosophila eye development. Cell 1997; 91:881-891.
76. Abdel-Hak S, Kalatzis V, Heilig R et al. A human homologue of the Drosophila eyes absent gene underlies branchio-oto-renal (BOR) syndrome and identifies a novel gene family. Nat Genet 1997; 15:157-164.
77. Borsani G, DeGrandi A, Ballabio A et al. EYA4, a novel vertebrate gene related to Drosophila eyes absent. Hum Mol Genet 1999; 8:11-23.
78. Ohto H, Kamada S, Tago K et al. Cooperation of six and eya in activation of their target genes through nuclear translocation of Eya. Mol Cell Biol 1999; 19:6815-6824.
79. Zhu CC, Dyer MA, Uchikawa M et al. Six3-mediated auto repression and eye development requires its interaction with members of the Groucho-related family of corepressors. Development 2002; 129:2835-2849.
80. Kobayashi M, Nishikawa K, Suzuki T et al. The homeobox protein six3 interacts with the groucho corepressor and acts as a transcriptional repressor in eye and forebrain formation. Dev Biol 2001; 232:315-326.
81. Kirby RJ, Hamilton GH, Finnegan DJ et al. The Drosophila homologue of the myotonic dystrophy associated gene, SIX5, is required for muscle and gonad development. Current Biology 2001; 11:1044-1049.
82. Goncalves I, Duret L, Mouchiroud D. Nature and structure of human genes that generate retropseudogenes. Genome Res 2000; 10:672-678.
83. Hirotsune S, Yoshida N, Chen A et al. An expressed pseudogene regulates the messenger-RNA stability of its homologous coding gene. Nature 2003; 423:91-96.
84. Marchese A, Beischlag TV, Nguyen T et al. Two gene duplication events in the human and primate dopamine D5 receptor gene family. Gene 1995; 154:153-158.
85. Figeuroa A, Cuadrado A, Fan J et al. Role of HuR in skeletal myogenesis through coordinate regulation of muscle differentiation genes. Mol Cell Biol 2003; 23:4991-5004.
86. Jansen G, Groenen PJ, Bachner D et al. Abnormal myotonic dystrophy protein kinase levels produce only mild myopathy in mice. Nat Genet 1996; 13:316-324.
87. Benders AAGM, Groenen PJTA, Oerlmans FTJJ et al. Myotonic dystrophy protein kinase is involved in the modulation of the Ca(2+) homeostasis in skeletal muscle cells. J Clin Invest 1997; 100:1440-1447.
88. Sarkar PS, Appukuttan B, Han J et al. Heterozygous loss of Six5 in mice is sufficient to cause ocular cataracts. Nat Genet 2000; 25:110-114.
89. Sarkar PS, Han J, Reddy S. Six5 loss results in lens, and muscle defects. Am J Hum Genet 2000; 67(suppl):366(A2049).
90. Winchester CL, Ferrier RK, Sermoni A et al. Characterization of the expression of DMPK and SIX5 in the human eye and implications for pathogenesis in myotonic dystrophy. Hum Mol Genet 1999; 8:481-492.
91. Monckton DG, Coolbaugh MI, Ashizawa KT et al. Hypermutable myotonic dystrophy CTG repeats in transgenic mice. Nat Genet 1997; 15:193-196.
92. Gourdon G, Radvanyi F, Lia AS et al. Moderate intergenerational and somatic instability of a 55-CTG repeat in transgenic mice. Nat Genet 1997; 15:190-192.
93. Mankodi A, Logigian E, Callahan L et al. Myotonic dystrophy in transgenic mice expressing an expanded CUG repeat. Science 2000; 289:1769-1772.
94. Kanadia RN, Johnstone K, Mankodi M et al. A Muscleblind knockout model for myotonic dystrophy. Science 2003; 392:1978-1980.
95. Seo HC, Curtiss J, Mlodzik M et al. Six class homeobox genes in Drosophila belong to three distinct families and are involved in head development. Mech Dev 1999; 83:127-139.
96. Gomes-Pereira M, Fortune TM, Monckton DG. Mouse tissue culture models of unstable triplet repeats: In vitro selection for larger alleles, mutational expansion bias and tissue specificity, but no association with cell division rates. Hum Mol Genet 2001; 10:845-854.
97. Cleary JD, Nichol K, Wang Y-H et al. Evidence of cis-acting factors in replication-mediated trinucleotide repeat instability in primate cells. Nat Genet 2002; 31:37-46.
98. Gehring WJ, Affolter M, Burglin T. Homeodomain proteins. Ann Rev Biochem 1994; 63:487-526.

CHAPTER 14

Spinal Muscular Atrophy

Robert Olaso, Jérémie Vitte, Nouzha Salah and Judith Melki

Abstract

Spinal muscular atrophies (SMA) are characterized by degeneration of lower motor neurons and occasionally bulbar motor neurons leading to progressive limb and trunk paralysis associated with muscular atrophy. SMA is a recessive autosomal disorder with an incidence of 1 out of 5000 newborns and represents one of the most frequent genetic causes of death in childhood. Identification of the survival of motor neuron gene (*SMN1*) and mutations found in SMA patients greatly improved the clinical management and family-planning options of SMA patients and their parents. The last years have seen major advances in the biochemistry of SMN although the molecular pathway linking SMN defect to the SMA phenotype remains to be elucidated. Animal models of SMA have been generated providing valuable tools to clarify SMA pathogenesis and for designing therapeutic approaches of this devastating neurodegenerative disease for which no curative treatment is known so far.

Clinical Aspects

Based on age of onset of symptoms, achievement of motor milestones and age at death, childhood SMA has been subdivided into three clinical types.[1] The acute form or Werdnig-Hoffmann disease (type I) is characterized by severe, generalized muscle weakness at birth or within the first 6 months. Feeding and breathing difficulties are usually responsible for death within two years of age. Type II SMA is a more slowly progressive generalized disease with a variable prognosis. Infants are able to sit unsupported but not able to stand or walk unaided. Clinical progression is slow or appears even to arrest. All children will develop, if untreated, severe scoliosis and respiratory ventilation defect. Life expectancy is highly variable, ranging up to adult life in some cases. In juvenile SMA or type III or Kugelberg-Welander disease, symptoms start from 18 months of life to the age of 20 years. All patients are able to walk and proximal muscle weakness is progressive. Subsequently, the patients show difficulties in climbing stairs, gait become waddling. About one fourth of SMA type III patients exhibit a hypertrophy of the calves, a feature similar to that observed in Duchenne or Becker muscular dystrophies. Finally, adult SMA is defined by an age of onset of symptoms beyond 20 years of age, the diagnostic criteria being identical to the other forms.

Whatever the age of onset, the clinical features consist of symmetrical muscle weakness (more proximal than distal) associated with muscle atrophy, absence or marked decrease of deep tendon reflexes, fasciculations of the tongue, and tremor of hands. There is no evidence of sensory or upper motor neuron involvement. Electromyographic studies show a pattern of denervation with neither sensory involvement nor marked decrease of motor nerve conduction velocities. Finally, muscle biopsy, which has been recently replaced by genetic testing for diagnosis, provides evidence of skeletal muscle denervation with groups of atrophic and hypertrophic fibers or fiber type grouping most often found in chronic cases (Types II and III).

The most striking neuropathological feature found in autopsy material of SMA patients is a loss of the large anterior horn cells of the spinal cord. In the remaining surviving motor

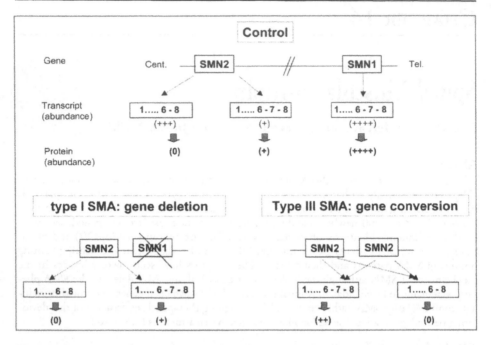

Figure 1. Genomic organization of the SMA locus in human control and SMA individuals. Transcripts derived from *SMN1* or *SMN2* genes are indicated in boxes. Full-length or truncated transcripts are indicated as (1...6-7-8) or (1...6-8), respectively. Deletion or gene conversion events of *SMN1* gene are responsible for severe (type I) or mild (type III) SMA, respectively. Only one allele is presented on each panel. Truncated transcripts are translated into unstable protein indicated as (0). The abundance of transcripts or protein is indicated as (0, +, ++ or ++++).

neurons, severe degree of central chromatolysis is visible. These cells appear as large ballooned cells without stored substances. Other anterior cells are pyknotic. In addition, there are occasional figures of neuronophagia associated with astrogliosis and the anterior roots are small.

Genetic Basis of SMA

The SMA locus on chromosome 5q13 is characterized by an inverted duplication, each element (about 500 kb) containing several genes. Deletions or conversion events including exon 7 of the telomeric copy of the survival of motor neuron gene (*SMN*, renamed *SMN1*) are found in 93% of SMA patients.[2] In the remaining patients, intragenic mutations of *SMN1* including missense, nonsense or splice site mutations have identified *SMN1* as the SMA disease gene.[2,3]

SMN1 is duplicated with a highly homologous copy called *SMN2* and both genes are transcribed (Fig. 1). The *SMN2* gene is present in all patients but is not able to compensate for the *SMN1* gene defects. At the genomic level, the presence of a gene dosage effect in type I but not in type III SMA has suggested that type I SMA was caused by deletion of *SMN1* whereas type III was associated with conversion event of *SMN1* into *SMN2* gene leading to an increased number of *SMN2* genes (Fig. 1).[2,4] This is in agreement with the presence of a tight inverted correlation between the amount of the protein encoded by the *SMN2* gene and the clinical severity of human SMA disease.[5] In patients, the SMN protein level depends on *SMN2* copy number making *SMN2* as a modifying gene in SMA. These data have ascribed SMA to a dose effect of SMN. Mutations of *SMN1* and *SMN2* on both chromosomes have not been reported. Such a genotype would be likely responsible for an extremely severe form of SMA or non viable fetus.

Why is *SMN2*, which remains present in SMA patients, not able to compensate for the *SMN1* defect? Five nucleotides distinguish *SMN2* from *SMN1* genes without any effect on the amino acid sequence.[2] One of these nucleotide substitutions located in exon 7 is responsible for the alternative splicing of exon 7 which is specific to the *SMN2* transcripts.[2,6] Full-length transcripts are almost exclusively produced by *SMN1*, whereas the predominant form encoded by *SMN2* is lacking exon 7. Full length transcript is also encoded by *SMN2* and translated into functional protein but its abundance is much lower than that encoded by *SMN1*. The truncated transcript lacking exon 7 encodes a protein lacking the last C-terminal 16 residues (SMN$^{\Delta 7}$). In vitro data strongly suggested a dominant negative effect of SMN$^{\Delta 7}$ on the full length SMN protein.[7-9] However, such effect was not observed in vivo in both mice and humans carrying heterozygous deletion of *SMN* exon 7.[10] Dominant negative effect SMN$^{\Delta 7}$ could be ascribed to over-expression of SMN$^{\Delta 7}$ induced by in vitro but not in vivo systems. Therefore, mutations of *SMN1* are responsible for a loss of SMN function.

SMN: A Multifunctional Protein

SMN is an ubiquitously expressed protein of 294 amino acids with a molecular weight of 38 kDa. Sequence analysis showed that SMN contains a phylogenetically conserved sequence called the Tudor domain which has been found in many proteins involved in RNA metabolism.[11] The SMN complex is found both in the cytoplasm and in the nucleus where it is concentrated in a structure called gems (for "gemini of coiled bodies") most often associated with or identical to Cajal bodies (coiled bodies) depending on the cell type or tissue analyzed.[12] Cajal bodies are nuclear structures enriched in U small nuclear ribonucleoprotein (snRNP) and U small nucleolar RNP proteins (snoRNP). SMN forms a large complex of approximately 1 MDa and most of components of this complex have been identified. SMN interacting proteins include Gemin 2 (formerly SIP1 for "SMN interacting protein 1"), Gemin 3 (a DEAD box putative RNA helicase previously known as dp103), Gemin 5, the spliceosomal snRNP Sm and Lsm protein, the snoRNP including fibrillarin and GAR1, heterogeneous nuclear RNP-Q and coilin.[13] The SMN complex is also composed of Gemin 4 and 6, two novel proteins of unknown function.[13] The interaction of SMN with components of this large complex is mediated directly by the Tudor domain and is enhanced by a SMN oligomerization domain corresponding to exon 6. The identification of SMN interacting proteins of known function strongly supports the view that SMN is involved and facilitates cytoplasmic assembly of snRNP into the spliceosome, a large RNA-protein complex that catalyzes the splicing reaction. In the nucleus, SMN appears to be directly involved in pre-mRNA splicing, transcription and metabolism of ribosomal RNA. SMN can be regarded as an assembly factor that mediates formation of the Sm core domain.

SMN has also been shown to interact directly with other proteins including Bcl-2 and p53, two proteins involved in apoptotic processes, although these interactions and the involvement of an apoptotic process in SMA remain to be clarified in vivo. Other direct or indirect partners of SMN include the FUSE binding protein, profilin II, a zinc-finger protein called ZRP1, RNA helicase A, RNA polymerase II, and RNA.[3] Therefore, SMN appears to be a multifunctional protein.

How does partial deficiency of SMN, but not complete absence, lead to the SMA phenotype? Among the various putative functions of SMN, which deficiency is responsible for the SMA phenotype? SMN is indeed ubiquitously expressed and the selective involvement of the neuromuscular system in SMA suggest that cells of different types do not require the same amount of SMN for survival unless SMN has an as yet unknown function specific to neurons or skeletal muscle. To answer these questions, animal models of SMA have been generated.

Animal Models of SMA

In various species including *M. musculus, D. melanogaster, C. elegans*, SMN orthologue is not duplicated and its deletion leads to early embryonic lethality.[14-17] These data hampered our ability to clarify the function of SMN and the pathway involved in SMA pathogenesis in vivo. To circumvent embryonic lethality of mice knocked out for the *Smn* gene, several transgenic approaches were undertaken to create mouse models. One strategy was based on the generation of mice carrying genomic organization similar to that of human SMA.[18,19] It consisted of the creation of two mouse lines, one carrying a deletion of *Smn* through homologous recombination and the other line carrying a transgene expressing the human *SMN2* gene. Mice carrying both homozygous deletion of *Smn* and the human *SMN2* gene developed a phenotype depending on the number of *SMN2* transgenes. Mice carrying low copy number of *SMN2* and harbouring a homozygous null *Smn* allele developed a severe phenotype leading to death either in utero or in the first days after birth. Mutant mice displayed abnormal motor defects associated with a moderate loss of motor neurons of the spinal cord (up to 35%) at the latest stage of disease. The absence of abnormal phenotype in mutant mice carrying a high copy number of *SMN2* transgene indicated that *SMN2* was able to prevent the embryonic lethality of the *Smn* knock out mice and confirmed that an increased copy number of *SMN2* reduced the severity of the phenotype, in agreement with data observed in human SMA.[18,19]

Another approach has been carried out by using the Cre-loxP recombination system. A mouse line carrying two loxP sequences flanking *Smn* exon 7 (*SmnF7*) has been established through homologous recombination. Cre-mediated deletion of *Smn* exon 7, the most frequent mutation found in SMA patients, has been directed to neurons by crossing *SmnF7* mice to transgenic mice expressing Cre recombinase in neurons ("neuronal" mutant).[17,20] Neuronal mutant mice in which full length *Smn* transcripts are lacking, displayed severe motor defect leading to complete paralysis and death at a mean age of 4 weeks. Analysis of skeletal muscle revealed a severe muscle denervation process, a dramatic and progressive loss of motor axons (up to 73% reduction) contrasting with mild reduction of motor neuron cell bodies (29%). In addition, abnormal synaptic terminals of neuromuscular junctions filled with neurofilaments, including phosphorylated forms, were observed in mutant mice associated with defects in axonal sprouting.[20] These findings are likely responsible for motor neuron dysfunction and suggest that loss of motor neurons occurs through a dying back axonopathy. These data demonstrated that motor neurons are a primary target of the *Smn* gene defect in SMA. Consistently, antisense morpholinos to reduce SMN levels in zebrafish causes defects in motor axon pathfinding suggesting that SMN has a role in motor axon development.[21] Moreover, point mutations in *SMN* similar to those found in SMA patients have been found in *Drosophila* resulting in defects of the neuromuscular junctions including disorganization of synaptic motor neuron boutons and reduction of post synaptic receptor subunits.[22] Therefore, defects in SMN in various organisms highlighted an essential role of SMN in motor axon and neuromuscular junction development or maintenance.

Is there an effect of *SMN* gene defect in other tissues? Deletion of *Smn* exon 7 has been directed to murine skeletal muscle using the same strategy than that described above ("muscular mutant").[10] Unexpectedly, mutant mice displayed a dystrophic phenotype leading to muscle paralysis and death at a mean age of 1 month. The dystrophic phenotype was associated with destabilization of the sarcolemma. Although sarcolemma destabilization is not sufficient to explain death of mutant mice, this study revealed that *Smn* gene defect leads to a degenerative process of skeletal muscle fibers. The severity of muscle phenotype suggested an impairment of muscle regeneration resulting from SMN defect in both skeletal muscle fibers and progenitors in SMN "muscular" mutant. This was supported by *Smn* targeting directed to myotubes but not muscle progenitor cells. Mutant mice develop similar myopathic process but exhibit mild phenotype with median survival of 8 months (instead of 1 month), motor performance similar to that of controls within the first six months of age and high proportion of regenerating myofibers. These data suggest that SMN has an important role in skeletal muscle differentiation and/or regeneration.[23]

To know whether SMN defects lead to deleterious effect in all mammalian cell types or is restricted to the neuromuscular system, which might suggest functions of SMN specific to this system, *Smn* mutation has been directed to liver, a tissue non affected in human SMA. Using a transgenic line that expressed the Cre recombinase in liver only, deletion of *Smn* exon 7 was restricted to that tissue. Homozygous deletion of *Smn* exon 7 leads to late embryonic lethality. Mutant phenotype was characterized by marked atrophy of liver associated with severe liver dysfunction and absence of regeneration (J.V. and J.M., unpublished data). Altogether, these data support the hypothesis that SMN plays an ubiquitous and essential role in all cells including motor neurons.

How to explain the difference of effect in response to SMN defect between human and mouse? Murine *Smn* gene is present as a single copy while in human, *SMN1* is duplicated in a highly homologous copy *SMN2*. The residual amount of SMN produced by full length transcripts of the *SMN2* gene, which remains present in patients, is likely sufficient to ensure normal functions of various organs including liver but not motor neurons.

Why are motor neurons more sensitive to reduced level of SMN? Motor axons can reach more than 1 m long, making motor neuron one of the largest cells in the body by both volume and surface area. Such large cellular unit requires molecular machinery able to produce and regulate molecules, including proteins and RNA, from the cell body to the neuromuscular junction through motor axons. Mild impairment in production or stability of molecules involved in these processes could have detrimental effect on axonal growth. Consistently, defects of axonal growth and neuromuscular junctions are the main features found in animal models. However, other neurons including sensory neurons are also very large in size. Recently, involvement of these neurons has been described in severe cases of SMA in agreement with such hypothesis.[24] Skeletal muscle has also been recently shown to be affected by severe SMN defect in mice,[10] features found in clinical reports suggesting a primary involvement of skeletal muscle, in addition to motor neurons, in human SMA.[25-28] The involvement of skeletal muscle which has a critical role in motor neuron maintenance could modulate the expression or progression of SMA disease. Altogether, these data suggest that involvement of other tissues may occur during prolonged SMA disease course or in very severe forms of SMA.

Ubiquitous expression of SMN and the deleterious effect of lack of SMN in various mammalian cell types strongly suggest an ubiquitous role of SMN rather than a function specific to the neuromuscular system. Recently, SMN has been localized in neurites and growth cones in which SMN appeared to be transported linked to components of the cytoskeleton.[29] Moreover, RNA and RNA binding proteins were shown to interact with microtubules in neurons or microfilaments in fibroblasts.[30] Therefore, it is tempting to hypothesize that defects in transport of mRNA complexes resulting from reduced levels of SMN could have detrimental effect in large cells, such as motor neurons. Alternatively, neurons express high numbers of genes with respect to other cell types and alternative splicing or stability of mRNA are important post-transcriptional mechanisms to create or regulate protein diversity.[31] Although a recent report did not reveal abnormal splicing pattern of a single gene in a heterozygous SMA mouse mutant,[32] we cannot however exclude the hypothesis of a defect in metabolism of some RNA sub-classes involved in the structural specificities of motor neurons. Search for transcript defects through microarray analysis should allow to test this hypothesis.

Therapeutic Strategies in SMA

From a better knowledge of genetic basis of SMA, upregulation of *SMN2* gene expression, preventing exon 7 skipping of *SMN2* transcripts or stabilizing SMN[47] should represent attractive therapeutic strategies in SMA (Fig. 2). Several groups have generated cell systems to identify compounds able to induce these modifications. Molecules including interferon, sodium butyrate, valproic acid or aclarubicin have been recently found to increase the amount of SMN protein encoded by the *SMN2* gene either by activating the *SMN2* gene promoter and/or by preventing the alternative splicing of exon 7 in vitro.[33-36] Their effects should be validated in vivo on animal models carrying the human *SMN2* gene and could represent candidate molecules for therapeutic trials in human SMA.

Identifying molecules involved in or associated with the SMA degenerative process should help in designing targeted therapeutics in SMA. Cellular or animal models should contribute

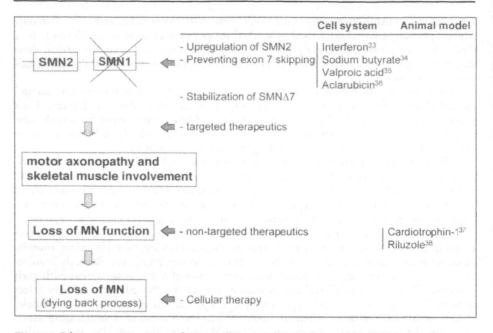

Figure 2. Schematic representation of the current knowledge on SMA pathophysiology and therapeutic research strategies. For each therapeutic strategy, molecules found to be active on either cell system or animal model are indicated. MN: motor neuron.

to elucidate the molecular events leading to SMA. Neuroprotective agents or neurotrophic factors may protect neurons against toxicity or promote axonal sprouting of motor neurons. The availability of animal models allows us to test the efficiency, however non specific, of molecules with such biological properties. Recently, cardiotrophin I and riluzole which exhibit neurotrophic activities have shown therapeutic benefits on neuronal mouse model of SMA indicating for the first time that SMA disease progression might be attenuated (Fig. 2).[37-38]

Finally, several reports have revealed that stem cells derived from either bone marrow or other sources have the capacity to differentiate into or to fuse to various cell types including skeletal muscle or neurons. The fantastic capacity of these cells could serve as shuttle for gene therapy. Mouse models of SMA offer hereafter the opportunity of testing the capacity of cells to migrate to damaged tissue, to divide and to differentiate into functional cells.

Conclusion

Refined characterization of the degenerative process in SMA and the identification of the defective molecular pathway downstream from the SMN defect will provide further exciting insight into this disease in the near future. They should contribute to clarify the pathophysiology of SMA, the function of SMN and should help in designing potential targeted or non targeted therapeutic molecules.

Acknowledgements

Work made in our laboratory was supported by INSERM, the Association Française contre les Myopathies, Families of SMA (U.S.A.), the Fondation Bettencourt Schueller and Genopole. We gratefully thank Natacha Roblot, Bénédicte Desforges, Gaelle Millet, Sabrina Courageot and Vandana Joshi for their invaluable assistance.

References

1. Munsat TL. Workshop report: International SMA Collaboration. Neuromuscul Disord 1991; 1:81.
2. Lefebvre S, Burglen L, Reboullet S et al. Identification and characterization of a spinal muscular atrophy-determining gene. Cell 1995; 80:155-165.
3. Frugier T, Nicole S, Cifuentes-Diaz C et al. The molecular bases of spinal muscular atrophy. Curr Opin Genet Dev 2002; 12:294-298.
4. Campbell L, Potter A, Ignatius J et al. Genomic variation and gene conversion in spinal muscular atrophy: Implications for disease process and clinical phenotype. Am J Hum Genet 1997; 61:40-50.
5. Lefebvre S, Burlet P, Liu Q et al. Correlation between severity and SMN protein level in spinal muscular atrophy. Nat Genet 1997; 16:265-269.
6. Lorson CL, Androphy EJ. An exonic enhancer is required for inclusion of an essential exon in the SMA-determining gene SMN. Hum Mol Genet 2000; 9:259-265.
7. Pellizzoni L, Kataoka N, Charroux B et al. A novel function for SMN, the spinal muscular atrophy disease gene product, in premRNA splicing. Cell 1998; 95:615-624.
8. Lorson CL, Strasswimmer J, Yao JM et al. SMN oligomerization defect correlates with spinal muscular atrophy severity. Nat Genet 1998; 19:63-66.
9. Vyas S, Bechade C, Riveau B et al. Involvement of survival motor neuron (SMN) protein in cell death. Hum Mol Genet 2002; 11:2751-2764.
10. Cifuentes-Diaz C, Frugier T, Tiziano FD et al. Deletion of murine SMN exon 7 directed to skeletal muscle leads to severe muscular dystrophy. J Cell Biol 2001; 152:1107-1114.
11. Talbot K, Ponting CP, Theodosiou AM et al. Missense mutation clustering in the survival motor neuron gene: A role for a conserved tyrosine and glycine rich region of the protein in RNA metabolism? Hum Mol Genet 1997; 6:497-500.
12. Liu Q, Dreyfuss G. A novel nuclear structure containing the survival of motor neurons protein. Embo J 1996; 15:3555-3565.
13. Paushkin S, Gubitz AK, Massenet S et al. The SMN complex, an assemblyosome of ribonucleoproteins. Curr Opin Cell Biol 2002; 14:305-312.
14. Miguel-Aliaga I, Culetto E, Walker DS et al. The Caenorhabditis elegans orthologue of the human gene responsible for spinal muscular atrophy is a maternal product critical for germline maturation and embryonic viability. Hum Mol Genet 1999; 8:2133-2143.
15. Miguel-Aliaga I, Chan YB, Davies KE et al. Disruption of SMN function by ectopic expression of the human SMN gene in Drosophila. FEBS Lett 2000; 486:99-102.
16. Schrank B, Gotz R, Gunnersen JM et al. Inactivation of the survival motor neuron gene, a candidate gene for human spinal muscular atrophy, leads to massive cell death in early mouse embryos. Proc Natl Acad Sci USA 1997; 94:9920-9925.
17. Frugier T, Tiziano FD, Cifuentes-Diaz C et al. Nuclear targeting defect of SMN lacking the C-terminus in a mouse model of spinal muscular atrophy. Hum Mol Genet 2000; 9:849-858.
18. Hsieh-Li HM, Chang JG, Jong YJ et al. A mouse model for spinal muscular atrophy. Nat Genet 2000; 24:66-70.
19. Monani UR, Sendtner M, Coovert DD et al. The human centromeric survival motor neuron gene (SMN2) rescues embryonic lethality in Smn(-/-) mice and results in a mouse with spinal muscular atrophy. Hum Mol Genet 2000; 9:333-339.
20. Cifuentes-Diaz C, Nicole S, Velasco ME et al. Neurofilament accumulation at the motor endplate and lack of axonal sprouting in a spinal muscular atrophy mouse model. Hum Mol Genet 2002; 11:1439-1447.
21. McWhorter ML, Monani UR, Burghes AH et al. Knockdown of the survival motor neuron (Smn) protein in zebrafish causes defects in motor axon outgrowth and pathfinding. J Cell Biol 2003; 162:919-931.
22. Chan YB, Miguel-Aliaga I, Franks C et al. Neuromuscular defects in a Drosophila survival motor neuron gene mutant. Hum Mol Genet 2003; 12:1367-1376.
23. Nicole S, Desforges B, Millet G et al. Intact satellite cells lead to remarkable protection against Smn gene defect in differentiated skeletal muscle. J Cell Biol 2003; 161:571-582.
24. Rudnik-Schoneborn S, Goebel HH, Schlote W et al. Classical infantile spinal muscular atrophy with SMN deficiency causes sensory neuronopathy. Neurology 2003; 60:983-987.
25. Kugelberg E, Welander V. Heredofamilial juvenile muscular atrophy simulating muscular dystrophy. Acta Neurol Psychiatr 1956; 75:500-509.
26. Namba T, Aberfeld DC, Grob D. Chronic spinal muscular atrophy. J Neurol Sci 1970; 11:401-423.
27. Mastaglia FL, Walton JN. Histological and histochemical changes from cases of chronic juvenile and early adult spinal muscular atrophy (the Kugelberg-Welander syndrome). J Neurol Sci 1971; 12:15-44.

28. Bouwsma G, Vanwijngaarden GK. Spinal muscular atrophy and hypertrophy of the calves. J Neurol Sci 1980; 44:275-279.
29. Zhang HL, Pan F, Hong D et al. Active transport of the survival motor neuron protein and the role of exon-7 in cytoplasmic localization. J Neurosci 2003; 23:6627-6637.
30. Bassell GJ, Singer RH. Neuronal mRNA localization and the cytoskeleton. In: Richter D, ed. Cell polarity and subcellular localization. Springer, Berlin: 2001:41-56.
31. Stamm S, Zhang MQ, Marr TG et al. A sequence compilation and comparison of exons that are alternatively spliced in neurons. Nucleic Acids Res 1994; 22:1515-1526.
32. Jablonka S, Schrank B, Kralewski M et al. Reduced survival motor neuron (Smn) gene dose in mice leads to motor neuron degeneration: An animal model for spinal muscular atrophy type III. Hum Mol Genet 2000; 9:341-346.
33. Baron-Delage S, Abadie A, Echaniz-Laguna A et al. Interferons and IRF-1 induce expression of the survival motor neuron (SMN) genes. Mol Med 2000; 6:957-968.
34. Chang JG, Hsieh-Li HM, Jong YJ et al. Treatment of spinal muscular atrophy by sodium butyrate. Proc Natl Acad Sci USA 2001; 98:9808-9813.
35. Brichta L, Hofmann Y, Hahnen E et al. Valproic acid increases the SMN2 protein level: A well-known drug as a potential therapy for spinal muscular atrophy. Hum Mol Genet 2003; 12:2481-2489.
36. Andreassi C, Jarecki J, Zhou J et al. Aclarubicin treatment restores SMN levels to cells derived from type I spinal muscular atrophy patients. Hum Mol Genet 2001; 10:2841-2849.
37. Lesbordes JC, Cifuentes-Diaz C, Miroglio A et al. Therapeutic benefits of cardiotrophin-1 gene transfer in a mouse model of spinal muscular atrophy. Hum Mol Genet 2003; 12:1233-1239.
38. Haddad H, Cifuentes-Diaz C, Miroglio A et al. Riluzole attenuates spinal muscular atrophy disease progression in a mouse model. Muscle and Nerve 2003; 28:432-437.

CHAPTER 15

The Pathophysiological Role of Impaired Calcium Handling in Muscular Dystrophy

Kay Ohlendieck

Abstract

Although the primary deficiency in dystrophin and the concomitant reduction in surface glycoproteins are well established factors in the molecular pathogenesis of Duchenne muscular dystrophy, the pathophysiological events that render a muscle fibre more susceptible to necrosis are not well understood. One proposed mechanism involves abnormal calcium homeostasis in mechanically stressed fibres that eventually leads to skeletal muscle weakness. This chapter examines the calcium hypothesis of muscular dystrophy and outlines how unbalanced ion cycling through the sarcolemma and the sarcoplasmic reticulum may contribute to enhanced degradation of muscle proteins. Studying calcium handling in muscular dystrophy is not only important for increasing our knowledge on the multifaceted process of muscle degeneration, but might also have implications for the future design of therapeutic approaches to treating x-linked muscular dystrophy. In this respect, it is encouraging that the pharmacological elimination of Ca^{2+}-dependent proteolysis counteracts dystrophic changes and that the removal of excess cytosolic Ca^{2+} appears to convey natural protection to dystrophin-deficient fibres.

Calcium Handling Proteins

Calcium ions represent an ubiquitous second messenger molecule involved in the regulation of various metabolic and physiological processes.[1] In skeletal muscle, cytosolic Ca^{2+}-levels dictate the overall contractile status, and Ca^{2+}-cycling through intracellular compartments is maintained under precise spatial and temporal control.[2] Signal transduction during excitation-contraction coupling and fibre relaxation is mediated by complex interactions between voltage sensors, Ca^{2+}-channels, Ca^{2+}-binding proteins, Ca^{2+}-transporters, ion exchangers and ion pumps.[3-5] Table 1 lists established elements involved in ion handling that are potentially impaired in muscular dystrophy. Almost all Ca^{2+}-regulatory elements exist as several isoforms[6] reflecting the heterogeneity of skeletal muscle fibres.[7] Due to the enormous complexity of the Ca^{2+}-handling apparatus, small changes in individual steps involved in modulating Ca^{2+}-signals might result in major pathophysiological consequences. Although it is not yet established how many auxiliary proteins act as intrinsic regulators of Ca^{2+}-release complexes, Ca^{2+}-uptake units and the luminal Ca^{2+}-reservoir complex, abnormal expression patterns of ion-regulatory components are clearly involved in muscular dystrophy.[8-12] The exact degree of abnormal Ca^{2+}-handling is difficult to judge,[13] since it is not fully understood whether key Ca^{2+}-handling proteins operate at their full capacity under normal conditions.[3,14-16]

Table 1. Ion-regulatory proteins of the excitation-contraction-relaxation cycle

Component	Physiological/Pathophysiological Role
Dihydropyridine receptor (α_{1S}-subunit)	Non-junctional L-type Ca^{2+}-channel
Dihydropyridine receptor (α_{1S}-subunit)	Junctional voltage sensor
Dihydropyridine receptor (α_2/δ-β-γ subunits)	Regulation of t-tubular Ca^{2+}-channel
Sarcolemmal Ca^{2+}-leak channel	Surface Ca^{2+}-flux
Nicotinic acetylcholine receptor	Surface Ca^{2+}-leakage
Transient receptor potential channel (TRPC-1, TRPC-4, TRPC-6)	Surface cation channel
Growth factor-regulated GRC channel	Ca^{2+}-permeable surface cation channel
Sarcolemmal Ca^{2+}-ATPase (PMCA)	Surface Ca^{2+}-pump
Na^{2+}/Ca^{2+}-exchanger / Na^+/K^+-ATPase	Surface Ca^{2+}-removal system
Ryanodine receptor (RyR1 isoform)	Sarcoplasmic reticulum Ca^{2+}-release channel
FKBP12 (12 kDa binding protein)	Regulation of Ca^{2+}-release channel
Annexin VI	Regulation of Ca^{2+}-release channel
Calmodulin	Regulation of Ca^{2+}-channels / Ca^{2+}-pumps
SR ADP/ATP translocator	Regulation of Ca^{2+}-release channel
Triadin	Triadic mediator between calsequestrin and RyR1 Ca^{2+}-release channel
Juntional face proteins (JP-90, JP-45)	Regulators of Ca^{2+}-homeostasis
Calsequestrin	Luminal Ca^{2+}-storage / Regulation of Ca^{2+}-release channel
Calsequestrin-like proteins (CLPs)	Luminal Ca^{2+}-storage
Junctin	Structural CSQ-binding protein
Calreticulin	Luminal Ca^{2+}-storage
Histidine–rich Ca^{2+}-binding protein	Luminal Ca^{2+}-storage
Sarcalumenin	Luminal Ca^{2+}-transport/storage
Fast Ca^{2+}-ATPase (SERCA1)	Sarcoplasmic reticulum Ca^{2+}-uptake
Sarcolipin	Regulation of SERCA1 Ca^{2+}-pump
Slow Ca^{2+}-ATPase (SERCA2)	Sarcoplasmic reticulum Ca^{2+}-uptake
Phospholamban	Regulation of SERCA2 Ca^{2+}-pump
Parvalbumin	Cytosolic Ca^{2+}-binding protein

Proteins involved in voltage sensing, receptor coupling, sarcolemmal ion fluxes, luminal and cytosolic calcium buffering and ion pumping, are potentially affected in down-stream events of the molecular pathogenesis of muscular dystrophy.

Calcium Hypothesis of Muscular Dystrophy

Historically, the formulation of the calcium hypothesis of muscular dystrophy[17] preceded the pioneering discovery of the genetic defect responsible for Duchenne muscular dystrophy.[18] A decade before Kunkel and coworkers described the primary abnormality in x-linked muscular dystrophy using a positional cloning strategy,[18,19] anomalous intracellular Ca^{2+}-accumulation was described in fibres from patients afflicted with x-linked muscular dystrophy.[20] In conjunction with the electron microscopical demonstration of membrane damage in dystrophic fibres,[21] this finding led to the idea that chronic Ca^{2+}-overload might play a central role in fibre degeneration.[17] Since dystrophinopathies are primarily diseases of the membrane cytoskeleton,[22,23] it is

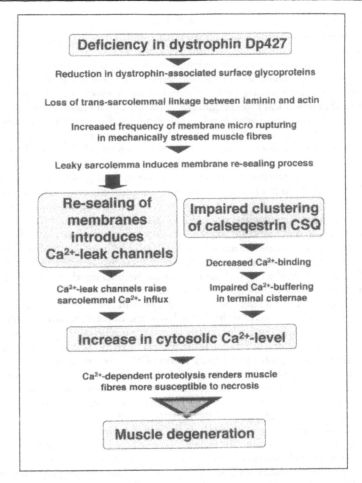

Figure 1. Flow chart of the mechanical calcium hypothesis of muscular dystrophy. Outlined are two major factors involved in the elevated cytosolic Ca^{2+}-levels observed in muscular dystrophy. Deficiency in the membrane cytoskeletal component dystrophin causes a drastic reduction in a specific subset of surface glycoproteins, such as the dystroglycans and sarcoglycans. This in turn results in the loss of sarcolemmal integrity and causes an increase in membrane micro-rupturing. During the natural resealing process, Ca^{2+}-leak channels are introduced into the muscle periphery and cause an increased Ca^{2+}-influx. It is not yet established what exact complement of surface Ca^{2+}-channels is involved in this process. In addition, impaired luminal Ca^{2+}-buffering in the sarcoplasmic reticulum exacerbates this pathophysiological condition. Chronically elevated Ca^{2+}-levels initiate Ca^{2+}-dependent proteolysis and thereby render dystrophin-deficient muscle fibres more susceptible to necrosis.

important to consider that disruption of the cytoskeletal network has a direct effect on Ca^{2+}-fluxes.[24,25] This suggests a direct involvement of the membrane cytoskeleton in the maintenance of subsarcolemmal Ca^{2+}-levels and Ca^{2+}-mediated signal transduction pathways.

Molecular Defects in Calcium Homeostasis

The current mechanical calcium hypothesis of muscular dystrophy assumes that a causal connection exists between the disintegration of sarcolemmal integrity and Ca^{2+}-dependent proteolysis of muscle proteins,[8-12] as outlined in the flow chart of Figure 1. Various aspects of individual steps involved in this destructive process have been discussed in recent reviews.[8-12,26,27]

Deficiency in the dystrophin isoform Dp427[19] results in a drastic reduction in various sarcolemma-associated proteins,[28] such as the sarcoglycans[29] and dystroglycans.[30] The resulting destabilisation of the linkage between the actin cytoskeleton and the extracellular matrix component laminin renders the muscle periphery more susceptible to micro-rupturing.[31] This in turn, triggers a Ca^{2+}-dependent membrane resealing process at the sites of sarcolemmal disintegration. The influx of extracellular Ca^{2+}-ions initiates an exocytotic event that adds plasma membrane patches to seal leaky micro domains of the disrupted muscle periphery.[8] Ca^{2+}-leak channels present in the newly introduced sarcolemmal patches cause an increased influx of Ca^{2+}-ions into the cytoplasm of dystrophic fibres. The exact extent of perturbation of Ca^{2+}-handling in dystrophin-deficient fibres is still controversial.[8-12] Contradictory findings may be partially due to the usage of different physiological techniques and standardisation in the measurement of intracellular Ca^{2+}-levels. However, it is generally accepted that the Ca^{2+}-overload is not global but restricted to cytosolic domains adjacent to the sarcolemma.[32,33] The recent characterisation of dysferlin-null mice has highlighted the importance of proper membrane resealing in muscular disorders. Although dysferlin-deficient fibres maintain a functional dystrophin-glycoprotein complex, they develop a progressive muscular dystrophy.[34] In skeletal muscle, surface membrane repair is an active process involving dysferlin whereby disruption of the sarcolemma repair apparatus triggers muscle degeneration.[34] This clearly demonstrates the importance of a fast Ca^{2+}-dependent sarcolemma resealing process to efficiently counter-act injuries to the muscle plasma membrane.

The pathological consequence of increased cytosolic Ca^{2+}-levels is an activation of Ca^{2+}-dependent proteases, such as skeletal muscle-specific calpains, resulting in the net degradation of muscle proteins in dystrophic muscle cells.[35-38] Localised Ca^{2+}-elevations are believed to contribute to a destructive cycle of enhanced protease activity and Ca^{2+}-leak channel activation.[36] Because increased proteolytic activity renders Ca^{2+}-leak channels constitutively active, insertion of membrane vesicles for resealing, following exercise-induced membrane rupturing, results in persistent Ca^{2+} influx at affected sarcolemma domains. Hence, more Ca^{2+} influx causes increased proteolysis, and more proteolysis causes increased Ca^{2+} influx, giving rise to a pathophysiological cycle.[8] The theory of the initiating role of mechano-sensitive Ca^{2+}-channels in triggering fibre degradation has recently been challenged by the study of transgenic mdx muscle that over-expresses utrophin.[39] It therefore can not be excluded that the Ca^{2+}-leak channel activation in mechanically stressed fibres is preceded by modifications in other Ca^{2+}-handling proteins. For example, surface membrane leakage of Ca^{2+} into the cytoplasm may also be due to abnormalities in the nicotinic acetylcholine receptor.[40,41] This fact, however, does not challenge the overall concept of deranged intracellular Ca^{2+}-handling being involved in muscular dystrophy. Both, Ca^{2+}-permeable growth factor-regulated channels[42] (GRC) and transient receptor potential channels[33] (TRPC) may facilitate abnormal Ca^{2+}-influxes through the dystrophin-deficient muscle periphery. The expression of Ca^{2+}-permeable nonselective cation channels of the GRC type were shown to be elevated in the sarcolemma of mdx fibres[42] and isoforms TRPC-1, 4 and 6 of the transient receptor potential channels are present in mdx membrane.[33] In contrast, surface L-type Ca^{2+}-channels of the dihydropyridine receptor class do not appear to contribute to elevated cytosolic Ca^{2+}-levels in muscular dystrophy.[43]

Impaired Excitation-Contraction Coupling

In addition to abnormal sarcolemma physiology, changes in the Ca^{2+}-cycling patterns through the lumen of the sarcoplasmic reticulum[44] and mitochondria[45] have also been shown to occur in dystrophin-deficient skeletal muscle. The excitation-contraction-relaxation cycle is regulated by the physiological interplay between the voltage-sensing dihydropyridine receptor of the transverse tubules, the ryanodine receptor Ca^{2+}-release channel of the junctional sarcoplasmic reticulum and the Ca^{2+}-ATPases of the longitudinal tubules.[3-5] It would not be surprising if these ion-regulatory elements were secondarily involved in muscular dystrophy, since numerous inherited muscle diseases exhibit defects in excitation-contraction coupling or muscle relaxation. This includes malignant hyperthermia, central core disease, hypokalemic periodic paralysis and Brody's disease.[46] In normal muscle fibres, the release of intracellular Ca^{2+} in response to sarcolemmal depolarization is one of the key steps of excitation-contraction coupling.[3] Following voltage-sensing by the α_{1S}-subunit of the dihydropyridine receptor, physical coupling between the II-III loop domain of the transverse-tubular tetrad complex and the cytosolic foot region of the junctional ryanodine receptor tetramer triggers transient opening of the Ca^{2+}-release channel.[5,16] Thus, at least initially, a direct physical receptor interaction process couples membrane depolarization to the release of Ca^{2+}-ions into the cytosol. In a subsequent step, comparable to the cardiac Ca^{2+}-induced Ca^{2+}-release mechanism, sarcoplasmic reticulum Ca^{2+}-channels are also locally activated by excess Ca^{2+} that greatly amplifies the overall Ca^{2+}-signal.[3-5,16] Fibre relaxation is initiated by the energy-dependent reversal of the cytosolic Ca^{2+}-signal. Reuptake of Ca^{2+}-ions in the lumen of the sarcoplasmic reticulum is achieved by the SERCA type Ca^{2+}-pumps.[14] In contrast to heart muscle, Ca^{2+}-removal via the surface Na^+/Ca^{2+}-exchanger does not play a major role in skeletal muscle fibres. An important physiological regulator that mediates between Ca^{2+}-release and Ca^{2+}-uptake is the luminal Ca^{2+}-reservoir complex of the sarcoplasmic reticulum.[15] Besides the histidine-rich Ca^{2+}-binding protein, calreticulin, sarcalumenin and calsequestrin-like proteins, the most abundant Ca^{2+}-binding element of the terminal cisternae are clusters of 63 kDa calsequestrin molecules.[4]

It has not been determined how many auxiliary proteins are involved in the fine regulation of receptor coupling, Ca^{2+}-release, Ca^{2+}-buffering and Ca^{2+}-uptake and remains to be elucidated which triad-associated components prevent passive disintegration of supramolecular membrane assemblies. Recently identified auxiliary proteins such as triadin, junctin, JP-45 and JP-90 only represent part of the total protein complement of the junctional couplings involved in excitation-contraction coupling.[47-50] Current muscle proteome projects are addressing this question and aim at the identification of all sarcoplamic reticulum proteins involved in Ca^{2+}-cycling. Until these results become available, we lack an overall understanding of the complexity of the Ca^{2+}-handling apparatus. Nevertheless, the initial analysis of the main players involved in excitation-contraction coupling revealed that calsequestrin-like proteins (CLP) are greatly reduced in muscular dystrophy,[44] as illustrated in the immunoblot analysis shown in Figure 2. While the expression of the junctional ryanodine receptor Ca^{2+}-release channel, the voltage-sensing dihydropyridine receptor, the Ca^{2+}-ATPase and calsequestrin was not affected, a drastic decline in CLP-150, CLP-170 and CLP-220 was observed in dystrophic microsomes.[44] Although immunoblotting could clearly demonstrate the reduction in these terminal cisternae markers, this technique could not address the question of whether the reduction in high-molecular-mass isoforms of calsequestrin is due to decreased expression levels or because of an impaired oligomersiation pattern.

Comparative chemical crosslinking analysis of normal and dystrophic microsomal membranes revealed a crosslinker-induced reappearance of CLP species in dystrophin-deficient preparations.[51] Hence, CLP-150, CLP-170 and CLP-220 appear to represent clusters of the calsequestrin monomers and not distinct high-molecular-mass isoforms. Since the introduction of short crosslinker probes could restore the appearance of CLPs in dystrophic membranes, their drastically reduced expression in muscular dystrophy is due to impaired oligomerisation of calsequestrin monomers, and not based on a loss in individual isoforms of this Ca^{2+}-binding protein. Equilibrium dialysis experiments have shown that the overall Ca^{2+}-binding capacity was reduced by approximately 20% in dystrophic sarcoplasmic reticulum.[44] Thus, impaired protein coupling between calsequestrin units in the terminal cisternae region might be directly involved in triggering impaired Ca^{2+}-sequestration within the lumen of the sarcoplasmic reticulum. In muscular dystrophy, disturbed sarcolemmal Ca^{2+}-fluxes appear to influence the complex Ca^{2+}-cycling apparatus causing distinct changes in the oligomerisation pattern of a subset of Ca^{2+}-handling proteins (Fig. 3).

With respect to cytosolic Ca^{2+}-binding elements, parvalbumin plays an ion-buffering role in fast muscle fibres.[52] Chronic low-frequency stimulation suppresses parvalbumin expression

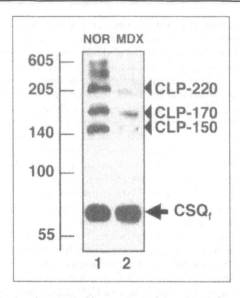

Figure 2. Drastic reduction in calsequestrin-like proteins in dystrophin-deficient skeletal muscle. Shown is an immunoblot of microsomal proteins from normal (lane 1) and dystrophic mdx (lane 2) muscle fibres labelled with monoclonal antibody VIIID1$_2$ to the fast isoform of the major Ca^{2+}-binding protein calsequestrin (CSQ$_f$) and its high-molecular-mass aggregates, the calsequestrin-like proteins (CLP) termed CLP-150, CLP-170 and CLP-220. Isolation of membrane proteins and immunoblotting was performed by standard methodology.[44] The position of the immuno-decorated CSQ band is indicated by an arrow and CLPs are marked by arrow heads. The relative position of molecular mass standards (x10^{-3}) is indicated on the left. In muscular dystrophy, the oligomerisation of calsequestrin appears to be impaired resulting in an approximately 20% reduced Ca^{2+}-binding capacity of the sarcoplasmic reticulum.[44]

in fast muscles demonstrating that a slow motoneuron-like impulse pattern rapidly silences the parvalbumin gene.[53] An increased turnover of parvalbumin has been described in dystrophic fast-twitching mdx fibres.[54] Elevated serum parvalbumin levels appear to be indicative of the severity of the disease status in dystrophic mouse muscle.[54] Compared to the severe dystrophic phenotype of Duchenne patients, the *mdx* mouse model exhibits a much milder dystrophy. Since parvalbumin is expressed at only very low levels in human fibres, but is abundant in rodent IIB fibres,[55] the presence of parvalbumin might be responsible for this difference in the severity of the dystrophinopathy. This relatively species-specific Ca^{2+}-binding element could buffer the increased Ca^{2+}-influx through the *mdx* sarcolemma thereby maintaining a lower cytosolic Ca^{2+}-level as compared to dystrophic human muscle fibres. However, double mutant mice that are deficient in both dystrophin and parvalbumin exhibit a cytosolic Ca^{2+}-content similar to that of mdx fibres.[56] Thus, parvalbumin does not appear to play a central role in eliminating abnormal Ca^{2+}-cycling through the cytosol of mdx muscle.

Therapeutic Implications

In addition to helping in understanding the molecular pathways underlying muscular dystrophy, studying Ca^{2+}-regulatory proteins might also lead to the discovery of new therapeutic targets. One potential way forward is the analysis of naturally protected muscle phenotypes that lack dystrophin but do not exhibit severe symptoms. The sparing of dystrophin-deficient extraocular fibres is attributed to the special protective properties of small-diameter fibres.[57] Recently, it was shown that in contrast to leg muscles, toe fibres from the dystrophic animal model *mdx* also show a very low degree of muscle degeneration.[58,59] This is histochemically evident by a low percentage of centrally located nuclei. Toe fibres from the mdx mouse exhibit

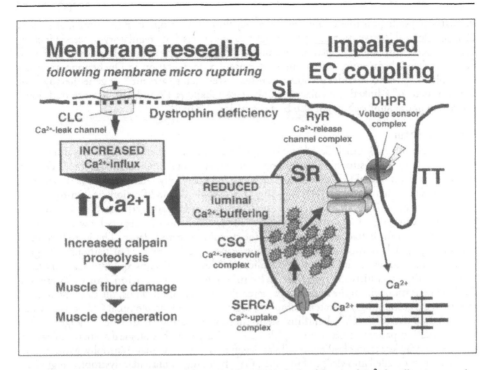

Figure 3. Diagrammatic presentation of the pathophysiological role of abnormal Ca^{2+}-handling in muscular dystrophy. In adult skeletal muscle, excitation-contraction (EC) coupling is mediated by direct physical interactions between the voltage-sensing α_{1S}-subunit of the dihydropyridine receptor (DHPR) of the transverse tubules (TT) and the junctional ryanodine receptor (RyR) Ca^{2+}-release channel complex of the sarcoplasmic reticulum (SR). Following Ca^{2+}-triggered fibre contraction, muscle relaxation is induced by the energy-dependent reuptake of Ca^{2+}-ions via the SERCA Ca^{2+}-pump units. Besides various minor Ca^{2+}-binding proteins, the major luminal Ca^{2+}-reservoir complex is represented by calsequestrin (CSQ). Reduction in the dystrophin-glycoprotein complex results in the loss of the linkage between the actin membrane cystoskeleton and the extracellular matrix component laminin, thereby rendering the sarcolemma (SL) more susceptible to contraction-induced membrane micro-rupturing. Naturally occuring resealing in dystrophic fibres introduces Ca^{2+}-leak channels (CLC) causing elevated Ca^{2+}-levels in the sub-sarcolemmal cytosol. Both, the surface Ca^{2+}-leak channel activity and the reduced Ca^{2+}-buffering capacity of the sarcoplasmic reticulum trigger a chronically elevated cytosolic Ca^{2+}-concentration. This results in the Ca^{2+}-dependent proteolysis of muscle proteins and consequently fibre degeneration.

a preserved expression of the critical trans-sarcolemmal linker α-dystroglycan[58] and an up-regulation and extra-junctional localization of the autosomal dystrophin homologue utrophin.[58,59] Possibly utrophin acts as a substitute for the deficient Dp427 isoform and thereby prevents the disintegration of the surface-associated glycoprotein complex.[60] Interestingly, an increase in both the expression of the fast SERCA1 isoform of the sarcoplasmic reticulum Ca^{2+}-pump and the total Ca^{2+}-ATPase activity was described in protected mdx toe fibres.[58] Possibly, the up-regulation of Ca^{2+}-ATPase units causes an increased removal of cytosolic Ca^{2+}-ions from mdx toe fibres thereby significantly reducing the degree of Ca^{2+}-induced myonecrosis in dystrophin-deficient fibres.[58] That a decrease in cytosolic Ca^{2+}-levels has a protective effect was shown by experiments with carnitine-linked leupeptin.[61,62] The targeted introduction of protease inhibitors appears to specifically inactivate Ca^{2+}-dependent calpain activity.[62] Calpain inhibition could be correlated with rentention of myofibre size.[61] Overexpression of the calpain inhibitor calpastatin in dystrophin-deficient fibres reduced the dystrophic phenotype[63] suggesting that specific calpains are primarily involved in muscle necrosis. Based on the pathophysiological role of abnormal Ca^{2+}-cycling in dystrophic fibres, potential sites of pharmacological interventions present themselves at various Ca^{2+} regulatory processes including the sarcolemma and sarcoplasmic reticulum.[64-66] Especially calpain inhibition promises to be an excellent therapeutic target to prevent muscular degeneration in Duchennne muscular dystrophy.[61-63] Hence, besides myoblast transfer,[67] stem cell treatment[68] or gene therapy[69] for the treatment of x-linked muscular dystrophy (see Chapters 17 and 18), a more traditional pharmacological approach might also be a promising option.[70]

Conclusions

In conclusion, abnormal Ca^{2+}-handling might be an important factor in the progressive functional decline of dystrophic muscle fibres. Although the total extent and exact micro-domain localisation of the initial Ca^{2+}-disturbance has not yet been fully established, it is understood that even small changes in Ca^{2+}-cycling trigger a cascade of modifications in ion-regulatory muscle proteins. The deficiency in dystrophin leads primarily to a reduction in dystrophin-associated proteins which in turn destroys sarcolemmal integrity by weakening the linkage between the extracellular matrix and the membrane cytoskeleton. The introduction of Ca^{2+}-leak channels during membrane resealing appears to be the central pathophysiological incident that causes an increase in cytosolic Ca^{2+}-levels and consequently Ca^{2+}-induced myonecrosis. In addition, impaired calsequestrin oligomerisation causes decreased Ca^{2+}-sequestration within the lumen of the sarcoplasmic reticulum. Thus, the disintegration of the surface dystrophin-glycoprotein complex triggers disturbed sarcolemmal Ca^{2+}-fluxes which in turn impairs other Ca^{2+}-handling steps thereby resulting in distinct changes in the oligomerisation pattern of Ca^{2+}-binding elements. This pathophysiological scenario might explain one of the key routes leading to cellular degeneration in muscular dystrophy. Most importantly, a more detailed knowledge of impaired Ca^{2+} handling in muscular dystrophy might lead to the design of novel approaches to counter-act fibre degeneration in genetic muscle diseases.

Acknowledgments

Research in the author's laboratory on calcium handling in neuromuscular disorders was supported by project grants from the Irish Health Research Board (HRB-95/95, HRB-01/98, HRB-01/99, HRB-01/01), Enterprise Ireland (SC/2000/386) and Muscular Dystrophy Ireland (MDI-95, MDI-02), as well as network grants from the European Commission (FMRX-CT960032, QLRT-1999-02034, RTN2-2001-00337). I would like to thank Dr. Kevin Culligan for the immunoblot analysis of dystrophic fibres.

References

1. Berridge MJ, Bootman MD, Roderick HL. Calcium signalling: Dynamics, homeostasis and remodelling. Nat Rev Mol Cell Biol 2003; 4:517-529.
2. Berchtold MW, Brinkmeier H, Muntener M. Calcium ion in skeletal muscle: Its crucial role for muscle function, plasticity, and disease. Physiol Rev 2000; 80:1215-1265.
3. Melzer W, Herrmann-Frank A, Luttgau HC. The role of Ca^{2+} ions in excitation-contraction coupling of skeletal muscle fibres. Biochim Biophys Acta 1995; 1241:59-116.
4. Murray BE, Froemming GR, Maguire PB et al. Excitation-contraction-relaxation cycle: Role of Ca^{2+}-regulatory membrane proteins in normal, stimulated and pathological skeletal muscle fibres (review). Int J Mol Med 1998; 1:677-697.
5. Leong P, MacLennan DH. Complex interactions between skeletal muscle ryanodine receptor and dihydropyridine receptor proteins. Biochem Cell Biol 1998; 76:681-694.
6. Ohlendieck K. Changes in Ca^{2+}-regulatory membrane proteins during the stimulation-induced fast-to-slow transition process. Bas Appl Myol 2000; 10:99-106.
7. Pette D, Staron RS. Transitions of muscle fiber phenotypic profiles. Histochem Cell Biol 2001; 115:359-372.
8. Alderton JM, Steinhardt RA. How calcium influx through calcium leak channels is responsible for the elevated levels of calcium-dependent proteolysis in dystrophic myotubes. Trends Cardiovasc Med 2000; 10:268-272.
9. Gailly P. New aspects of calcium signaling in skeletal muscle cells: Implications in Duchenne muscular dystrophy. Biochim Biophys Acta 2002; 1600:38-44.

10. Gillis JM. Membrane abnormalities and Ca homeostasis in muscles of the mdx mouse, an animal model of the Duchenne muscular dystrophy: A review. Acta Physiol Scand 1996; 156:397-406.
11. Ruegg UT, Nicolas-Metral V, Challet C et al. Pharmacological control of cellular calcium handling in dystrophic skeletal muscle. Neuromuscul Disord 2002; 12:S155-S161.
12. Culligan K, Ohlendieck K. Abnormal calcium handling in muscular dystrophy. Bas Appl Myol 2002; 12:147-157.
13. Divet A, Huchet-Cadiou C. Sarcoplasmic reticulum function in slow- and fast-twitch skeletal muscles from mdx mice. Pflugers Arch 2002; 444:634-643.
14. Stokes DL, Wagenknecht T. Calcium transport across the sarcoplasmic reticulum: Structure and function of Ca^{2+}-ATPase and the ryanodine receptor. Eur J Biochem 2000; 267:5274-9.
15. MacLennan DH, Reithmeier RA. Ion tamers. Nat Struct Biol 1998; 5:409-411.
16. Franzini-Armstrong C, Protasi F. Ryanodine receptors of striated muscles: A complex channel capable of multiple interactions. Physiol Rev 1997; 77:699-729.
17. Duncan CJ. Role of intracellular calcium in promoting muscle damage: A strategy for controlling the dystrophic condition. Experientia 1978; 34:1531-1535.
18. Monaco AP, Neve RL, Colletti-Feener C et al. Isolation of candidate cDNAs for portions of the Duchenne muscular dystrophy gene. Nature 1986; 323:646-650.
19. Ahn AH, Kunkel LM. The structural and functional diversity of dystrophin. Nat Genet 1993; 3:283-291.
20. Bodensteiner JB, Engel AG. Intracellular calcium accumulation in Duchenne dystrophy and other myopathies: A study of 567,000 muscle fibers in 114 biopsies. Neurology 1978; 28:439-446.
21. Cullen MJ, Fulthorpe JJ. Stages in fibre breakdown in Duchenne muscular dystrophy. An electron-microscopic study. J Neurol Sci 1975; 24:179-200.
22. Campbell KP. Three muscular dystrophies: Loss of cytoskeleton-extracellular matrix linkage. Cell 1995; 80:675-679.
23. Ohlendieck K. Towards an understanding of the dystrophin-glycoprotein complex: Linkage between the extracellular matrix and the subsarcolemmal membrane cytoskeleton. Eur J Cell Biol 1996; 69:1-10.
24. Menke A, Jockusch H. Extent of shock-induced membrane leakage in human and mouse myotubes depends on dystrophin. J Cell Sci 1995; 108:727-33.
25. Hajnoczky G, Lin C, Thomas AP. Luminal communication between intracellular calcium stores modulated by GTP and the cytoskeleton. J Biol Chem 1994; 269:10280-10287.
26. Petrof BJ. The molecular basis of activity-induced muscle injury in Duchenne muscular dystrophy. Mol Cell Biochem 1998; 179:111-23.
27. Blake DJ, Weir A, Newey SE et al. Function and genetics of dystrophin and dystrophin-related proteins in muscle. Physiol Rev 2002; 82:291-329.
28. Culligan K, Mackey A, Finn D et al. Role of dystrophin isoforms and associated glycoproteins in muscular dystrophy (review). Int J Mol Med 1998; 2:639-648.
29. Hack AA, Groh ME, McNally EM. Sarcoglycans in muscular dystrophy. Microsc Res Tech 2000; 48:167-80.
30. Winder SJ. The complexities of dystroglycan. Trends Biochem Sci 2001; 26:118-124.
31. Clarke MS, Khakee R, McNeil PL. Loss of cytoplasmic basic fibroblast growth factor from physiologically wounded myofibers of normal and dystrophic muscle. J Cell Sci 1993; 106:121-133.
32. Mallouk N, Jacquemond V, Allard B. Elevated subsarcolemmal Ca^{2+} in mdx mouse skeletal muscle fibres detected with Ca^{2+}-activated K^+ channels. Proc Natl Acad Sci USA 2000; 97:4950-4955.
33. Vandebrouck C, Martin D, Colson-Van Schoor M et al. Involvement of TRPC in the abnormal calcium influx observed in dystrophic (mdx) mouse skeletal muscle fibers. J Cell Biol 2002; 158:1089-96.
34. Bansal D, Miyake K, Vogel SS et al. Defective membrane repair in dysferlin-deficient muscular dystrophy. Nature 2003; 423:168-172.
35. MacLennan PA, McArdle A, Edwards RH. Effects of calcium on protein turnover of incubated muscles from mdx mice. Am J Physiol 1991; 260:E594-E598.
36. Turner PR, Fong PY, Denetclaw WF et al. Increased calcium influx in dystrophic muscle. J Cell Biol 1991; 115:1701-1712.
37. Turner PR, Schultz R, Ganguly B et al. Proteolysis results in altered leak channel kinetics and elevated free calcium in mdx muscle. J Membr Biol 1993; 133:243-251.
38. Alderton JM, Steinhardt RA. Calcium influx through calcium leak channels is responsible for the elevated levels of calcium-dependent proteolysis in dystrophic myotubes. J Biol Chem 2000; 275:9452-9460.
39. Squire S, Raymackers JM, Vandebrouck C et al. Prevention of pathology in mdx mice by expression of utrophin: Analysis using inducible transgenic expression system. Hum Mol Genet 2002; 11:3333-3344.
40. Carlson CG. The dystrophinopathies: An alternative to the structural hypothesis. Neurobiol Dis 1998; 5:3-15.

41. Carlson CG. Acetylcholine receptor and calcium leakage activity in nondystrophic and dystrophic myotubes (MDX). Muscle Nerve 1996; 19:1258-1267.
42. Iwata Y, Katanosaka Y, Arai Y et al. A novel mechanism of myocyte degeneration involving the Ca^{2+}-permeable growth factor-regulated channel. J Cell Biol 2003; 161:957-967.
43. Collet C, Csernoch L, Jacquemond V. Intramembrane charge movement and L-type calcium current in skeletal muscle fibers isolated from control and mdx mice. Biophys J 2003; 84:251-265.
44. Culligan K, Banville N, Dowling P et al. Drastic reduction of calsequestrin-like proteins and impaired calcium binding in dystrophic mdx muscle. J Appl Physiol 2002; 92:435-445.
45. Robert V, Massimino ML, Tosello V et al. Alteration in calcium handling at the subcellular level in mdx myotubes. J Biol Chem 2001; 276:4647-4651.
46. Froemming GR, Ohlendieck K. The role of ion-regulatory membrane proteins of excitation-contraction coupling and relaxation in inherited muscle diseases. Front Biosci 2001; 6:D65-D74.
47. Froemming GR, Murray BE, Ohlendieck K. Self-aggregation of triadin in the sarcoplasmic reticulum of rabbit skeletal muscle. Biochim Biophys Acta 1999; 1418:197-205.
48. Froemming GR, Pette D, Ohlendieck K. The 90 kDa junctional sarcoplasmic reticulum protein forms an integral part of a supramolecular triad complex in skeletal muscle. Biochem Biophys Res Comm 1999; 261:603-609.
49. Zorzato F, Anderson AA, Ohlendieck K et al. Identification of a novel 45 kDa protein (JP-45) from rabbit sarcoplasmic-reticulum junctional-face membrane. Biochem J 2000; 351:537-543.
50. Jones LR, Zhang L, Sanborn K et al. Purification, primary structure, and immunological characterization of the 26-kDa calsequestrin binding protein (junctin) from cardiac junctional sarcoplasmic reticulum. J Biol Chem 1995; 270:30787-30796.
51. Dowling P, Lohan J, Ohlendieck K. Comparative analysis of Dp427-deficient mdx tissues shows that the milder dystrophic phenotype of extraocular and toe muscle fibres is associated with a persistent expression of beta-dystroglycan. Eur J Cell Biol 2003; 82:222-230.
52. Schleef M, Zuhlke C, Jockusch H et al. The structure of the mouse parvalbumin gene. Mamm Genome 1992; 3:217-225.
53. Huber B, Pette D. Dynamics of parvalbumin expression in low-frequency-stimulated fast-twitch rat muscle. Eur J Biochem 1996; 236:814-819.
54. Jockusch H, Friedrich G, Zippel M. Serum parvalbumin, an indicator of muscle disease in murine dystrophy and myotonia. Muscle Nerve 1990; 13:551-555.
55. Schmitt TL, Pette D. Fiber type-specific distribution of parvalbumin in rabbit skeletal muscle. A quantitative microbiochemical and immunohistochemical study. Histochemistry 1991; 96:459-465.
56. Raymackers JM, Debaix H, Colson-Van Schoor M et al. Consequence of parvalbumin deficiency in the mdx mouse: Histological, biochemical and mechanical phenotype of a new double mutant. Neuromuscul Disord 2003; 13:376-387.
57. Porter JD. Extraocular muscle sparing in muscular dystrophy: A critical evaluation of potential protective mechanisms. Neuromusc Disord 1998; 8:198-203.
58. Dowling P, Culligan K, Ohlendieck K. Distal mdx muscle groups exhibiting up-regulation of utrophin and rescue of dystrophin-associated glycoproteins exemplify a protected phenotype in muscular dystrophy. Naturwiss 2002; 89:75-88.
59. Matsumura K, Ervasti JM, Ohlendieck K et al. Association of dystrophin-related protein with dystrophin-associated proteins in mdx mouse muscle. Nature 1992; 360:588-591.
60. Rafael JA, Tinsley JM, Potter AC et al. Skeletal muscle-specific expression of utrophin transgene rescues utrophin-dystrophin deficient mice. Nature Genet 1998; 19:79-82.
61. Badalamente WA, Stracher A. Delay of muscle degeneration and necrosis in mdx mice by calpain inhibition. Muscle Nerve 2000; 23:106-111.
62. Stracher A. Calpain inhibitors as therapeutic agents in nerve and muscle degeneration. Ann NY Acad Sci 1999; 884:52-59.
63. Spencer MJ, Mellgren RL. Overexpression of a calpastatin transgene in mdx muscle reduces dystrophic pathology. Hum Mol Genet 2002; 11:2645-2655.
64. Leijendekker WJ, Passaquin AC, Metzinger L et al. Regulation of cytosolic calcium in skeletal muscle cells of the mdx mouse under conditions of stress. Br J Pharmacol 1996; 118:611-616.
65. Tutdibi O, Brinkmeier H, Rudel R et al. Increased calcium entry into dystrophin-deficient muscle fibres of MDX and ADR-MDX mice is reduced by ion channel blockers. J Physiol 1999; 515:859-68.
66. DeLuca A, Pierno S, Liantonio A et al. Alteration of excitation-contraction coupling mechanism in extensor digitorum longus muscle fibres of dystrophic mdx mouse and potential efficacy of taurine. Br J Pharmacol 2001; 132:1047-1054.
67. Partridge TA. Myoblast transplantation. Neuromuscul Disord 2002; 12:S3-S6.
68. Partridge TA. Stem cell route to neuromuscular therapies. Muscle Nerve 2003; 27:133-141.
69. Chamberlain JS. Gene therapy of muscular dystrophy. Hum Mol Genet 2002; 11:2355-2362.
70. Khurana TS, Davies KE. Pharmacological strategies for muscular dystrophy. Nat Rev Drug Discov 2003; 2:379-390.

CHAPTER 16

Cell Adhesion and Signalling in the Muscular Dystrophies

Steve J. Winder

Abstract

Many of the muscular dystrophies are caused by defects in proteins involved in maintaining connections between the cytoskeleton and extracellular matrix-cell adhesion complexes. Cell adhesion complexes in many other systems are associated with major signalling pathways involved in regulating cell architecture, morphology and survival. This chapter explores the links between defects in cell adhesion attributable to muscular dystrophy and downstream signalling pathways that may offer novel and more accessible routes for therapy.

Introduction

The loss of cell-matrix adhesion as a unifying theme in 3 of the muscular dystrophies for which the molecular defect was known at that time, namely; Duchenne muscular dystrophy (DMD), congenital muscular dystrophy (CMD) and a limb-girdle muscular dystrophy (LGMD 2C) was highlighted by Kevin Campbell in a review in 1995.[1] Not only can this theme now be applied to many more muscular dystrophies, but possibly also to include signalling cascades associated with cell-matrix adhesion. Ever since the identification of mutations in the DMD gene as the cause of Duchenne muscular dystrophy[2] and the complete sequence of the DMD gene product dystrophin was elucidated,[3-5] much emphasis has been placed on the role of this large cytoskeletal linker protein in the maintenance of muscle cell integrity, for example see.[6-8] The majority view has been that dystrophin performs a mechanical role in the maintenance of sarcolemmal integrity, connecting the actin cytoskeleton[9,10] primarily to laminin in the extracellular matrix[11] via dystroglycan which is anchored in the muscle cell membrane.[12] Whilst a simple role in physically connecting cytoskeleton to extracellular matrix is undoubtedly part of the story, there is increasing evidence to suggest that perturbations in signalling pathways also play a significant role in the aetiology of the muscular dystrophies. Moreover, with the acceptance of the dystrophin glycoprotein complex (DGC) as a specialised cell adhesion complex, more emphasis has been placed on searching for adhesion-mediated signalling cascades associated with the DGC, based largely on the paradigm set by the integrins. Indeed, as we shall see, there is even the potential for cross-talk or functional replacement between the dystroglycan and integrin adhesion systems.

Cell Adhesion: Dystroglycans vs. Integrins

In adult skeletal muscle the primary constituents of the basement membrane are laminin, collagen IV, perlecan and nidogen-1.[13] Laminin is most likely to be the major cell adhesive protein found in basement membranes, and the principal laminin receptors in muscle are the DGC and integrin $\alpha7\beta1$.[14] The role of dystroglycan has been discussed extensively elsewhere in this book (Chapters 1, 6, 7). It is important to mention here, however, the contribution of dystroglycan to muscle fibre adhesion. Loss of dystroglycan function in muscle leads to a dystrophic phenotype.[15-18] This is assumed to arise because of loss of connectivity between laminin or other laminin G (LG) domain proteins in the extracellular matrix and dystrophin in the subsarcolemmal cytoskeleton leads to sarcolemmal instability and dystrophy. However, a more detailed deletion of the cytoplasmic domain of β-dystroglycan gave rise to a slightly different conclusion.[19] Mutant mice lacking the last 15 amino acids in β-dystroglycan, the dystrophin binding site,[20,21] did not result in a dystrophic phenotype,[19] suggesting either a contribution from other proteins such as plectin or ezrin in maintaining the connection to the actin cytoskeleton (see below), or alternative binding sites for dystrophin in dystroglycan as has been suggested previously.[22] However, when the entire cytoplasmic domain was deleted, a severe dystrophic phenotype was observed[19], presumably as this would eliminate any cytoskeletal connection or intracellular signalling role for dystroglycan. On the other hand however, and given the success in overexpressing proteins to compensate for loss of function in dystrophic muscle, it is somewhat surprising that overexpression of dystroglycan in *mdx* mouse muscle did not improve the dystrophic phenotype in these animals.[23] As the basement membrane forms correctly in muscle lacking dystroglycan,[16,17] this would tend to suggest that there is no endogenous protein(s), or enough endogenous protein(s) that can efficiently bind to the extra dystroglycan to make the connection to the actin cytoskeleton. Alternatively, simple overexpression of dystroglycan, which is central to the adhesion role of the DGC and therefore any signalling associated with it, may not be sufficient to restore that signalling function. Recent evidence also suggests cooperativity between extracellular matrix protein receptors and organisation of the extracellular matrix. Integrins act in concert with dystroglycan in the organisation of the extracellular matrix in muscle (reviewed in ref. 14) and in particular formation of the muscle basal lamina.[24,25] Dystroglycan appears to bind initially to laminin followed by laminin association with β1 integrins. A network of laminin, dystroglycan and integrins then undergoes rearrangements with associated reorganisation of the actin cytoskeleton.[24,25] As a result, a reciprocal distribution of dystroglycan and integrin arises with corresponding reciprocal arrangements in the cytoskeletal elements that bind to these transmembrane receptors. Perlecan, another LG domain-containing ligand for dystroglycan, has also been implicated in the process. β1 integrins and perlecan appear to be required for the overall assembly of the laminin matrix after it has been bound to the cell by dystroglycan.[25] All these studies clearly point to a role for cell adhesion in maintenance of muscle integrity. Furthermore, depending upon the particular cell adhesion receptors employed, then different aspects of the cytoskeleton are also involved in binding to these receptors. Some of the molecular mechanisms that underlie the interactions between dystroglycan and laminin and between laminin and integrins are gradually becoming better understood. As suggested above, perlecan co-operates with laminin in the establishment of a dystroglycan and integrin containing lattice on the surface of muscle cells.[25] The failure of the formation of this primary laminin scaffold is thought to be the main cause of congenital muscular dystrophy (CMD), which arises through mutations in the laminin α2 chain (see Chapter 6). There also appears to be a reciprocal and partially compensatory upregulation of α7β1 integrins or laminin chains where either one is absent or reduced (see Chapter 6 and ref. 26 for review). The localisation of integrin subunits to regions of cell adhesion in muscle, i.e., the myotendinous junction (MTJ) and costameres as well as to the neuromuscular junction (NMJ), strongly support a role for integrins in maintaining adhesion links in muscle. It comes as no surprise therefore that deletion of individual integrin subunits in mice leads to dystrophic phenotypes. From experiments conducted on chimeric mice or in embryonic stem cells, it is clear that there is a role for integrins in the organisation of both basement membrane and cytoskeleton in muscle cells. Integrin α5-deficient mice are embryonic lethal, whereas mice chimeric for integrin α5, with a high level of α5 integrin-deficient muscle fibres, develop a very early muscular dystrophy phenotype.[27] Mutations in the muscle-specific α7 integrin in humans are the cause of a congenital myopathy.[28] The disruption of the integrin α7 subunit in mice, contrary to expectation, did not cause lethality, however mice did exhibit a novel form of muscular dystrophy restricted to soleus and diaphragm muscle,[29] and a more general defect in MTJ architecture.[30,31] Conversely overexpression of α7 integrin in the severely dystrophic *mdx*/UTRN-/- mouse,[32,33] unlike dystroglycan overexpression in the *mdx* mouse,[23] improved viability and considerably improved the overall phenotype of these mice.[34] The success of this approach is likely to be due to the different nature of the cytoskeletal linkages employed by the two receptor types. It is clear that there is cooperation between extracellular matrix receptors at the level of extracellular matrix assembly and organisation, but to what extent there is crosstalk at the signalling level and how signalling events contribute to muscular dystrophy phenotypes is less clear. These will be discussed in more detail below.

Tyrosine Kinase Signalling

If one follows the integrin paradigm, then a key event in both the assembly and disassembly of adhesion structures is the tyrosine phosphorylation of proteins involved in mediating the link between adhesion receptor and cytoskeleton, see refs. 35-38 for reviews. Accordingly, various components of the DGC and associated and related proteins have been demonstrated to be phosphorylated on tyrosine residues, including; dystrobrevin,[39-41] dystroglycan,[42-44] dystrophin,[45] sarcoglycans,[46] syntrophin,[45] and utrophin.[42] However, the majority of these studies have simply documented the presence of phospho-tyrosine in the various proteins and provide little mechanistic insight into the functional consequences of the phosphorylation nor do they address the particular kinase or signalling cascade associated with the regulatory event. The presence of dystrobrevin isoforms at NMJ and MTJ that possess a carboxy-terminal tail that can be phosphorylated on tyrosine, has lead to the suggestion that tyrosine phosphorylation of dystrobrevin is important for NMJ assembly,[39,40,47,48] see Chapter 5. The kinase responsible remains to be identified however, as agrin activation of MuSK does not appear to lead to dystrobrevin phosphorylation.[40,48] Perhaps the best documented regulatory event is the tyrosine phosphorylation of β-dystroglycan on tyrosine either in muscle[43] or non-muscle cells.[42] In either case adhesion-dependent phosphorylation of tyrosine 892 within the WW domain interaction motif of the cytoplasmic tail of β-dystroglycan was demonstrated to negatively regulate the interaction between β-dystroglycan and dystrophin or utrophin. Interestingly, phosphorylation at the same site had no effect on the ability of caveolin-3 to interact with β-dystroglycan,[49,50] suggesting the potential for a tyrosine phosphorylation-dependent switch between caveolin-3 and dystrophin binding to β-dystroglycan.[49] Whilst in non-muscle cells at least, there appeared to be a laminin-dependent adhesion-mediated component to dystroglycan tyrosine phosphorylation.[42] However, apart from apparently eliminating focal adhesion kinase (FAK), the precise signalling pathways involved were not further delineated. Lisanti and co-workers went on to demonstrate that phosphorylation of tyrosine 892 in dystroglycan was able to recruit other SH2 domain-containing proteins including a number of adaptor proteins such as NCK and SHC and tyrosine kinases including Src and Fyn.[44] Furthermore, β-dystroglycan was shown also to be phosphorylated on tyrosine by Src, both in vitro and in cells expressing active Src, whereas FAK and Src family kinases Yes and Fyn were not able to phosphorylate β-dystroglycan.[44] Rather surprisingly, unlike the non-phosphorylated protein which can be found localised to sites of cell adhesion,[51-54] in both non-muscle and muscle cells, tyrosine-phosphorylated β-dystroglycan localised to an endosomal compartment,[55] possibly as part of a mechanism involved in recycling the (de)phosphorylated dystroglycan back to the cell surface.

Studies in rat myotubes revealed an association between the DGC and α5β1 integrin and other focal adhesion proteins, in this case *including* FAK. α5β1 integrins were shown to localise with dystrophin during muscle development.[56] Additionally, an integrin- and adhesion-dependent tyrosine phosphorylation of the DGC components α- and γ-sarcoglycan was observed in this system, thought to be mediated by Src or FAK.[46] It has also been demonstrated that a dystroglycan complex isolated from bovine brain synaptosomes contained FAK and that DGC components were phosphorylated on tyrosine.[57] Furthermore, targeted deletion of FAK in brain tissues resulted in abnormalities similar to type II cobblestone lisencephally,[58] a phenotype observed in congenital muscular dystrophy. This and other observations provide evidence for an inside-out signalling role for FAK in basement membrane assembly and further suggest that FAK signalling may be altered in muscular dystrophies involving perturbations of the dystroglycan-laminin axis. A connection between the DGC and signalling is also suggested by the observation that the adaptor protein Grb2 is able to associate directly with β-dystroglycan via an SH3-mediated interaction[59] which in turn could interact with phosphorylated FAK via its SH2 domain. A ternary complex comprising β-dystroglycan:Grb2:FAK has yet to be identified however,[57] as has any other binding partner for a β-dystroglycan:Grb2 complex in this context (but see section on GTPase signalling below). Studies in myotubes examining the integrin- and dystroglycan-dependent polymerisation of laminin also revealed a dependency on tyrosine phosphorylation,[24] but whether this was associated with the specific phosphorylation of integrin or dystroglycan complexes, or both, was not determined.

Serine/Threonine Kinase Signalling

The possibility that components of the DGC are post-translationally modified by phosphorylation was first raised by Ozawa and colleagues as a result of the identification of multiple charged species in two-dimensional gels of purified DGC.[60] Subsequent analyses however, have been mainly confined to the demonstration of the in vitro phosphorylation of this or that DGC component by one or other kinase with little functional or in vivo evidence for the existence or consequences of pathway activation in muscle. The majority of effort has been expended on demonstrating the phosphorylation of dystrophin itself. Most reports have at least attempted to identify or demonstrate the presence of co-purifying kinases with the DGC, or at least in a heavy microsome or WGA purified fraction of muscle, but many have fallen short of defining a functional role for the pathway in muscle. One of the earliest reports suggested that several S/T kinases including cAMP- cGMP- and Ca^{2+}/calmodulin-dependent kinases were present in microsomal fractions of muscle and could phosphorylate dystrophin.[61] Given the apparent importance of calcium handling in dystrophic muscle (see Chapter 15) Ca^{2+}/calmodulin-dependent kinase has been studied in somewhat more detail.[62,63] As such the carboxy-terminus of dystrophin has been defined as a region that is phosphorylated by Ca^{2+}/calmodulin-dependent kinase and also dephosphorylated by the Ca^{2+}/calmodulin-dependent phosphatase calcineurin.[63] In a series of papers from the Michalak group, the carboxy-terminal region of dystrophin was also shown to be phosphorylated by p34^{cdc2} protein kinase, p44 MAP kinase, and sequentially phosphorylated by casein kinase 2 and GSK-3,[64-66] but functional consequences of any of these phosphorylation events remain to be elucidated. Other approaches such as yeast two-hybrid screening have also been used to identify novel signalling molecules associated with DGC components. In particular the syntrophins, or more specifically the PDZ domains of syntrophins, have been shown to interact with a number of signalling proteins including NOS, voltage-gated sodium channel, aquaporin and kinases (see Chapter 4). Of the kinases found to be associated with syntrophin are a family of S/T kinases of unknown function termed microtubule-associated or syntrophin-associated S/T kinases,[67] and perhaps more intriguingly stress-activated protein kinase 3 (also known as ERK6).[68] The latter finding is interesting for two reasons, on one hand the PDZ-mediated interaction between syntrophin α1 and SAPK3 is novel in that it is the syntrophin PDZ-SAPK3 interaction that is required for the binding and subsequent phosphorylation of syntrophin by SAPK3[68] and also because SAPK3 (ERK6) is a major regulator of muscle cell differentiation.[69-71] ERK6 has apparently little role in myoblast proliferation but is required specifically for myoblast fusion and differentiation into myotubes. Two other members of the MAP kinase family, MEK and its downstream kinase ERK1, were also shown to interact with β-dystroglycan,[54] and were differentially targeted to either membrane ruffles or adhesion structures respectively. Dystroglycan was not a substrate for either kinase, and did not affect the activity of the kinases, suggesting a simple scaffold or targeting role for dystroglycan. The possible role of MAP kinases in muscular dystrophy will be discussed in more detail below.

GTPase Signalling

Rho family GTPases are best known for their roles in the regulation of the actin cytoskeleton,[72] but increasingly have been found to be involved in more diverse areas of cell function including regulation of kinase, cell polarity and cell survival pathways to name but a few, reviewed in.[73] Evidence is also emerging of a role for Rho GTPases in skeletal muscle signalling, and in particular, in relation to the DGC and muscular dystrophy. Data from studies on dystroglycan and utrophin in non-muscle cells suggests roles for these proteins in localising activated Cdc42 to membrane sites where it exerts effects on actin organisation and cell motility respectively[74,75] (and Holzfeind, Bittner and Winder unpublished). Deletion of another large cytoskeletal protein – plectin, which is associated with epidermolysis bullosa simplex with muscular dystrophy[76] and can bind to dystroglycan, dystrophin and utrophin (Winder and Wiche unpublished), also leads to aberrant GTPase signalling in fibroblast cells.[77] To date the role of dystroglycan, dystrophin, utrophin and plectin in GTPase signalling in muscle have not been elucidated. However, experimentally induced muscle atrophy in rats, not only leads to a reduction in DGC components including dystroglycan and dystrophin, but also to a perturbation in Rho GTPase activity/signalling[78] suggesting a causative link between dystroglycan depletion and altered GTPase signalling. As discussed above, dystroglycan has been demonstrated to interact with Grb2.[59] In association with integrins or receptor tyrosine kinases, Grb2 is also able to associate with the GTP exchange factor (GEF) known as SOS[79-83] which in turn can activate small GTPases, principally Ras. But as suggested above, no partner for a dystroglycan:Grb2 complex has so far been found. Jarret and colleagues however, identified syntrophin as a Grb2 binding protein,[84] and in a further study this group went on to propose a signalling cascade acting through syntrophin, Grb2 and SOS resulting in Rac activation leading to an increase in jun kinase (JNK) phosphorylation.[85] These data however, seem to be at odds with their own earlier findings where they previously demonstrated a loss of DGC components including dystrophin and dystroglycan, but not syntrophin, as a consequence of experimentally induced atrophy.[78] This might be expected to lead to a reduction in Rac activity if the pathways described above were functional, instead they found that Rac activity was not altered, possibly because syntrophin levels were not reduced, whereas Ras and Cdc42 activity were reduced. In this case it is difficult to rationalise how syntrophin can still maintain Rac activity if it no longer has a connection to laminin through dystrophin and dystroglycan.

From our own work it is clear that dystroglycan and utrophin are able to recruit activated Cdc42 to peripheral sites in non-muscle cells,[75,86] and dominant-negative Cdc42, but not Rho or Rac, prevents dystroglycan-mediated regulation of the actin cytoskeleton.[74] Furthermore, we have identified downstream components of the laminin-dystroglycan-Cdc42 signalling pathway including the p21-activated kinase PAK1, implicated in the JNK signalling pathways (discussed below and Fig. 1), and ezrin a member of the Ezrin-Radixin-Moesin (ERM) family of proteins. ERM proteins are known to also recruit components of the Rho GTPase signalling machinery including a RhoGEF and Rho GDP dissociation inhibitor (Rho GDI)[87,88] suggesting a possible mechanism for the dystroglycan-induced activation of Cdc42. Furthermore these signalling pathways can be activated in the functional absence of dystrophin or utrophin, i.e., in a utrophin null fibroblast,[74] suggesting that they could operate in the context of a DMD muscle fibre and could also explain maintenance or Rac activity in atrophied muscle where DGC components are reduced.[78]

MAP Kinase Pathways

The role of the mitogen-activated protein (MAP) kinase pathways in dystrophic muscle have so far received relatively little attention. Given the importance of the MAP kinases in cell survival signalling, this shortcoming is likely to soon be rectified. Despite their name, mitogen-activated protein kinases can be activated not only by mitogens, but by a host of stimuli including cell adhesion, growth factors, cytokines and cellular stress, see Figure 1 and[89,90] for reviews. Equally they have a plethora of roles in the cell, from activation of transcription,

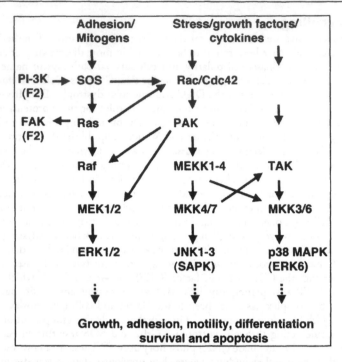

Figure 1. Crosstalk between mitogen-activated protein kinase cascades. Not all components are shown and an arrow doesn't necessarily imply direct activation. Horizontal arrows with numbers in parentheses refer to links to Figure 2 via the indicated component.

regulation of proliferation, modulation of cell motility and not least, cell survival. This latter point may be of most relevance given that apoptosis is associated with many muscular dystrophies.[91,92] It is worth noting however, that the mechanisms by which apoptosis is induced in the muscular dystrophies vary. The sarcoglycans being a case in point: sarcoglycan deficiency in mice gives rise to muscular dystrophy and associated apoptosis, but without loss of the physical link between actin and extracellular matrix mediated by dystrophin and dystroglycan,[92] moreover by virtue of the presence of dystrophin and dystroglycan, sarcoglycan deficient muscles are not susceptible to mechanical injury.[93] This raises two possible mechanisms. Sarcoglycan (which is also lost from the sarcolemma in DMD) may be a key regulator of survival signals and its loss either as a consequence of dystrophin mutation in DMD or directly as in the case of LGMD results in aberrant signalling and apoptosis. Alternatively, there are distinct or overlapping pathways mediated via different mechanisms involved in each muscular dystrophy. The latter, though not impossible, seems less likely. So what can we discern from what is known about MAP kinase signalling in muscular dystrophy? Experimentally, in vitro disruption of the α-dystroglycan interaction with laminin in myotubes in culture results in loss of survival signals mediated via the phosphinositide 3-kinase (PI3K)/ protein kinase B (AKT) pathway[94] (Fig. 2) and induction of apoptosis.[15,94] In studies in *mdx* mice, the more severely dystrophic *mdx*: MyoD-/-[95] and *mdx*/UTRN-/-[32] mice and the caveolin-3 null mouse,[96] there appears to be a range of observed differences in activation of MAP kinase activities in skeletal, diaphragm and cardiac muscles. In skeletal and diaphragm muscles of both *mdx* and *mdx*:MyoD-/-, there appears to be an increase in JNK1 activation, with little or no change in ERK or p38 MAP kinase activity.[97] Furthermore the authors were able to demonstrate an improvement in myofiber survival on expression of a JNK inhibitory protein. The JNK activator MKK7 led to reduced myotube differentiation in transfected myoblasts substantiating a role for JNK in the dystrophic

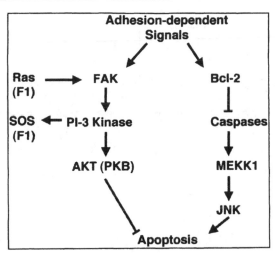

Figure 2. Adhesion-dependent signals to cell survival pathways. Arrows indicate activating steps, flat ended lines inhibitory steps. Consequently a continuous adhesion signal would maintain a constant inhibitory effect on apoptosis. Loss of adhesion signals leads to anchorage-dependent apoptosis or anoikis. Horizontal arrows relate to connections with the indicated components in Figure 1.

phenotype. Somewhat at odds with these findings however, is the observation that there was a muscle-specific *activation* of JNK associated with the DGC in normal myotubes.[85] The proposed pathway involved laminin engagement by dystroglycan with recruitment of a Grb2:Sos1 complex with activation of Rac1, and via the Cdc42/Rac effector p21 activated kinase 1 (PAK1) to JNK. Though it is worth noting that not all elements of this cascade were delineated in myotubes, some were only identified in myoblasts, and the existence of parts of the pathway were inferred rather than directly demonstrated. Nonetheless, if as suggested there is a DGC-mediated activation of Rac signalling leading to the phosphorylation of JNK, one would expect a decrease, rather than increase in JNK activity, if components of the signalling pathway were reduced or missing as the DGC is in the *mdx* mouse. In defence of this hypothesis however, activation of JNK could occur through different pathways, e.g., via crosstalk from the p38 pathway (see Fig. 1). The situation in cardiac muscle appears more complex: reports suggest little or no increase in JNK activation in *mdx* mouse heart,[98-100] but a more robust activation of JNK in the more severely affected *mdx*:MyoD-/- and *mdx*/UTRN-/- mice.[98,100] By contrast, p38 MAPK activity was not found to be altered in *mdx* but dramatically increased in *mdx*:MyoD-/- heart,[98] whereas Nakamura and colleagues found significantly elevated levels of activated p38 MAPK in *mdx*, exercised *mdx* and *mdx*/UTRN-/- heart.[99,100] The reasons for these differences are not entirely clear, though as one might expect, there does appear to be a trend toward increased activation of JNK and p38 MAPK with increasing severity of dystrophy and increasing muscle damage through exercise.

The situation regarding ERK activation is also unclear. Kolodziejczyk et al[97] claim no difference in ERK activity in either cardiac or skeletal muscle of *mdx* or *mdx*:MyoD-/- mice, despite the clear presence of a slower migrating ERK1 band (likely to be active ERK1) in the skeletal muscle sample from 5 month old *mdx* mouse (Fig. 1B in Ref. 97). Furthermore, significantly elevated levels of active ERK were found in the hearts of non-exercised and exercised hearts of *mdx* mice as compared to either non-exercised or exercised controls.[100] ERK activity was also found to be increased in the hearts of caveolin-3 null mice[101] probably because caveolin is able to interact with ERK and suppress its activation.[102] The interaction of caveolin-3 with β-dystroglycan[50] and the observed ability of β-dystroglycan to interact with, but not regulate, MEK and ERK[54] provide potential clues as to how DGC signalling to ERK might be mediated

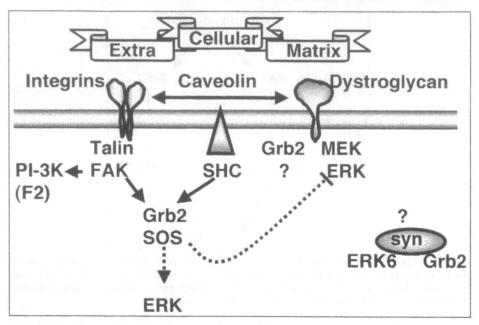

Figure 3. Possible interrelationships between integrins, caveolin and dystroglycan involving signalling to and/or binding of ERK MAP kinase cascade components. Caveolin can interact with integrins or dystroglycan, and in so doing may effectively act to segregate the ERK cascade into signalling and non-signalling domains. Dystroglycan may also interact and sequester ERK components directly. ? represent unknown or incomplete pathways/interactions, syn=syntrophin.

(Fig. 3). ERK activation downstream of MEK, has previously been associated with integrin engagement, but dystroglycan has also been shown to be found localised to integrin-containing focal adhesion structures in a number of cell types.[51,52] Moreover, cell adhesion induces the tyrosine phosphorylation of β-dystroglycan,[42] which in turn leads to the recruitment of SH2 domain–containing signalling and adaptor proteins such as Src, FAK and Shc.[44,57] Thus tyrosine phosphorylation of dystroglycan recruits components of several kinase cascades making it a multifunctional transducer of adhesion-dependent signalling. It has recently been suggested that in certain cells dystroglycan can act antagonistically to other adhesion molecules, such as integrins, and suppress the activation of downstream kinase cascades.[103] Dystroglycan was able to suppress the activation of the ERK-MAP kinase cascade induced by the interaction of integrin α6β1 with laminin,[103] further suggesting the possibility of an interaction between β-dystroglycan and components of the ERK-MAP kinase cascade. The direct interaction of dystroglycan with MEK might be sufficient to sequester and inactivate or prevent the activation of MEK and therefore reduce the activity of its downstream kinase ERK.[54] Consistent with this hypothesis is the observation that ERK activity is increased in the hearts of *mdx*[100] and caveolin-3 null[101] mice suggesting that alterations in dystroglycan level or function may have an uncoupling effect on MEK deregulation leading to increased ERK signalling. Furthermore it has previously been reported that the SH2/SH3 adaptor protein Grb2 also associates with β-dystroglycan.[59,104] Grb2 is capable of linking SOS, a GEF for Ras and the upstream activator of the Raf-MEK-ERK-MAP kinase cascade, to membrane receptors.[79-81]

An essential part of dystroglycan function is its interaction with laminin, where it binds to the LG domains of the E3 fragment.[105] However, the dystroglycan-laminin interaction requires the presence of correctly glycosylated dystroglycan.[11] Mutations in glycosyltransferase genes such as LARGE[106] and Fukutin and related proteins[107,108] affect dystroglycan glycosylation and hence its ability to interact with laminin which in turn leads to muscular dystrophy (see Chapter 7). The ability of dystroglycan to inhibit laminin-induced $\alpha6\beta1$ integrin activation of the ERK MAP kinase cascade[103] and the association of dystroglycan with components of the ERK MAP kinase cascade, namely MEK and ERK, may play an important role in normal muscle homeostasis. Perturbation of dystroglycan-laminin interactions by aberrant glycosylation could therefore lead to an imbalance in ERK signalling that could affect cell cycle progression and gene expression thus contributing to the muscular dystrophy phenotype.

Conclusions

Our understanding of the role of cell adhesion and signalling in the muscular dystrophies is at a relatively early stage. But given the enormous advances that have been made in recent years in understanding the molecular genetics of muscular dystrophies and with increasing use of animal and cell models, it is likely that in the near future we will see an explosion of information in relation to understanding alterations in signalling cascades associated with muscular dystrophy. Hopefully a better understanding of aberrant signalling in muscular dystrophy will provide us with more easily accessible therapeutic targets for this devastating and difficult to treat group of diseases.

Acknowledgements

I am grateful to members of my laboratory for critical reading of the manuscript. Work from my group cited in this article was funded by the MRC, Muscular Dystrophy Campaign and Wellcome Trust.

References

1. Campbell KP. Three muscular dystrophies: loss of cytoskeleton-extracellular matrix linkage. Cell 1995; 80:675-679.
2. Kunkel LM, Monaco AP, Middlesworth W et al. Specific cloning of DNA fragments absent from the DNA of a male patient with an X chromosome deletion. Proc Natl Soc Sci USA 1985; 82:4778-4782.
3. Hoffman EP, Jr. RHB, Kunkel LM. Dystrophin: the protein product of the Duchenne muscular dystrophy locus. Cell 1987; 51:919-28.
4. Koenig M, Hoffman EP, Bertelson CJ et al. Complete cloning of the Duchenne muscular-dystrophy (DMD) cDNA and preliminary genomic organization of the DMD gene in normal and affected individuals. Cell 1987; 50:509-517.
5. Koenig M, Monaco AP, Kunkel LM. The complete sequence of dystrophin predicts a rod-shaped cytoskeletal protein. Cell 1988; 53:219-226.
6. Ervasti JM. Costameres: the Achilles' Heel of Herculean Muscle. J Biol Chem 2003; 278:13591-13594.
7. Spence HJ, Chen Y-J, Winder SJ. Muscular dystrophies, the cytoskeleton and cell adhesion. Bioessays 2002; 24:542-552.
8. Winder SJ. The membrane-cytoskeleton interface: the role of dystrophin and utrophin. J Muscle Res Cell Motil 1997; 18:617-629.
9. Hemmings L, Kuhlman PA, Critchley DR. Analysis of the actin-binding domain of α-actinin by mutagenesis and demonstration that dystrophin contains a functionally homologous domain. J Cell Biol 1992; 116:1369-1380.
10. Way M, Pope B, Cross RA et al. Expression of the N-terminal domain of dystrophin in E. coli and demonstration of binding to F-actin. FEBS Lett 1992; 301:243-245.
11. Ervasti JM, Campbell KP. A role for the dystrophin glycoprotein complex as a transmembrane linker between laminin and actin. J Cell Biol 1993; 112:809-823.
12. Ibraghimov-Beskrovnaya O, Ervasti JM, Leveille CJ et al. Primary structure of dystrophin-associated glycoproteins linking dystrophin to the extracellular matrix. Nature 1992; 355:696-702.
13. Timpl R, Brown JC. Supramolecular assembly of basement membranes. Bioessays 1996; 18:123-132.
14. Mayer U. Integrins: Redundant or Important Players in Skeletal Muscle? J Biol Chem 2003; 278:14587-14590.
15. Brown SC, Fassati A, Popplewell L et al. Dystrophic phenotype induced in vitro by blockade of muscle α-dystroglycan-laminin interaction. J Cell Sci 1999; 112:209-216.

16. Cote PD, Moukhles H, Lindenbaum M et al. Chimeric mice deficient in dystroglycans develop muscular dystrophy and have disrupted myoneural synapses. Nature Genet 1999; 23:338-342.
17. Parsons MJ, Campos I, Hirst EMA et al. Removal of dystroglycan causes severe muscular dystrophy in zebrafish embryos. Development 2002; 129:3505-3512.
18. Cohn RD, Henry MD, Michele DE et al. Disruption of DAG1 in differentiated skeletal muscle reveals a role for dystroglycan in muscle regeneration. Cell 2002; 110:639-648.
19. Kusano H, Lee J, Flanagan J et al. The Cytoplasmic Domain of Dystroglycan Is Essential for the Cytoskeleton-muscle Membrane Linkage and Prevention of Muscular Dystrophy. Mol Biol Cell 2003; 14:44a.
20. Jung D, Yang B, Meyer J et al. Identification and characterization of the dystrophin anchoring site on β-dystroglycan. J Biol Chem 1995; 270:27305-27310.
21. Rosa G, Ceccarini M, Cavaldesi M et al. Localisation of the dystrophin binding site at the carboxyl terminus of β-dystroglycan. Biochem Biophys Res Commun 1996; 223:272-277.
22. Rentschler S, Linn H, Deininger K, Bedford MT et al. The WW domain of dystrophin requires EF-hands region to interact with β-dystroglycan. Biological Chemistry 1999; 380:431-442.
23. Hoyte K, Jayasinha V, Xia B, Martin PT. Transgenic Overexpression of Dystroglycan Does Not Inhibit Muscular Dystrophy in mdx Mice. Am J Pathol 2004; 164:711-718.
24. Colognato H, Winkelmann DA, Yurchenco PD. Laminin polymerisation induces a receptor-cytoskeleton network. J Cell Biol 1999; 145:619-631.
25. Henry M, Satz J, Brakebusch C et al. Distinct roles for dystroglycan, β1 integrin and perlecan in cell surface laminin organisation. J Cell Sci 2001; 114:1137-1144.
26. Cohn R, Campbell K. Molecualr basis of muscualr dystrophies. Muscle Nerve 2000; 23:1456-1471.
27. Taverna D, Disatnik M-H, Rayburn H et al. Dystrophic muscle in mice chimeric for expression of α5 integrin. J Cell Biol 1998; 143:849-859.
28. Hayashi YK, Chou F-L, Engvall E et al. Mutations in the integrin alpha 7 gene cause congenital myopathy. Nature Gemet 1998; 19:94-97.
29. Mayer U, Saher G, Fassler R et al. Absence of integrin alpha-7 causes a novel form of muscular dystrophy. Nature Genet 1999; 17:318-323.
30. Miosge N, Klenczar C, Herken R et al. Organization of the myotendinous junction is dependent on the presence of alpha7beta1 integrin. Lab Invest 1999; 79:1591-1599.
31. Nawrotzki R, Willem M, Miosge N et al. Defective integrin switch and matrix composition at alpha 7-deficient myotendinous junctions precede the onset of muscular dystrophy in mice. Hum Mol Genet 2003;12:483-495.
32. Deconinck AE, Rafael JA, Skinner JA et al. Utrophin-dystrophin-deficient mice as a model for Duchenne muscular dystrophy. Cell 1997; 90:717-727.
33. Grady RM, Teng H, Nicholl MC et al. Skeletal and cardiac myopathies in mice lacking utrophin and dystrophin: a model for Duchenne muscular dystrophy. Cell 1997; 90:729-738.
34. Burkin DJ, Wallace GQ, Nicol KJ et al. Enhanced Expression of the α7β1 Integrin Reduces Muscular Dystrophy and Restores Viability in Dystrophic Mice. J Cell Biol 2001; 152:1207-1218.
35. Brakebusch C, Fassler R. New EMBO Member's Review: The integrin-actin connection, an eternal love affair. EMBO J 2003; 22:2324-2333.
36. Flier Avd, Sonnenberg A. Function and interactions of integrins. Cell Tissue Res 2001; 305:285-298.
37. Hynes RO. Integrins: versatility, modulation and signaling in cell adhesion. Cell 1992; 69:11-25.
38. Schlaepfer DD, Hunter T. Integrin signalling and tyrosine phosphorylation: just th e FAKs? Trends Cell Biol 1998; 8:151-157.
39. Balasubramanian S, Fung ET, Huganir RL. Characterization of the tyrosine phosphorylation and distribution of dystrobrevin isoforms. FEBS Lett 1998; 432:1330140.
40. Nawrotzki R, Loh NY, Ruegg MA et al. Characterisation of alpha-dystrobrevin in muscle. J Cell Sci 1998; 111:2595-2605.
41. Wagner KR, Cohen JB, Huganir RL. The 87K postsynaptic membrane protein from Torpedo is a protein-tyrosine kinase substrate homologous to dystrophin. Neuron 1993; 10:511-522.
42. James M, Nuttall A, Ilsley JL et al. Adhesion-dependent tyrosine phosphorylation of β-dystroglycan regulates its interaction with utrophin. J Cell Sci 2000; 113:1717-1726.
43. Ilsley JL, Sudol M, Winder SJ. The interaction of dystrophin with β-dystroglycan is regulated by tyrosine phosphorylation. Cell Signal 2001; 13:625-632.
44. Sotgia F, Lee H, Bedford M et al. Tyrosine phosphoryaltion of b-dystroglycan at its WW domain binding motif, PPxY, recruits SH2 domain containing proteins. Biochem 2001; 40:14585-14592.
45. Wagner KR, Huganir RL. Tyrosine and serine phosphorylation of dystrophin and the 58-kDa protein in the postsynaptic membrane of Torpedo electric organ. J Neurochem 1994; 62:1947-1952.
46. Yoshida T, Pan Y, Hanada H et al. Bidirectional signaling between sarcoglycans and the integrin adhesion system in cultured L6 myocytes. J Biol Chem 1998; 273:1583-1590.

47. Blake DJ, Nawrotzki R, Peters MF et al. Isoform Diversity of Dystrobrevin, the Murine 87-kDa Postsynaptic Protein. J Biol Chem 1996; 271:7802-7810.
48. Grady RM, Akaaboune M, Cohen AL et al. Tyrosine-phosphorylated and nonphosphorylated isoforms of {alpha}-dystrobrevin: roles in skeletal muscle and its neuromuscular and myotendinous junctions. J Cell Biol 2003; 160:741-752.
49. Ilsley JL, Sudol M, Winder SJ. The WW domain: linking cell signalling to the membrane cytoskeleton. Cell Signal 2002; 14:183-189.
50. Sotgia F, Lee JK, Das K et al. Caveolin-3 directly interacts with the c-terminal tail of β-dystroglycan: identification of a central WW-like domain within caveolin family members. J Biol Chem 2000; 275:38048-38058.
51. Belkin AM, Smallheiser NR. Localization of cranin (dystroglycan) at sites of call-matrix and cell-cell contact: recruitment to focal adhesions is dependent upon extracellular ligands. Cell Adhesion Commun 1996; 4:281-296.
52. James M, Man Nt, Wise CJ et al. Utrophin-dystroglycan complex in membranes of adherent cultured cells. Cell Motil Cytoskel 1996; 33:163-174.
53. Khurana TS, Kunkel LM, Frederickson AD et al. Interaction of chromosome 6-encoded dystrophin related protein with the extracellular matrix. J Cell Sci 1995; 108:173-185.
54. Spence HJ, Dhillon AS, James M et al. Dystroglycan a scaffold for the ERK-MAP kinase cascade. EMBO Rep 2004; 5:484-489.
55. Sotgia F, Bonuccelli G, Bedford M et al. Localization of Phospho-β-dystroglycan (pY892) to an Intracellular Vesicular Compartment in Cultured Cells and Skeletal Muscle Fibers in Vivo. Biochem 2003; 42:7110-7123.
56. Lakonishok M, Muschler J, Horwitz AF. The α5β1 integrin associates with a dystrophin-containing lattice during muscle development. Devel Biol 1992; 152:209-220.
57. Cavaldesi M, Macchia G, Barca S et al. Association of the dystroglycan complex isolated from bovine brain synaptosomes with proteins involved in signal transduction. J Neurochem 1999; 72:1648-1655.
58. Beggs HE, Schahin-Reed D, Zang K et al. FAK Deficiency in Cells Contributing to the Basal Lamina Results in Cortical Abnormalities Resembling Congenital Muscular Dystrophies. Neuron 2003; 40:501-514.
59. Yang B, Jung D, Motto D et al. SH3 domain-mediated interaction of dystroglycan and Grb2. J Biol Chem 1995; 270:11711-11714.
60. Yamamoto H, Hagiwara Y, Mizuno Y et al. Heterogeneity of dystrophin-associated proteins. J Biochem 1993; 114:132-139.
61. Luise M, Presotto C, Senter L, et al. Dystrophin is phosphorylated by endogenous protein kinases. Biochem J 1993; 293:243-247.
62. Madhavan R, Jarrett HW. Calmodulin-activated phosphorylation of dystrophin. Biochem 1994; 33:5797-5804.
63. Walsh MP, Busaan JL, Fraser ED et al. Characterisation of the recombinant C-terminal domain of dystrophin: phosphorylation by calmodulin-dependent protein kinase II and dephosphorylation by type 2B protein phosphatase. Biochem 1995; 34:5561-5568.
64. Michalak M, Fu SY, Milner RE et al. Phosphorylation of the carboxyl-terminal region of dystrophin. Biochem Cell Biol 1996; 74:431-437.
65. Milner RE, Bussan JL, Holmes CFB et al. Phosphorylation of Dystrophin: The carboxyl-terminal region of dystrophin is a substrate for in vitro phosphorylation by p34^{cdc2} protein kinase. J Biol Chem 1993; 268:21901-21905.
66. Shemanko C, Sanghera JS, Milner RE et al. Phosphorylation of the carboxyl terminal region of dystrophin by mitogen-activated (MAP) kinase. Mol Cell Biochem 1995; 152:63-70.
67. Lumeng C, Phelps S, Crawford G et al. Interactions between beta 2-syntrophin and a family of microtubule-associated serine/threonine kinases. Nature Neurosci 1999; 2:611-617.
68. Hasegawa M, Cuenda A, Spillantini M et al. Stress-activated protein kinase-3 interacts with the PZ domain of α1-syntrophin. J Biol Chem 1999; 274:12626-12631.
69. Bennett AM, Tonks NK. Regulation of Distinct Stages of Skeletal Muscle Differentiation by Mitogen-Activated Protein Kinases. Science 1997; 278:1288-1291.
70. Lechner C, Zahalka MA, Giot J-F et al. ERK6, a mitogen-activated protein kinase involved in C2C12 myoblast differentiation. PNAS 1996; 93:4355-4359.
71. Gredinger E, Gerber AN, Tamir Y et al. Mitogen-activated Protein Kinase Pathway Is Involved in the Differentiation of Muscle Cells. J Biol Chem 1998; 273:10436-10444.
72. Hall A. Rho GTPases and the actin cytoskeleton. Science 1998; 279:509-514.
73. Etienne-Manneville S, Hall A. Rho GTPases in cell biology. Nature 2002; 420:629-635.

74. Chen Y-J, Spence HJ, Cameron JM et al. Direct interaction of β-dystroglycan with F-actin. Biochem J 2003; 375:329-337.
75. Spence HJ, Chen Y-J, Batchelor CL et al. Ezrin-dependent regulation of the actin cytoskeleton by β-dystroglycan. Hum Mol Genet 2004; 13:1657-1668.
76. Smith FJD, Eady RAJ, Leigh IM, et al. Plectin deficiency results in muscular dystrophy with epidermolysis bullosa. Nature Genet 1996; 13:450-457.
77. Andrä K, Nikolic B, Stöcher M, et al. Not just scaffolding: plectin regulates actin dynamics in cultured cells. Genes Dev 1998; 12:3442-3451.
78. Chockalingham PS, Cholera R, Oak SA et al. Dystrophin-glycoprotein complex and Ras and Rho-GTPase signaling are altered in muscle atrophy. Am J Physiol 2002; 283:C500-C511.
79. Lowenstein EJ, Daly RJ, Batzer AG et al. The SH2 and SH3 domain containing protein GRB2 links receptor tyrosine kinases to ras signaling. Cell 1992; 70:431-442.
80. Buday L, Downward J. Epidermal growth factor regulates p21(Ras) through the formation of a complex of receptor, GRB2 adapter protein, and SOS nucleotide exchange factor. Cell 1993; 73:611-620.
81. Egan SE, Geddings BW, Brooks MW et al. Association of SOS Ras exchange protein with GRB2 is implicated in tyrosine kinase signal transduction and transformation. Nature 1993; 363:45-51.
82. Schlaepfer DD, Jones KC, Hunter T. Multiple Grb2-mediated integrin-stimulated signaling pathways to ERK2/mitogen-activated protein kinase: summation of both c-Src- and focal adhesion kinase-initiated tyrosine phosphorylation events. Mol Cell Biol 1998; 18:2571-2585.
83. Schlaepfer DD, Hanks S, Hunter T et al. Integrin-mediated signal transduction linked to Ras pathway by GRB2 binding to focal adhesion kinase. Nature 1994; 372:786-791.
84. Oak SA, Russo K, Petrucci TC et al. Mouse alpha1-syntrophin binding to Grb2: further evidence of a role for syntrophin in cell signaling. Biochem 2001; 40:11270-11278.
85. Oak SA, Zhou YW, Jarrett HW. Skeletal muscle signaling pathway through the dystrophin glycoprotein complex and Rac1. J. Biol. Chem. 2003;278:39287-39295.
86. Holzfeind PJ, Winder SJ, Spence HJ, Davies KE, Bittner RE. Utrophin promotes actin-based cell motility. Proc. Natl Acad Sci USA (submitted).
87. Takahashi K, Sasaki T, Mammoto A et al. Direct interaction of the Rho GDP dissociation inhibitor with ezrin/radixin/moesin initiates the activation of Rho small G protein. J Biol Chem 1997; 272:23371-23375.
88. Takahashi K, Sasaki T, Mammoto A et al. Interaction of radixin with Rho small G protein GDP/GTP exchange protein Dbl. Oncogene 1998; 16:3279-3284.
89. Aplin AE, Howe A, Alahari SK et al. Signal transduction and cell modulation by cell adhesion receptors: the role of integrins, cadherins, immunoglobulin-cell adhesion molecules and selectins. Pharmacol Rev 1998; 50:197-262.
90. Giancotti FG, Tarone G. Positional control of cell fate through joint integrin/receptor protein kinase signaling. Annu Rev Cell Dev Biol 2003; 19:173-206.
91. Matsuda R, Nishikawa A, Tanaka H. Visualisation of dystrophic muscle fibers in mdx mouse by vital staining with Evans blue: evidence of apoptosis in dystrophin-deficient muscle. J Biochem 1995; 118:959-964.
92. Hack AA, Ly CT, Jiang F et al. γ-Sarcoglycan Deficiency Leads to Muscle Membrane Defects and Apoptosis Independent of Dystrophin. J Cell Biol 1998; 142:1279-1287.
93. Hack AA, Cordier L, Shoturma DI et al. Muscle degeneration without mechanical injury in sarcoglycan deficiency. PNAS 1999; 96:10723-10728.
94. Langenbach KJ, Rando TA. Inhibition of dystroglycan binding to laminin disrupts the PI3K/AKT pathway and survival signaling in muscle cells. Muscle Nerve 2002; 26:644-653.
95. Megeney LA, Kablar B, Garrett K et al. MyoD is required for myogenic stem cell function in adult skeletal muscle. Genes Dev 1996; 15:1173-1183.
96. Galbiati F, Engelman JA, Volonte D et al. Caveolin-3 Null Mice Show a Loss of Caveolae, Changes in the Microdomain Distribution of the Dystrophin-Glycoprotein Complex, and T-tubule Abnormalities. J Biol.Chem 2001; 276:21425-21433.
97. Kolodziejczyk SM, Walsh GS, Balazsi K et al. Activation of JNK1 contributes to dystrophic muscle pathogenesis. Curr.Biol 2001; 11:1278-1282.
98. Megeney LA, Kablar B, Perry RLS et al. Severe cardiomyopathy in mice lacking dystrophin and MyoD. PNAS 1999; 96:220-225.
99. Nakamura A, Harrod GV, Davies KE. Activation of calcineurin and stress activated protein kinase/p38-mitogen activated protein kinase in hearts of utrophin-dystrophin knockout mice. Neuromusc Disord 2001; 11:251-259.

100. Nakamura A, Yoshida K, Takeda S et al. Progression of dystrophic features and activation of mitogen-activated protein kinases and calcineurin by physical exercise, in hearts of mdx mice. FEBS Lett 2002; 520:18-24.
101. Woodman SE, Park DS, Cohen AW et al. Caveolin-3 Knock-out Mice Develop a Progressive Cardiomyopathy and Show Hyperactivation of the p42/44 MAPK Cascade. J Biol Chem 2002; 277:38988-38997.
102. Engelman JA, Chu C, Lin A et al. Caveolin-mediated regulation of signaling along the p42/44 MAP kinase cascade in vivo: A role for the caveolin-scaffolding domain. FEBS Lett 1998; 428:205-211.
103. Ferletta M, Kikkawa Y, Yu H et al. Opposing Roles of Integrin α6Aβ1 and Dystroglycan in Laminin-mediated Extracellular Signal-regulated Kinase Activation. Mol Biol Cell 2003; 14:2088-2103.
104. Russo K, Stasio ED, Macchia G et al. Characterisation of the β-dystroglycan-growth factor receptor 2 (Grb2) interaction. Biochemical and Biophysical Research Communications 2000; 274:93-98.
105. Talts JF, Andac Z, Gohring W et al. Binding of the G domains of laminin alpha 1 and alpha 2 chains and perlecan to heparin, sulfatides, alpha -dystroglycan and several extracellular matrix proteins. EMBO J 1999; 18:863-870.
106. Grewal PK, Holzfeind PJ, Bittner RE et al. Mutant glycosytrnsferase and altered glycosylation of a-dystroglycan in the myodystrophy mouse. Nature Genet 2001; 28:151-154.
107. Brockington M, Blake DJ, Prandini P et al. Mutations in the Fukutin-Related Protein Gene (FKRP) Cause a Form of Congenital Muscular Dystrophy with Secondary Laminin 2 Deficiency and Abnormal Glycosylation of -Dystroglycan. Am. J. Hum. Genet 2001; 69:1198-1209.
108. Michele DE, Barresi R, Kanagawa M, et al. Post-translational disruption of dystroglycan-ligand interactions in congenital muscular dystrophies. Nature 2002; 418:417-422.

CHAPTER 17

Gene Therapies for Muscular Dystrophies

Dominic J. Wells

Abstract

Gene therapy for the muscular dystrophies aims to restore the normal biochemistry by either modifying the damaged gene (or mRNA) or by expression of a therapeutic transgene. Animal models, mostly mice, are available for many of the muscular dystrophies and so can be used to test gene therapy strategies. Germ-line gene therapy (transgenic) experiments in mouse models have been valuable in assessing the potential of different therapeutic genes. The majority of gene therapy studies have been conducted in the *mdx* mouse model of the commonest muscular dystrophy, Duchenne muscular dystrophy. Most somatic gene therapy experiments have studied the effects of local administration but for clinical benefit it will be essential to develop suitable systems for treating multiple muscles at once. Immune system responses to the gene vectors and the therapeutic protein are concerns for human clinical trials. Currently the use of antisense oligonucleotides to modify splicing to produce a translatable mRNA, adeno-associated virus gene transfer and regional vascular delivery of plasmid DNA appear to be the most promising gene therapy approaches to the treatment of muscular dystrophies.

Introduction

The development of therapies for muscular dystrophies is a daunting task. There are over 600 muscles in the human body and the majority of these are affected in the various muscular dystrophies, although the extent and pattern of the pathological process varies between the different disorders. In addition, not only are the skeletal muscles affected, but the cardiac and smooth muscle can also be involved and in some patients there is clear evidence of developmental abnormalities in the brain that lead to nonprogressive cognitive defects. Our knowledge of the underlying genetic causes of the various muscular dystrophies has increased rapidly since the identification in 1987 of dystrophin, the protein whose absence or abnormality is responsible for Duchenne muscular dystrophy, DMD, and the related Becker muscular dystrophy, BMD.[1-3] A list of the different muscular dystrophies and the mutant proteins is given in Table 1.[4]

A wide range of therapeutic approaches have been proposed for the treatment of these devastating neuromuscular disorders and they can be divided into three categories: physiotherapy/mechanical assistance, pharmacological and genetic. Developments in respiratory and cardiac management together with the judicious use of corticosteroids have already substantially increased the average lifespan of patients with DMD.[5,6] However, these treatments are palliative and do not correct the underlying biochemical deficit that causes the muscle wasting in these disorders. The aim of gene therapy is to restore the normal biochemistry by either modifying the damaged gene or mRNA or by expression of a therapeutic transgene (Table 2). To date all gene therapy studies have concentrated on the treatment of skeletal muscle as this is the most marked manifestation of these disorders. Although preliminary studies can be conducted in cells in culture, the availability of animal models of the different muscular

Table 1. Animal models of muscular dystrophy

Disease	Inheritance	Gene Locus	Gene Product	Animal Model
Duchenne/Becker MD	XR	Xp21	Dystrophin	*mdx* mouse, GRMD dog
Emery-Dreifuss MD	XR	Xp28	Emerin	-
Emery-Dreifuss MD	AD	1qll	Lamin A/C	-
LGMD 1A	AD	5q31	Myotilin	-
LGMD 1B	AD	1q11	Lamin A/C	Lmna $^{-/-}$ mouse
LGMD 1C	AD	3p25	Caveolin-3	Cav3 $^{-/-}$ mouse
LGMD 1D	AD	6q23	?	-
LGMD 1E	AD	7q32	?	-
LGMD 1F	AD	5q31	?	-
LGMD 2A	AR	15q15	Calpain-3	Capn3 $^{-/-}$ mouse
LGMD 2B	AR	2p13	Dysferlin	SJL mouse
LGMD 2C	AR	13q12	γ-sarcoglycan	Sgcg $^{-/-}$ mouse
LGMD 2D	AR	17q12	α-sarcoglycan	Sgca $^{-/-}$ mouse
LGDM 2E	AR	4q12	β-sarcoglycan	Sgcb $^{-/-}$ mouse
LGMD 2F	AR	5q33	δ-sarcoglycan	Sgcd $^{-/-}$ mouse BIO14.6 hamster
LGMD 2G	AR	17q11	Telethonin	-
LGMD 2H	AR	9q31	TRIM31	-
LGMD 2I	AR	19q13	Fukutin-related protein	-
Miyoshi myopathy	AR	2p13	Dysferlin	SJL mouse
Tibial MD	AD	2q31	?	-
Classical CMD	AR	6q22	Laminin α2	dy/dy mouse
Fukuyama CMD	AR	9q31	Fukutin	-
MDC1C	AR	19q13	Fukutin-related protein	-
α7 integrin CMD	AR	12q13	α7 integrin	Itga $^{-/-}$ mouse
Ulrich CMD	AR	?	Collagen VI α2	-
Walker-Warburg syndrome	AR	?	?	-
Rigid spine CMD	AR	1p35	Selenoprotein N	-
Muscle-eye-brain disease	AR	1p32	POMGnT1	-
Bethlem myopathy	AD	21q22	Collagen VI α1	Col6α1 $^{-/-}$ mouse
Bethlem myopathy	AD	21q22	Collagen VI α2	-
Bethlem myopathy	AD	2q37	Collagen VI α3	-
EB and MD	AR	8q24	Plectin	Plectin $^{-/-}$ mouse
Facioscapulohumeral MD	AD	4q35	?	-
Scapuloperoneal MD	AD	12q21	?	-
Oculopharyngeal	AD	14q11.2	Poly A binding	-

dystrophies has been essential in the development of different therapeutic strategies. Animal models are not available for all the muscular dystrophies and the majority of those that are available are mice (see Table 1). Some of the models, such as the *mdx* and SJL mice, are natural mutants but increasingly gene knockout mice are being developed where natural mutants have not been found.

The majority of gene therapy studies have concentrated on DMD as this is the most common of the muscular dystrophies. Most of the proposed genetic therapies have aimed to restore expression of a functional version of dystrophin. The development of strategies to achieve this goal has relied on the use of the dystrophic *mdx* mouse. The *mdx* mouse is a natural dystrophin

Table 2. A summary of gene therapy approaches and methods

General Approach	Method	Details
Gene repair (DNA)	Chimeroplasts	
	Short fragment homo logous recombination	
mRNA modification	antisense oligonucleotides ribozymes	
Gene replacement	Cell transplants	(see Chapter 19)
	Viral vectors	Adenovirus
		Oncoretrovirus (MMLV)
		Lentivirus
		Adeno-associated virus
		Herpesvirus
	Non-viral vectors	Naked plasmid DNA
		Complexes with plasmid DNA

mutant that was first detected in 1984[7] and was confirmed as the biochemical model of DMD in 1987.[2] The *mdx* mouse lacks dystrophin due to a point mutation in exon 23 that leads to a premature stop codon and the production of an unstable dystrophin peptide.[8] As a result of this mutation, the muscle fibres are easily damaged and the mouse undergoes cycles of muscle fibre necrosis and regeneration from about 2 weeks of age.[9] The *mdx* mouse thus provides a convenient animal model that allows testing of a variety of gene therapy strategies.

Another important animal model is the GRMD (also known as the CXMD) dog and several colonies exist around the world. These dogs have a splice site mutation in intron 6 of the canine DMD gene and so fail to produce dystrophin.[10,11] They suffer a more dramatic muscular dystrophy than the *mdx* mouse and exhibit severe symptoms at 6 months of age.[12]

Gene Repair

In theory it should be possible to induce the natural DNA repair mechanisms to correct certain mutations, such as point mutations. Synthetic oligonucleotides containing DNA and RNA sequences (chimeroplasts) designed to intercalate between the two strands of genomic DNA have been used to try and stimulate this phenomenon. Initial successes in cells in culture lead to attempts to perform the similar corrections in vivo. Although apparent gene repair was observed in the *mdx* mouse[13,14] and the GRMD dog[15] the effect was very modest and the major hurdle of delivery of sufficient oligonucleotide has to be overcome. An alternative strategy for gene repair, short fragment homologous recombination has been proposed by Kapsa and colleagues[16] but again appears to work at very low efficiency. It is possible that such strategies could also be used to modify larger deletions to restore the open-reading frame and thus generate expression of an internally deleted dystrophin protein.[17]

Message Modification

Antisense oligonucleotides can be used to modify splicing of the primary transcript to skip exons containing mutations that alter the reading frame. In this way internally truncated proteins will be produced and the resultant protein should retain some degree of function. This strategy is based on the observation that there are rare dystrophin positive fibers that can be detected both in the *mdx* mouse[18] and in many DMD patients.[19-22] These fibers are termed revertants and in the *mdx* mouse these revertant fibres arise due to exon skipping of the mutant exon 23 (see Fig. 1). By targeting the splicing sites of exon 23 with complementary antisense oligonucleotides it is possible to alter splicing of the primary transcript to exclude exon 23. This leads to the generation of an in-frame mRNA and a variety of studies have shown that this can lead to transient in vivo expression of a near full-length dystrophin protein.[24-26]

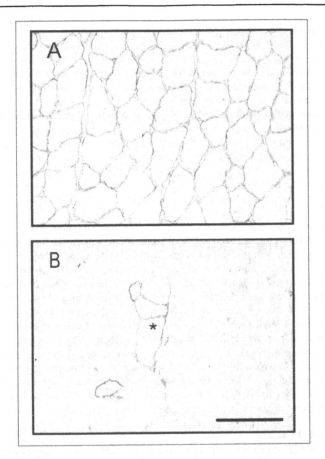

Figure 1. Images of transverse sections of normal muscle (from C57BL10 mouse) and dystrophic muscle (from an *mdx* mouse) stained with a rabbit polyclonal antibody directed against the C-terminus of dystrophic (DysC3750, 23). The normal muscle shows sarcolemmal staining in all muscle fibres whereas the majority of the dystrophic muscle shows no staining except for rare dystrophin positive "revertant" fibres (marked with an asterisk). Scale bar equals 100 micrometres.

Importantly, the resultant protein restores muscle function in the *mdx* mouse regardless of the age of the mouse.[27] Similar studies have been performed successfully with DMD patient derived muscle cells in culture and it has been estimated that this approach may be suitable for more than 65% of DMD cases.[28]

Not all gene mutations lead to a loss of function. In some cases the mutation may lead to an undesirable gain of function where the gene product exhibits a "dominant negative" effect. This appears to be the case in myotonic dystrophy, in which the large expansion of the CTG repeat in the 3' untranslated region of the DMPK gene leads to sequestration of the mutant RNA in the nucleus. The presence of the large repeat region in the mutant mRNA binds proteins involved in the correct splicing of primary transcript leading to abnormal functions in a number of other genes (reviewed in ref. 29). This can be prevented by the introduction of antisense RNA that can prevent the binding of proteins to the repeat expansions and hence restore RNA homeostatsis.[30] An alternative strategy is the use of ribozymes to target and modify or destroy those RNA molecules with expanded repeat sequences[31,32] (see Chapter 13).

Germ-Line Gene Therapy

The whole body consequences of the transfer of recombinant dystrophin cDNAs have been assessed in transgenic *mdx* mice, effectively a form of germ-line gene therapy. A number of studies have demonstrated that expression of full-length or a range of internally truncated dystrophin proteins can prevent the normal development of dystrophic pathology. The transgenic studies also show that there is a wide therapeutic window for the expression of dystrophin with no pathology evident with as little as 20% of endogenous dystrophin at the membrane or with as much as 50 times the normal levels of expression (reviewed in ref. 33). However, the transgenic studies have all involved the expression of dystrophin prior to the onset of pathology in *mdx* mice and so are not informative as to the therapeutic potential of such recombinant minigenes in muscle that has established dystrophic pathology. One study attempted to resolve that question using an inducible dystrophin transgene and, although troubled by nonuniform expression of the transgene, appears to show that administration of dystrophin can be beneficial in damaged muscle.[34]

Another closely related protein, utrophin, has also been tested for its ability to substitute for dystrophin using transgenic *mdx* mice. Several studies have shown that overexpression of recombinant utrophins can prevent the development of muscular dystrophy in transgenic *mdx* mice.[35,36] This is not particularly surprising as utrophin is a developmental precursor of dystrophin being present at the muscle membrane prior to the expression of dystrophin in late gestation, after which utrophin is restricted to the neuromuscular and myotendinous junctions in skeletal muscles.[37] As utrophin is encoded by a gene located on an autosome (chromosome 6) and is normally expressed in DMD patients, there are hopes that the early developmental pattern of expression can be restimulated pharmacologically and thus compensate for the lack of dystrophin. A recent experiment using an inducible utrophin transgene suggests that utrophin may not be as effective as dystrophin in correcting the muscle pathology once dystrophy develops.[38] However, a number of studies have also considered utrophin gene transfer as a therapeutic option as, unlike dystrophin, the immune system of DMD patients will be familiar with utrophin as a self protein (see later section on immune response, also Chapter 3).

Somatic Gene Therapy

Gene therapy for DMD patients will not involve germ line gene therapy but rather requires the specific treatment of affected muscle (somatic gene therapy) and for ethical reasons must avoid the transmission of genetic changes to subsequent generations. The transfer of a functional dystrophin gene into skeletal muscle requires a vector to deliver the DNA to the myofibre. Such vectors can be broadly divided into two categories, viral and nonviral.

Viral Vectors

A number of viruses have been used to introduce genes involved in muscular dystrophy into skeletal muscle in vivo. These include adenovirus e.g.,[39-41] adeno-associated virus e.g.,[42-45] herpesvirus,[46] lentivirus[47,48] retrovirus[49,50] and hybrids e.g., adenovirus carrying retrovirus.[51] A summary of the properties of the different viruses is shown in Table 3. Of these the adenovirus and adeno-associated viruses have been most widely used.

All vectors have some or all of the viral genome deleted thus rendering them replication deficient and providing space for the insertion of exogenous transgenes. However, all of the viral vectors possess coat proteins and these often provoke humoral immune responses that can limit repeated delivery of the vector.

Adenoviral Vectors

The most commonly used adenoviral vector is based on the human adenovirus serotype 5. Initial versions of this vector were constructed as replication deficient by removal of the first early gene that is required for viral replication. Subsequent deletion of parts of the third early gene and fourth early gene provided additional cloning capacity as well as further reducing the possibility of producing replication competent virus.[52,53] These vectors contained sufficient

Table 3. A summary of the different properties of the main viral vectors used for gene transfer into skeletal muscle

Vector	Genome	Packaging Capacity	Inflammatory Potential	Status of Genome in Cell	Main Limitations	Main Advantages
Adenovirus	dsDNA	8kb[1] 30kb[2]	High	Episomal	Capsid proteins induce a strong inflammatory response	Efficient, particularly in immature muscle
Adeno-associated virus	ssDNA	<5kb	Low	>90% Episomal <10% Integrated	Small capacity for exogenous DNA	Very efficient and non-pathogenic
Herpes-virus	dsDNA	40kb[1] 150kb[3]	High	Episomal	Inflammatory and inefficient passage across connective tissue barriers	Large packaging capacity
Lentivirus	RNA	8kb	Low	Integrated	May cause insertional oncogenesis	Transduces non-dividing cells so suitable for muscle fibres
Retrovirus	RNA	8kb	Low	Integrated	Only transduces dividing cells. May cause insertional oncogenesis	Persists in dividing cells. Can modify the satellite cell pool

[1] Replication defective, [2] Extensively deleted viral genome (Gutted virus), [3] Amplicon, dsDNA is double stranded DNA, ssDNA is single stranded DNA.

capacity to accommodate the minidystrophin construct and were extensively used to study the consequences of somatic gene transfer in the *mdx* mouse (reviewed in ref. 33). However, such studies were limited by the strong cytotoxic immune responses evoked by leaky expression of the residual viral genes. Adenoviral gene transfer into neonatal mice with an immature immune system allowed long-term expression of the transgene. Results thus resembled the transgenic studies as expression was initiated prior to the onset of pathology. These early generation adenoviral vectors could only be used in older mice if accompanied by strong immunosuppression.[54,55] The same requirement for immunosuppression when using these vectors was evident in the GRMD dog.[56]

Work in several laboratories lead to the generation of high capacity "gutted" adenoviral vectors that had all functional viral genes deleted.[57-59] The deletion of the remaining viral genes also enlarged the cloning capacity enabling these new vectors to be used for the transfer of cDNAs encoding the full-length dystrophin protein. A number of problems were associated with these high capacity adenoviral vectors; they required a helper virus for replication that could contaminate the final preparation and final viral titres were generally lower than that achieved by the earlier generation adenoviruses. These problems have been largely overcome,[60] and recent experiments have shown significant gene transfer into adult mice with amelioration of the pathology and functional correction of the treated muscle.[61-63]

Adeno-Associated Viral Vectors

The human parvovirus, first described as an adenoviral contaminant and hence known as an adeno-associated virus (AAV), is a common nonpathogenic small human virus that exhibits a complex life cycle involving integration into a specific site on chromosome 19. AAV appears not to transduce antigen presenting cells and this may explain the much reduced immune responses to transgenes introduced into muscle using an AAV vector.[64]

AAV vectors have a limited capacity for exogenous DNA, accomodating a maximum of 4.5kb even after removal of the entire viral genome. For use in gene therapy for DMD this size limitation has required the development of microdystrophin constructs with deletion of the majority of the central rod domain and parts of the amino and carboxy terminii. Remarkably, transgenic studies have shown that these highly deleted dystrophins prevent the development of muscular dystrophy in transgenic *mdx* mice.[65,66] Microdystrophin gene transfer using AAV vectors has demonstrated a prevention of pathology in young *mdx* mice.[44,67] More importantly, AAV based microdystrophin transfer has partially corrected existing pathology and improved muscle function in adult *mdx* with established muscle pathology.[44,65,68] No immune responses to microdystrophin gene expression have been noted in these studies despite the report that immune responses to an AAV transferred reporter gene were generated in dystrophic muscle.[69]

AAV appears ideal for gene transfer of sarcoglycans in the treatment of LGMD 2C-2F as the vector can easily accommodate the 2kb cDNAs for each gene. Several groups have shown efficient AAV based sarcoglycan delivery[70,42,43,45] but it should be noted that the range of expression that can be therapeutic may be limited in some cases. Dressman and colleagues have demonstrated that overexpression of α-sarcoglycan lead to toxicity but overexpression of β-sarcogylcan or δ-sarcoglycan did not. This result was consistent with the predicted order of assembly of the sarcoglycan complex.[71] Interestingly, Cordier and colleagues noted problems when expression of γ-sarcoglycan was driven by the strong cytomegalovirus promoter but not when driven by a muscle specific creatine kinase promoter and they attributed this to an immune response triggered by the high levels and nonspecific expression.[72] However, Zhu et al reported a severe muscular dystrophy in transgenic mice overexpressing γ-sarcoglycan and, based on their experiments, suggested that care will need to be taken in regulating the extent of expression of this molecule.[73]

Nonviral Vectors

Nonviral vectors are all based on the use of plasmid DNA. Plasmid DNA is a double stranded covalently closed circular form of DNA found in bacteria. This form of DNA can be taken up and expressed in skeletal muscle as shown by the pioneering work of Jon Wolff and colleagues.[74] However, the efficiency of gene transfer using intramuscular injection of naked plasmid DNA is very low in rodents and even less efficient in larger animals.[75] A number of studies have attempted to increase the efficiency of this process by changing the solvent for the plasmid,[76] selecting the sex and age of the recipient[77] or initiating muscle regeneration prior to injection of the plasmid.[78-82] The improvements seen in the aforementioned studies have generally been only 2-10 fold and have not been shown to substantially improve the efficiency in larger species.

Plasmid can also be complexed to other agents to protect it against degradation by DNAses and to increase the efficiency of uptake by cells. Most of these methods, such as mixing plasmid with cationic lipids, which works well with cells in culture, are inefficient when applied to intact skeletal muscle, presumably due to their inability to penetrate the extracellular matrix (endomysium) of the muscle fibres. Recent developments have improved this methodology and Lu et al[83] have shown that the use of a polyoxamer can substantially improve the efficiency of plasmid gene transfer and expression in skeletal muscle.

Two recent developments have dramatically improved the efficiency of naked plasmid DNA delivery to skeletal muscle. The application of electrical fields to muscle immediately following intramuscular injection dramatically improves the uptake and expression of plasmid DNA by

10-1,000 fold.[84-87] This method also works in the larger species[88] and the applications, methodology and limitations of this method has been recently reviewed.[89] Transfection of the majority of muscle fibres with minimal procedure associated damage can be achieved by pretreatment of the muscle with a dilute solution of hyaluronidase.[87]

The other improvement is the delivery of plasmid DNA through the vascular system. By using high volume rapid intra-arterial injection into a temporarily isolated limb treated with agents that vasodilate and increase capillary permeability, multiple muscles can be shown to take up and express plasmid DNA at levels 5-20 fold that achieved with optimal intramuscular injection.[90] Importantly, this method also works well in larger species and has been shown to efficiently treat multiple muscles in primate limbs.[91]

Overexpression of Compensating Genes

A number of other genes can be transferred with the technology discussed above that, though they do not lead to the correct expression of the mutant protein, may offer functional protection against the consequences of the original gene mutation. Utrophin can be considered the foremost example of this category, where the mouse experiments show that it is a functional substitute for the lack of dystrophin (Chapter 3). Other genes do not show the same degree of structural and functional homology but, at least in the mouse model, are capable of partially protecting the muscles from developing marked muscular dystrophy. For example, overexpression of a nitric oxide synthase (NOS) transgene in the *mdx* mouse reduced muscle pathology, possibly through reduction of the macrophage infiltration and activity that is associated with the initial phase of myonecrosis in the *mdx*.[92] Expression of an insulin-like growth factor 1 transgene (IGF-1) leads to increased muscle bulk and reduced pathology in the *mdx* mouse.[93] Overexpression of ADAM12 leads to post-transcriptional changes in protein expression that alleviates the pathology in the *mdx* mouse.[94,95] A calpastatin transgene also reduced muscle pathology in young *mdx* mice, possibly through decreased calpain activation.[96] More dramatically, a synaptic cytotoxic T cell GalNAc transferase transgene[97] substantially reduced muscle pathology in the *mdx*, possibly by creating a functional dystrophin associated protein complex. Similarly, overexpression of alpha 7 beta one intregin reduced pathology in the dys-/-utr-/- double knockout mouse.[98] However, apart from IGF-1[99] none of these observations have yet to be translated into any form of somatic gene therapy.

Immune Responses to Gene Therapy

For human clinical trials it will be necessary to consider not only the potential immune responses to the vector(s) employed but also to the therapeutic transgene. DMD patients commonly have little or no dystrophin expression and thus expression of recombinant dystrophin following gene transfer may lead to a specific immune response directed against the "foreign" protein. A number of studies have shown immune responses to exogenous dystrophin in both human and experimental animals (reviewed in ref. 100). In many cases the presence of dystrophin specific antibodies have been demonstrated, however these are probably of little importance to a cytoskeletal protein which will not be accessible to antibody in intact muscle fibres. Of more concern is the presence of cytotoxic T cells that have the potential to destroy genetically modified muscle fibres, a situation that would be highly undesirable in DMD patients who already have very few intact muscle fibres. The analysis of the cytotoxic immune responses in experimental animal models of DMD has been complicated by the addition of other foreign proteins when using viral or cell based methods of gene transfer (see ref. 100). Nonviral gene transfer using naked plasmid DNA does not in itself generate cytotoxic T cell responses as no proteins are associated with the vector system. Thus, studies of immune responses using plasmid DNA have been able to demonstrate that specific immune responses to human dystrophin can be generated in *mdx* mice[101,102] but no immune responses are directed against expression of full-length murine dystrophin.[102] The latter study suggests that the *mdx* mice are tolerant of full-length dystrophin, despite their failure to produce dystrophin in their skeletal muscles due to the exon 23 stop mutation. This tolerance may be due to one or both of the following

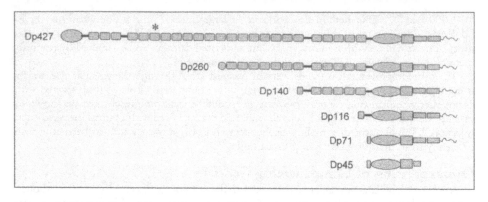

Figure 2. A diagram illustrating the different isoforms of dystrophin and the regions of identity. The figure is based on a previously published figure.[103] The asterisk marks the position of exon 23 relative to the protein structure.

events. The stop mutation in exon 23, whilst preventing expression of the 427kDa muscle, cardiac and Purkinjie isoforms of dystrophin, does not prevent expression of the smaller isoforms of dystrophin (Fig. 2, ref. 103). Although these isoforms lack the amino terminus of the full length molecules, they ensure that the immune system recognises part of the rod domain and C-terminal epitopes as self. In addition, as previously noted, there are rare dystrophin positive revertant fibres that can be detected both in the *mdx* mouse and DMD patients. Although these exon skipping events are individually idiosyncratic and may involve skipping of a large number of exons,[104] it is likely that with sufficient revertant fibres the vast majority of exons are expressed and thus presented to the immune system for recognition. Thus the immune system in the mouse may have encountered the vast majority, if not all, potential immunogenic epitopes of dystrophin. Unfortunately the situation in many patients is somewhat different. The majority of DMD patients have large deletions within the dystrophin gene (65% of cases reported in ref. 105) and, although exon skipping may allow the production of dystrophin positive revertant fibres by restoring the reading frame, the immune system of patients with such deletions will not encounter the potential immunogenic epitopes in the deleted region of the molecule. However, for the patients with duplications, small deletions and point mutations (35% of cases reported in ref. 105), the production of revertant fibres may be sufficient to induce tolerance to the expression of recombinant dystrophin following gene transfer. The potential importance of large deletions has been modelled by examining anti-dystrophin immune responses following human dystrophin gene transfer into transgenic *mdx* mice expressing various internally truncated human dystrophin transgenes. Despite deletion of the majority of the rod domain (exons 13 to 47) there was no evidence of cytotoxic immune responses to expression of full-length human dystrophin.[106] Thus many patients with large deletions may also not produce cytotoxic immune responses following dystrophin gene therapy. In order to model specific human haplotype responses to dystrophin gene transfer, Ginhoux and colleagues have transferred a full-length dystrophin cDNA using plasmid gene transfer into H-2-negative HLA-A*0201 transgenic mice. They noted a strong cytotoxic immune response to an epitope in spectrin-like repeat 9, a region not present in the minidystrophin construct.[107]

As previously mentioned, the transfer of recombinant utrophin genes offers a safe alternative to recombinant dystrophins as the immune system should be tolerant to the expression of recombinant utrophin. However, although both germ-line and somatic gene transfer studies have shown that utrophin can prevent the development of dystrophic pathology in the *mdx* mouse (reviewed in ref. 33), it is still not certain that utrophin can successfully substitute for dystrophin in damaged muscle, particularly in humans. Utrophin has differences compared to dystrophin in the manner of binding to the actin cytoskeleton[108] and does not relocalise nNOS to the muscle sarcolemma.[109] Although these differences do not appear significant in the mouse form of muscular dystrophy, it is possible that these differences may be more important in man.

Immune responses to other genes used in the treatment of muscular dystrophies have been less intensively studied. However, as noted earlier, a decrease in γ-sarcoglycan expression over time from a cytomegalovirus driven γ-sarcoglycan construct in Sgcg$^{-/-}$ mice was associated with a humoral and cytotoxic immune response.[72] Several groups have suggested that the use of muscle-specific promoters will help to avoid or reduce the generation of immune responses following gene transfer into skeletal muscle.[110,111]

Potential Complications Awaiting Clinical Trials of Gene Therapy for DMD

There are a number of concerns regarding translation of experiments performed in animals into a practical clinical treatment for DMD.

The first concern relates to safety of the vector systems used for gene transfer. Many people have been previously exposed to adenovirus serotype 5 and 55% have neutralising antibodies to this virus. Likewise, much of the population has been exposed to adeno-associated virus serotype 2 and 32% have neutralising antibodies to this virus.[112] Plasmid DNA is recognised as prokaryotic by the innate immune system and provokes nonspecific inflammatory responses.[113] The route of administration of the vector may also be important. High dose adenovirus administration via the vascular route was associated with a fatality in a trial of gene therapy for ornithine transcarbamylase deficiency.[114]

The efficiency of the different vector systems may be markedly different between mouse and man, yet much of our understanding of these vectors is based on studies in the mouse. There have been two gene transfer trials to date involving patients with muscular dystrophy. The first used an adeno-associated virus vector to introduce sarcogylcan into one of the muscles of the foot in patients with LGMD.[115] After treating two patients the trial was halted due to problems arising in an unrelated trial with an adenoviral vector. Analysis of biopsies from the treated muscle compared to a similar biopsy from the untreated contralateral muscle failed to show evidence of gene transfer although both patients showed no side effects from the administration of the vector (MDA press release). The other gene transfer trial involved the intramuscular injection of naked plasmid DNA carrying a full-length dystrophin cDNA into DMD and BMD patients. This was a dose escalation trial to assess the safety of the procedure and the possible generation of immune responses.[116] A preliminary report notes that no adverse effects were recorded although the levels of expression of dystrophin were very low (Transgene press release).

The vast majority of the gene transfer studies in animals cited earlier have involved intramuscular injections of individual muscles. The diffusion of any of the gene transfer vectors is very limited,[117] and treatment of muscles larger than those of the mouse will require multiple injections to genetically modify sufficient fibres to significantly improve muscle function. While this may be a necessary step in showing safety and efficacy in clinical trials of any gene therapy approach, treatment of individual muscles is unlikely to provide a clear clinical benefit. Consequently it will be important to use approaches that ensure gene transfer into multiple muscles. Vascular delivery of therapeutic or reporter genes has been achieved in animal models using both viral[70,118] and nonviral vectors.[90,91,119] The vascular approach requires increased pressure and/or increased capillary permeability to deliver the vector to the muscle cell membrane and thus involves the generation of oedema as a consequence of the treatment. Such oedema might be life-threatening in the lungs or brain, but by temporarily isolating a limb for treatment these dangerous side-effects can be minimised. Treatment of the arm muscles with gene therapy to prevent excessive muscle loss could have a significant impact on the quality of life for patients with muscular dystrophy, allowing them to maintain an important degree of independence. Clearly, treatment must be applied while there is sufficient muscle and before such independence is lost.

Conclusion

Gene therapy is widely seen by patients and family members as the main hope for a cure for muscular dystrophies. Although there have been some notable developments in preclinical gene therapy studies, it still has to be proved that the vectors selected for human clinical trials will be as effective when administered to patients with muscular dystrophy. In addition, such gene therapy is unlikely to treat the nonmuscle manifestations of these disorders. Recent developments in antisense oligonucleotide induced splicing to convert the DMD to a BMD phenotype show therapeutic promise but the technique is currently limited to local delivery.[25-27] However, the small size of the oligonucleotides offers a realistic possibility of systemic delivery provided sufficient oligonucleotide can be targeted to the skeletal muscles. Such delivery would need to be atraumatic, as modification of the mRNA would require frequent repeat dosing.

The most promising approach for recombinant dystrophin gene transfer appears to be the regional delivery of viral or nonviral vectors to the muscles of the arms.[70,91] While this form of gene therapy may do little to increase the longevity of patients, if successful this treatment should substantially increase quality of life as it will enable the patient to maintain a degree of independence. Thus while a cure for muscular dystrophy is not yet within sight, a combination of mechanical, pharmacological and genetic therapies have the potential within the near future to offer an increased lifespan and quality of life for patients with these devastating neuromuscular disorders.

Note Added in Proof

Very recently there have been significant developments in the vascular delivery of viral and non-viral vectors to skeletal muscle. Delivery of dystrophin plasmids to the skeletal muscle of *mdx* mice has been demonstrated by both the arterial[120] and venous[121] routes. The latter approach involves lower volumes and pressure and can be performed via a peripheral vein, thus making it a much more clinically applicable route of administration compared to regional delivery via the arterial system. More dramatically, the group led by Jeff Chamberlain have recently demonstrated that body wide transduction of skeletal muscle can be achieved after intravenous delivery of a novel serotype AAV6.[122] With the use of VEGF and/or high viral titres, the vast majority of all of the skeletal muscle fibers were transduced and this was demonstrated with delivery of a microdystrophin construct in *mdx* mice. These recent developments with vascular delivery of genetic material to multiple muscle groups mean that a therapeutically useful gene therapy for muscular dystrophy is within sight.

Acknowledgements

I would like to thank all members of my laboratory past and present who have contributed to studies of gene transfer into muscle and immune responses to transgene expression and in particular Kim Wells for critical reading and help in final formatting of this chapter. This work has been funded by grants from the Muscular Dystrophy Campaign of Great Britain and Northern Ireland, the Muscular Dystrophy Association of America, the Association Francaise contre les Myopathies, the United Kingdom Medical Research Council and the Wellcome Trust.

References

1. Koenig M, Hoffman EP, Bertelson CJ et al. Complete cloning of the Duchenne muscular dystrophy (DMD) cDNA and preliminary genomic organization of the DMD gene in normal and affected individuals. Cell 1987; 50:509-17.
2. Hoffman EP, Brown Jr RH, Kunkel LM. Dystrophin: The protein product of the Duchenne muscular dystrophy locus. Cell 1987; 51:919-28.
3. Monaco AP, Bertelson S, Liechti-Gallati S et al. An explanation for the phenotypic differences between patients bearing partial deletions of the DMD locus. Genomics 1988; 2:90-95.
4. Durbeej M, Campbell KP. Muscular dystrophies involving the dystrophin-glycoprotein complex: An overview of current mouse models. Curr Opin Genet Dev 2002; 12:349-61.

5. Eagle M, Baudouin SV, Chandler C et al. Survival in Duchenne muscular dystrophy: Improvements in life expectancy since 1967 and the impact of home nocturnal ventilation. Neuromuscul Disord 2002; 12:926-9.
6. Wong BL, Christopher C. Corticosteroids in Duchenne muscular dystrophy: A reappraisal. J Child Neurol 2002; 17:183-90.
7. Bulfield G, Siller WG, Wight PA et al. X chromosome-linked muscular dystrophy (mdx) in the mouse. Proc Natl Acad Sci USA 1984; 81:1189-92.
8. Sicinski P, Geng Y, Ryder-Cook AS et al. The molecular basis of muscular dystrophy in the mdx mouse: A point mutation. Science 1989; 244:1578-80.
9. Coulton GR, Morgan JE, Partridge TA et al. The mdx mouse skeletal muscle myopathy: I. A histological, morphometric and biochemical investigation. Neuropathol Appl Neurobiol 1988; 14:53-70.
10. Cooper BJ, Winand NJ, Stedman H et al. The homologue of the Duchenne locus is defective in X-linked muscular dystrophy of dogs. Nature 1988; 334:154-6.
11. Sharp NJ, Kornegay JN, Van Camp SD et al. An error in dystrophin mRNA processing in golden retriever muscular dystrophy, an animal homologue of Duchenne muscular dystrophy. Genomics 1992; 13:115-21.
12. Valentine BA, Cooper BJ, de Lahunta A et al. Canine X-linked muscular dystrophy. An animal model of Duchenne muscular dystrophy: Clinical studies. J Neurol Sci 1988; 88:69-81.
13. Rando TA, Disatnik MH, Zhou LZ. Rescue of dystrophin expression in mdx mouse muscle by RNA/DNA oligonucleotides. Proc Natl Acad Sci USA 2000; 97:5363-8.
14. Bertoni C, Rando TA. Dystrophin gene repair in mdx muscle precursor cells in vitro and in vivo mediated by RNA-DNA chimeric oligonucleotides. Hum Gene Ther 2002; 13:707-18.
15. Bartlett RJ, Stockinger S, Denis MM et al. In vivo targeted repair of a point mutation in the canine dystrophin gene by a chimeric RNA/DNA oligonucleotide. Nat Biotechnol 2000; 18:615-22.
16. Kapsa R, Quigley A, Lynch GS et al. In vivo and in vitro correction of the mdx dystrophin gene nonsense mutation by short-fragment homologous replacement. Hum Gene Ther 2001; 12:629-42.
17. Rando TA. Oligonucleotide-mediated gene therapy for muscular dystrophies. Neuromuscul Disord 2002; 12:S55-60.
18. Hoffman EP, Morgan JE, Watkins SC et al. Somatic reversion/suppression of the mouse mdx phenotype in vivo. J Neurol Sci 1990; 99:9-25.
19. Burrow KL, Coovert DD, Klein CJ et al. Dystrophin expression and somatic reversion in prednisone-treated and untreated Duchenne dystrophy. CIDD Study Group. Neurology 1991; 11:661-6.
20. Nicholson LV, Johnson MA, Bushby KM et al. Integrated study of 100 patients with Xp21 linked muscular dystrophy using clinical, genetic, immunochemical, and histopathological data. Part 2. Correlations within individual patients. J Med Genet 1993; 30:737-44.
21. Fanin M, Danieli GA, Cadaldini M et al. Dystrophin-positive fibers in Duchenne dystrophy: Origin and correlation to clinical course. Muscle Nerve 1995; 18:1115-20.
22. Uchino M, Tokunaga M, Mita S et al. PCR and immunocytochemical analyses of dystrophin-positive fibers in Duchenne muscular dystrophy. J Neurol Sci 1995; 129:44-50.
23. Gollins H, McMahon J, Wells KE et al. High-efficiency plasmid gene transfer into dystrophic muscle. Gene Ther 2003; 10:504-12.
24. Mann CJ, Honeyman K, Cheng AJ et al. Antisense-induced exon skipping and synthesis of dystrophin in the mdx mouse. Proc Natl Acad Sci USA 2001; 98:42-7.
25. Gebski BL, Mann CJ, Fletcher S et al. Morpholino antisense oligonucleotide induced dystrophin exon 23 skipping in mdx mouse muscle. Hum Mol Genet 2003; 12:1801-11.
26. Wells KE, Fletcher S, Mann CJ et al. Enhanced in vivo delivery of antisense oligonucleotide to restore dystrophin expression in adult mdx mouse muscle. FEBS Lett 2003; 552:145-149.
27. Lu QL, Mann CJ, Lou F et al. Functional amounts of dystrophin produced by skipping the mutated exon in the mdx dystrophic mouse. Nat Med 2003; 9:1009-14.
28. van Deutekom JC, Bremmer-Bout M, Janson AA et al. Antisense-induced exon skipping restores dystrophin expression in DMD patient derived muscle cells. Hum Mol Genet 2001; 10:1547-54.
29. Mankodi A, Thornton CA. Myotonic syndromes. Curr Opin Neurol 2002; 15:545-52.
30. Furling D, Doucet G, Langlois MA et al. Viral vector producing antisense RNA restores myotonic dystrophy myoblast functions. Gene Ther 2003; 10:795-802.
31. Phylactou LA, Darrah C, Wood MJ. Ribozyme-mediated trans-splicing of a trinucleotide repeat. Nat Genet 1998; 18:378-81.
32. Langlois MA, Lee NS, Rossi JJ et al. Hammerhead ribozyme-mediated destruction of nuclear foci in myotonic dystrophy myoblasts. Mol Ther 2003; 7:670-80.

33. Wells DJ, Wells KE. Gene transfer studies in animals: What do they really tell us about the prospects for gene therapy in DMD? Neuromuscul Disord 2002; 12:S11-22.
34. Ahmad A, Brinson M, Hodges BL et al. Mdx mice inducibly expressing dystrophin provide insights into the potential of gene therapy for duchenne muscular dystrophy. Hum Mol Genet 2000; 9:2507-15.
35. Tinsley JM, Potter AC, Phelps SR et al. Amelioration of the dystrophic phenotype of mdx mice using a truncated utrophin transgene. Nature 1996; 384:349-53.
36. Tinsley J, Deconinck N, Fisher R et al. Expression of full-length utrophin prevents muscular dystrophy in mdx mice. Nat Med 1998; 4:1441-4.
37. Lin S, Burgunder JM. Utrophin may be a precursor of dystrophin during skeletal muscle development. Brain Res Dev Brain Res 2000; 119:289-95.
38. Squire S, Raymackers JM, Vandebrouck C et al. Prevention of pathology in mdx mice by expression of utrophin: Analysis using an inducible transgenic expression system. Hum Mol Genet 2002; 11:3333-44.
39. Vincent N, Ragot T, Gilgenkrantz H et al. Long-term correction of mouse dystrophic degeneration by adenovirus-mediated transfer of a minidystrophin gene. Nat Genet 1993; 5:130-4.
40. Petrof BJ, Acsadi G, Jani A et al. Efficiency and functional consequences of adenovirus-mediated in vivo gene transfer to normal and dystrophic (mdx) mouse diaphragm. Am J Respir Cell Mol Biol 1995; 13:508-17.
41. Clemens PR, Krause TL, Chan S et al. Recombinant truncated dystrophin minigenes: Construction, expression, and adenoviral delivery. Hum Gene Ther 1995; 6:1477-85.
42. Li J, Dressman D, Tsao YP et al. rAav vector-mediated sarcogylcan gene transfer in a hamster model for limb girdle muscular dystrophy. Gene Ther 1999; 6:74-82.
43. Cordier L, Hack AA, Scott MO et al. Rescue of skeletal muscles of gamma-sarcoglycan-deficient mice with adeno-associated virus-mediated gene transfer. Mol Ther 2000; 1:119-29.
44. Wang B, Li J, Xiao X. Adeno-associated virus vector carrying human minidystrophin genes effectively ameliorates muscular dystrophy in mdx mouse model. Proc Natl Acad Sci USA 2000; 97:13714-9.
45. Xiao X, Li J, Tsao YP et al. Full functional rescue of a complete muscle (TA) in dystrophic hamsters by adeno-associated virus vector-directed gene therapy. J Virol 2000; 74:1436-42.
46. Akkaraju GR, Huard J, Hoffman EP et al. Herpes simplex virus vector-mediated dystrophin gene transfer and expression in MDX mouse skeletal muscle. J Gene Med 1999; 1:280-9.
47. Scott JM, Li S, Harper SQ et al. Viral vectors for gene transfer of micro-, mini-, or full-length dystrophin. Neuromuscul Disord 2002; 12:S23-9.
48. Sampaolesi M, Torrente Y, Innocenzi A et al. Cell therapy of alpha-sarcoglycan null dystrophic mice through intra-arterial delivery of mesoangioblasts. Science 2003; 301:487-92.
49. Dunckley MG, Wells DJ, Walsh FS et al. Direct retroviral-mediated transfer of a dystrophin minigene into mdx mouse muscle in vivo. Hum Mol Genet 1993; 2:717-23.
50. Fassati A, Wells DJ, Sgro Serpente PA et al. Genetic correction of dystrophin deficiency and skeletal muscle remodeling in adult MDX mouse via transplantation of retroviral producer cells. J Clin Invest 1997; 100:620-8.
51. Roberts ML, Wells DJ, Graham IR et al. Stable micro-dystrophin gene transfer using an integrating adeno-retroviral hybrid vector ameliorates the dystrophic pathology in mdx mouse muscle. Hum Mol Genet 2002; 11:1719-30.
52. Haj-Ahmad Y, Graham FL. Development of a helper-independent human adenovirus vector and its use in the transfer of the herpes simplex virus thymidine kinase gene. J Virol 1986; 57:267-74.
53. Krougliak V, Graham FL. Development of cell lines capable of complementing E1, E4, and protein IX defective adenovirus type 5 mutants. Hum Gene Ther 1995; 6:1575-86.
54. Lochmuller H, Petrof BJ, Pari G et al. Transient immunosuppression by FK506 permits a sustained high-level dystrophin expression after adenovirus-mediated dystrophin minigene transfer to skeletal muscles of adult dystrophic (mdx) mice. Gene Ther 1996; 3:706-16.
55. Yang L, Lochmuller H, Luo J et al. Adenovirus-mediated dystrophin minigene transfer improves muscle strength in adult dystrophic (MDX) mice. Gene Ther 1998; 5:369-79.
56. Howell JM, Lochmuller H, OH A et al. High-level dystrophin expression after adenovirus-mediated dystrophin minigene transfer to skeletal muscle of dystrophic dogs: Prolongation of expression with immunosuppression. Hum Gene Ther 1998; 9:629-34.
57. Haecker SE, Stedman HH, Balice-Gordon RJ et al. In vivo expression of full-length human dystrophin from adenoviral vectors deleted of all viral genes. Hum Gene Ther 1996; 7:1907-14.
58. Kochanek S, Clemens PR, Mitani K et al. A new adenoviral vector: Replacement of all viral coding sequences with 28 kb of DNA independently expressing both full-length dystrophin and beta-galactosidase. Proc Natl Acad Sci USA 1996; 93:5731-6.

59. Kumar-Singh R, Chamberlain JS. Encapsidated adenovirus minichromosomes allow delivery and expression of a 14 kb dystrophin cDNA to muscle cells. Hum Mol Genet 1996; 5:913-21.
60. Barjot C, Hartigan-O'Connor D, Salvatori G et al. Gutted adenoviral vector growth using E1/E2b/E3-deleted helper viruses. J Gene Med 2002; 4:480-9.
61. DelloRusso C, Scott JM, Hartigan-O'Connor D et al. Functional correction of adult mdx mouse muscle using gutted adenoviral vectors expressing full-length dystrophin. Proc Natl Acad Sci USA 2002; 99:12979-84.
62. Gilbert R, Liu A, Petrof B et al. Improved performance of a fully gutted adenovirus vector containing two full-length dystrophin cDNAs regulated by a strong promoter. Mol Ther 2002; 6:501-9.
63. Gilbert R, Dudley RW, Liu AB et al. Prolonged dystrophin expression and functional correction of mdx mouse muscle following gene transfer with a helper-dependent (gutted) adenovirus-encoding murine dystrophin. Hum Mol Genet 2003; 12:1287-99.
64. Jooss K, Turka LA, Wilson JM. Blunting of immune responses to adenoviral vectors in mouse liver and lung with CTLA4Ig. Gene Ther 1998; 5:309-19.
65. Harper SQ, Hauser MA, DelloRusso C et al. Modular flexibility of dystrophin: Implications for gene therapy of Duchenne muscular dystrophy. Nat Med 2002; 8:253-61.
66. Sakamoto M, Yuasa K, Yoshimura M et al. Micro-dystrophin cDNA ameliorates dystrophic phenotypes when introduced into mdx mice as a transgene. Biochem Biophys Res Commun 2002; 293:1265-72.
67. Fabb SA, Wells DJ, Serpente P et al. Adeno-associated virus vector gene transfer and sarcolemmal expression of a 144 kDa micro-dystrophin effectively restores the dystrophin-associated protein complex and inhibits myofibre degeneration in nude/mdx mice. Hum Mol Genet 2002; 11:733-41.
68. Watchko J, O'Day T, Wang B et al. Adeno-associated virus vector-mediated minidystrophin gene therapy improves dystrophic muscle contractile function in mdx mice. Hum Gene Ther 2002; 13:1451-60.
69. Yuasa K, Sakamoto M, Miyagoe-Suzuki Y et al. Adeno-associated virus vector-mediated gene transfer into dystrophin-deficient skeletal muscles evokes enhanced immune response against the transgene product. Gene Ther 2002; 9:1576-88.
70. Greelish JP, Su LT, Lankford EB et al. Stable restoration of the sarcoglycan complex in dystrophic muscle perfused with histamine and a recombinant adeno-associated viral vector. Nat Med 1999; 5:439-43.
71. Dressman D, Araishi K, Imamura M et al. Delivery of alpha- and beta-sarcoglycan by recombinant adeno-associated virus: Efficient rescue of muscle, but differential toxicity. Hum Gene Ther 2002; 13:1631-46.
72. Cordier L, Gao GP, Hack AA et al. Muscle-specific promoters may be necessary for adeno-associated virus-mediated gene transfer in the treatment of muscular dystrophies. Hum Gene Ther 2001; 12:205-15.
73. Zhu X, Hadhazy M, Groh ME et al. Overexpression of gamma-sarcoglycan induces severe muscular dystrophy. Implications for the regulation of Sarcoglycan assembly. J Biol Chem 2001; 276:21785-90.
74. Wolff JA, Malone RW, Williams P et al. Direct gene transfer into mouse muscle in vivo. Science 1990; 247:1465-8.
75. Jiao S, Williams P, Berg RK et al. Direct gene transfer into nonhuman primate myofibers in vivo. Hum Gene Ther 1992; 3:21-33.
76. Wolff JA, Williams P, Acsadi G et al. Conditions affecting direct gene transfer into rodent muscle in vivo. Biotechniques 1991; 11:474-85.
77. Wells DJ, Goldspink G. Age and sex influence expression of plasmid DNA directly injected into mouse skeletal muscle. FEBS Lett 1992; 306:203-5.
78. Davis HL, Demeneix BA, Quantin B et al. Plasmid DNA is superior to viral vectors for direct gene transfer into adult mouse skeletal muscle. Hum Gene Ther 1993; 4:733-40.
79. Wells DJ. Improved gene transfer by direct plasmid injection associated with regeneration in mouse skeletal muscle. FEBS Lett 1993; 332:179-82.
80. Danko I, Fritz JD, Jiao S et al. Pharmacological enhancement of in vivo foreign gene expression in muscle. Gene Ther 1994; 1:114-21.
81. Vitadello M, Schiaffino MV, Picard A et al. Gene transfer in regenerating muscle. Hum Gene Ther 1994; 5:11-8.
82. Wells DJ, Maule J, McMahon J et al. Evaluation of plasmid DNA for in vivo gene therapy: Factors affecting the number of transfected fibers. J Pharm Sci 1998; 87:763-8.
83. Lu QL, Bou-Gharios G, Partridge TA. Nonviral gene delivery in skeletal muscle: A protein factory. Gene Ther 2003; 10:131-42.

84. Aihara H, Miyazaki J. Gene transfer into muscle by electroporation in vivo. Nat Biotechnol 1998; 16:867-70.
85. Mathiesen I. Electropermeabilization of skeletal muscle enhances gene transfer in vivo. Gene Ther 1999; 6:508-14.
86. Mir LM, Bureau MF, Gehl J et al. High-efficiency gene transfer into skeletal muscle mediated by electric pulses. Proc Natl Acad Sci USA 1999; 96:4262-7.
87. McMahon JM, Signori E, Wells KE et al. Optimisation of electrotransfer of plasmid into skeletal muscle by pretreatment with hyaluronidase — increased expression with reduced muscle damage. Gene Ther 2001; 8:1264-70.
88. Babiuk S, Baca-Estrada ME, Foldvari M et al. Electroporation improves the efficacy of DNA vaccines in large animals. Vaccine 2002; 20:3399-3408.
89. McMahon JM, Wells DJ. Electroporation for Gene Transfer to Skeletal Muscles: Current Status. BioDrugs 2004; 18:155-165.
90. Budker V, Zhang G, Danko I et al. The efficient expression of intravascularly delivered DNA in rat muscle. Gene Ther 1998; 5:272-6.
91. Zhang G, Budker V, Williams P et al. Efficient expression of naked DNA delivered intraarterially to limb muscles of nonhuman primates. Hum Gene Ther 2001; 12:427-38.
92. Wehling M, Spencer MJ, Tidball JG. A nitric oxide synthase transgene ameliorates muscular dystrophy in mdx mice. J Cell Biol 2001; 155:123-31.
93. Barton ER, Morris L, Musaro A et al. Muscle-specific expression of insulin-like growth factor I counters muscle decline in mdx mice. J Cell Biol 2002; 157:137-48.
94. Kronqvist P, Kawaguchi N, Albrechtsen R et al. ADAM12 alleviates the skeletal muscle pathology in mdx dystrophic mice. Am J Pathol 2002; 161:1535-40.
95. Moghadaszadeh B, Albrechtsen R, Guo L et al. Compensation for dystrophin-deficiency: ADAM12 overexpression in skeletal muscle results in increased {alpha}7 integrin, utrophin, and associated glycoproteins. Hum Mol Genet 2003.
96. Spencer MJ, Mellgren RL. Overexpression of a calpastatin transgene in mdx muscle reduces dystrophic pathology. Hum Mol Genet 2002; 11:2645-55.
97. Nguyen HH, Jayasinha V, Xia B et al. Overexpression of the cytotoxic T cell GalNAc transferase in skeletal muscle inhibits muscular dystrophy in mdx mice. Proc Natl Acad Sci USA 2002; 99:5616-21.
98. Burkin DJ, Wallace GQ, Nicol KJ et al. Enhanced expression of the alpha 7 beta 1 integrin reduces muscular dystrophy and restores viability in dystrophic mice. J Cell Biol 2001; 152:1207-18.
99. Barton-Davis ER, Shoturma DI, Musaro A et al. Viral mediated expression of insulin-like growth factor I blocks the aging-related loss of skeletal muscle function. Proc Natl Acad Sci USA 1998; 95:15603-7.
100. Wells DJ, Ferrer A, Wells KE. Immunological hurdles in the path to gene therapy for Duchenne muscular dystrophy. Expert Reviews in Molecular Medicine 2002, (http://www.expertreviews.org/0200515Xh.htm).
101. Braun S, Thioudellet C, Rodriguez P et al. Immune rejection of human dystrophin following intramuscular injections of naked DNA in mdx mice. Gene Ther 2000; 7:1447-57.
102. Ferrer A, Wells KE, Wells DJ. Immune responses to dystropin: Implications for gene therapy of Duchenne muscular dystrophy. Gene Ther 2000; 7:1439-46.
103. Winder SJ. The membrane-cytoskeleton interface: The role of dystrophin and utrophin. J Muscle Res Cell Motil 1997; 18:617-29.
104. Lu QL, Morris GE, Wilton SD et al. Massive idiosyncratic exon skipping corrects the nonsense mutation in dystrophic mouse muscle and produces functional revertant fibers by clonal expansion. J Cell Biol 2000; 148:985-96.
105. Roberts RG, Gardner RJ, Bobrow M. Searching for the 1 in 2,400,000: A review of dystrophin gene point mutations. Hum Mutat 1994; 4:1-11.
106. Ferrer A, Gollins H, Wells KE et al. Long term expression of full-length human dystrophin in transgenic mdx mice expressing truncated human dystrophins. Gene Therapy 2004; 884-893.
107. Ginhoux F, Doucet C, Leboeuf M et al. Identification of an HLA-A*0201-restricted epitopic peptide from human dystrophin: Application in duchenne muscular dystrophy gene therapy. Mol Ther 2003; 8:274-83.
108. Sutherland-Smith AJ, Moores CA, Norwood FL et al. An atomic model for actin binding by the CH domains and spectrin-repeat modules of utrophin and dystrophin. J Mol Biol 2003; 329:15-33.
109. Wells KE, Torelli S, Lu Q et al. Relocalization of neuronal nitric oxide synthase (nNOS) as a marker for complete restoration of the dystrophin associated protein complex in skeletal muscle. Neuromuscul Disord 2003; 13:21-31.

110. Hartigan-O'Connor D, Kirk CJ, Crawford R et al. Immune evasion by muscle-specific gene expression in dystrophic muscle. Mol Ther 2001; 4:525-33.
111. Weeratna RD, Wu T, Efler SM et al. Designing gene therapy vectors: Avoiding immune responses by using tissue-specific promoters. Gene Ther 2001; 8:1872-8.
112. Chirmule N, Propert K, Magosin S et al. Immune responses to adenovirus and adeno-associated virus in humans. Gene Ther 1999; 6:1574-83.
113. McMahon JM, Wells KE, Bamfo JE et al. Inflammatory responses following direct injection of plasmid DNA into skeletal muscle. Gene Ther 1998; 5:1283-90.
114. Carmen IH. A death in the laboratory: The politics of the Gelsinger aftermath. Mol Ther 2001; 3:425-8.
115. Stedman H, Wilson JM, Finke R et al. Phase I clinical trial utilizing gene therapy for limb girdle muscular dystrophy: Alpha-, beta-, gamma-, or delta-sarcoglycan gene delivered with intramuscular instillations of adeno-associated vectors. Hum Gene Ther 2000; 11:777-90.
116. Thioudellet C, Blot S, Squiban P et al. Current protocol of a research phase I clinical trial of full-length dystrophin plasmid DNA in Duchenne/Becker muscular dystrophies. Part I: Rationale. Neuromuscul Disord 2002; 12:S49-51.
117. O'Hara AJ, Howell JM, Taplin RH et al. The spread of transgene expression at the site of gene construct injection. Muscle Nerve 2001; 24:488-95.
118. Cho WK, Ebihara S, Nalbantoglu J et al. Modulation of starling forces and muscle fiber maturity permits adenovirus-mediated gene transfer to adult dystrophic (mdx) mice by the intravascular route. Hum Gene Ther 2000; 11:701-14.
119. Liu F, Nishikawa M, Clemens PR et al. Transfer of full-length Dmd to the diaphragm muscle of Dmd(mdx/mdx) mice through systemic administration of plasmid DNA. Mol Ther 2001; 4:45-51.
120. Zhang G, Ludtke JJ, Thioudellet C et al. Intra-arterial delivery of naked plasmid DNA expressing full-length mouse dystrophin in the mdx mouse model of Duchenne muscular dystrophy. Hum Gene Ther 2004; 15:770-782.
121. Hagstrom JE, Hegge J, Zhang G et al. A facile nonviral method for delivering genes and siRNAs to skeletal muscle of mammalian limbs. Mol Ther 2004; 10:386-398.
122. Gregorevic P, Blankenship MJ, Allen JM et al. Systemic delivery of genes to striated muscles using adeno-associated viral vectors. Nat Med 2004; 10:828-834.

CHAPTER 18

Cell Therapies for Muscular Dystrophy

Terence A. Partridge

Rationale

D estruction of muscle fibres accompanied by a reactive regenerative response is a major defining characteristic that distinguishes the muscular dystrophies from the more general category of myopathies caused by genetic defects that impact on skeletal muscle.[1] In the early stages of severe dystrophies, such as Duchenne muscular dystrophy (DMD), this pathology appears to hold a balance, the regenerative response effectively compensating the loss of muscle fibres. Eventually, however, there is a progressive loss of muscle mass and disruption of structure with consequent physiological dysfunction. This is generally attributed to exhaustion of the proliferative and myogenic competence of resident muscle precursor cells.[2,3] Hence, the notion of supplementing these resident cells by grafting of more potent myogenic cells. If these grafted cells were genetically normal or genetically corrected, they would achieve the dual objectives of replacing the failing endogenous myogenic capacity and, at the same time, of introducing normal copies of the defective gene into the newly formed muscle.[4-6] A particular advantage arises in the case of skeletal muscle from the fact that muscle fibres are syncytia formed by fusion of the individual myoblasts, for this raises the prospect of each functional copy of the introduced gene spreading its product along the length of the syncytium over several nuclear territories and thus of achieving a disproportionate beneficial effect.

Figure 1 summarizes the various options within this strategy.

Background

In the early 1960's two important advances in our understanding of skeletal muscle occurred almost simultaneously. First, it was demonstrated that the syncytial skeletal muscle fibre is formed by fusion together of mononucleated precursor cells, termed myoblasts.[7] Second, two independent reports described the satellite cell, lying beneath the basement membrane surrounding each muscle fibre both in the general population of muscle fibres that produce the contractile force[8] and the spindle fibres responsible for sensory information concerning length and tension within the muscle.[9] These swiftly became the almost unquestioned candidates as the endogenous source of myoblasts in skeletal muscle to the extent that the terms satellite cell and myoblast are commonly used synonymously. These two features have provided the accepted paradigm for myogenesis during postnatal muscle growth[10] and during regeneration following muscle injury[11] and have only been seriously challenged in recent times.

There has, however, always been room for heterodox voices, because the evidence for the link between the satellite cell and the histochemically recognizable myogenic cells that are seen in regions of muscle regeneration, or can be extracted and tissue cultured, is essentially circumstantial.

It has been suggested for example, that not all myogenic cells were derived from satellite cells but that some may arise from mononucleated buds from parts of damaged muscle fibres.[12] Such dedifferentiation and budding appears to occur in the regenerating limbs of urodele amphibians[13-15] in which cells are not found in the anatomical satellite cell niche.[16] Convincing

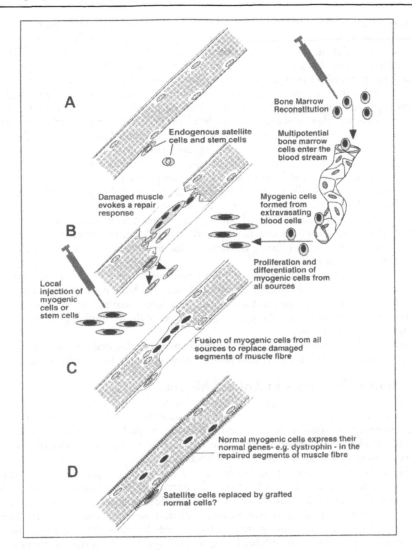

Figure 1. Principle of cell transplantation as a mode of therapy for muscular dystrophies. A) Muscle fibres carry myogenic precursors, called satellite cells on their surface. There is also some evidence that myogenic precursors exist in the tissue between muscle fibres. B) Damaged muscle fibres may die from end to end, but some sections commonly reseal and survive. Muscle damage activates the endogenous myogenic satellite cells to proliferate. It has been demonstrated that locally injected myogenic cells mostly die but the survivors are also activated to proliferate and to migrate for distances of up to a few millimetres. Muscle damage also attracts inflammatory cells, generated within the bone marrow into the region, including a few cells which are capable of differentiating into skeletal muscle. Thus cells derived from a normal bone marrow may participate in repair of skeletal muscle. C) Myogenic cells from all of the above sources have been demonstrated to contribute to muscle regeneration by fusion together to replace dead segments of muscle fibre and to form new muscle fibres in areas of complete death of the local muscle. D) In segments of fibre and in new fibres, genetically normal myonuclei within the fibres are able to expresss the normal gene that is lacking in the nuclei of the host, e.g., dystrophin. Within the syncytial muscle fibre, this gene product is able to diffuse some distance from the nucleus that contained the gene encoding it thus enhancing the beneficial effect of a given genomic contribution. In the case of dystrophin, this diffusion distance can be estimated at about 10 nuclear diameters, i.e., 100 or so μm.

evidence for such mechanisms in higher vertebrates has been hard to come by apart from recent demonstrations in vitro that myotubes derived from the C2 muscle cell line can be caused to split into mononucleated cells and to reenter the cell cycle, either by treatment with Myoseverin[17] or by over-expression of Msx1.[18] Another suggested nonsatellite cell source of myogenic cells, especially in regenerating muscles, has been the blood-born cells that enter myonecrotic lesions during the inflammatory phase. Such a source was suggested some years ago by findings that tritiated thymidine injected prior to injury of skeletal muscle was initially found predominantly in cells of the blood and bone marrow but later appeared in myonuclei of regenerated muscle fibres.[19] In retrospect, this result may well have represented reutilization of the thymidine label[20] but it was responsible for the first suggestion that stem-cell precursors of myogenic cells might reside in the bone marrow to be distributed via the blood vascular system, with the corollary that this route could be exploited to infiltrate genetically defective muscle with genetically normal or genetically corrected myogenic precursors. Subsequent studies, using enzyme polymorphisms as true lineage markers, did show that muscle precursors could move into regenerating muscles from the outside[21-23] but that they almost certainly came from neighbouring muscles[22] and no input from the circulation was detected.[24] This movement of myogenic cells between neighbouring muscles during regeneration has recently been convincingly confirmed by use of the GFP transgenic marker.[25]

Over the past few years, it has been demonstrated that there is indeed a contribution to regenerating muscle from circulating cells derived from the bone marrow.[26-28] This input is extremely small and has been detected only by use of sensitive markers, but it has again raised the prospect of using myogenic cells to carry good copies of the dystrophin gene into regenerating dystrophic muscle fibres.

The remainder of this article will discuss the prospects of using grafts of myogenic cells for therapy, the barriers to this approach and possible ways in which these might be surmounted.

Validation of Principles in Animal Models

Pending the identification of genuine animal models of genetic disease of skeletal muscle, experimental investigation of the idea of myoblast transplantation was limited to the use of various markers. Here again, until recent years, there was the problem of establishing the validity of most markers. Radio-labelled thymidine or BrdU for example are both toxic to cells at high concentrations and rapidly dilute out of the detectable range in proliferating cells. Moreover, they and most other labels applied to cells could, in principle, be passed to unlabelled cells in vivo, to an extent that could only be assessed by comparison with a known stable marker.[20] In the event the only reliable early markers were genetic polymorphisms between strains of animals, but these raised the problem of the potential immune response of the recipient animal to the grafted cells. It was found to be possible to graft between different strains of mice, matched at the major histocompatibility loci, but carrying different allotypes of enzymes such as malate dehydrogenase[29] and Glucose-6-phosphate dehydrogenase.[21] It was also clear however, that cell transplants were more successful in immunodeficient animals,[23] implying some level of immunological interference short of outright rejection. The one known genetic model of a muscular dystrophy in the mouse, the dy and dy[2J] mice, was clearly neuropathic as well as myopathic.[30] and has only in recent years been identified as a homologue of merosin-deficient congenital muscular dystrophy. Prior to this identification, clinical phenotypic correction of the disease was the only marker of success of myoblast transplantation, and a series of papers claiming dramatic rescue of mice suffering from this condition were published[31-33] and formed part of the basis for a trial of myoblast transplantation in man in the early 1990s. Myoblast transplant experiments conducted subsequent to identification of the underlying genetic defect[34] show replacement of merosin within the engrafted muscle but at far too low a level to sustain credibility as to the original claims of body-wide clinical rectification.[31,32]

Actual replacement of the product of a genetically defective gene in muscle was first accomplished in the mouse lacking phosphorylase-kinase,[35] in which there was no gross pathological consequence of the defect. More important for treatment of DMD, was the subsequent demonstration that normal myoblasts injected into muscles of the dystrophic mdx mouse were able to participate in regeneration of muscle in and to generate expression of dystrophin in the engrafted muscles.[5,36] However, as was found later in the merosin-deficient mouse, the bulk of the injected myoblasts dispersed only a few millimetres from the injection site: providing a far too limited expression of dystrophin to be of clinical value.

Subsequent animal experimentation has addressed 3 main problems. First, some considerable effort has been put into attempts to raise the efficiency of muscle formation from the injected myoblasts. Second, the problem of the limited dispersion of myoblasts from the injection site has been the subject of a number of diverse approaches. Last, the clear immune problems in the mouse have provoked continuing interest in the more general programme of research into suppression and obviation of the immune response to allografts and foreign transgenes.

Progress in all three areas has been limited. A major factor in the low efficiency of myoblast transplantation has turned out to be the massive and rapid death by necrosis of the vast majority of transplanted cells.[37-41] The cause of this early cell death remains mysterious. Where it is possible to make estimates of the number of donor myonuclei in the regenerated muscle, they correspond approximately the number of cells injected, or a fraction thereof[42] not the massive augmentation of yield that one might hope for from a stem cell. Despite a number of reports of regimes that reduce the very early loss of the grafted myoblasts,[43-46] no measures have succeeded in enhancing the final outcome by even one order of magnitude. Only two strategies appear to have led to significant improvement in the yield of muscle from a given number of injected cells. First, two types of stem-like cell have been isolated in mice. One is a cell isolated from muscle by virtue of its lack of adhesiveness to a collagen substrate[47] that gives rise to some 5-10 fold more nuclei in the resultant muscle fibres than was present in the cells that were injected. The second appears to be derived from an endothelial precursor[48] and its yield efficiency is less easy to assess but it too seems to generate large amounts of muscle from a relatively small injected dose.[49] It also has the virtue that it can be delivered relatively efficiently by intra-arterial injection. Up to now, no human equivalent of either of these cells has been reported. A second approach to improvement of the yield of muscle from transplanted myogenic cells is to provide a more stimulating environment. In mice, the most effective such measure is preirradiation of the graft site, which generates some 2-3 times as much muscle as would be formed from the same number of cells in a nonirradiated site,[42,50] not by improving myoblast survival but by driving the few surviving myogenic cells into extensive and rapid proliferation.[40] This appears to involve a genuine proliferative signal, since it drives myoblasts of the C2 cell line into rapid formation of invasive tumours.[51]

Although the early studies of muscle transplantation and of myoblast transplantation showed movement of muscle precursors between neighbouring muscles in the mouse leg, this was barely detectable with isoenzyme markers[22,23,52] and was not demonstrable in the rat.[53] More recently the histologically visualizable GFP and lacZ markers have confirmed that movement of muscle precursors between neighbouring muscles does occur in the mouse but is constrained close to the intermuscular interface.[25] Over the years a number of strategies have been applied to stimulate the migration of myoblasts from the injection site, but none have shifted the total range beyond a few millimetres.[54-56]

A ray of hope, on the problem of adequate dispersion was the discovery that bone marrow stem cells were capable of entering regenerating muscle and participating in the regeneration of damaged fibres.[26-28] These findings caused great excitement, holding promise of an almost ideal curative procedure; envisaged as a transplant of bone marrow from a normal donor, or of genetically corrected autologous bone marrow, that would form a reservoir of normal stem cells that would gradually feed into the dystrophic muscle lesions over time. Disappointingly, thus far, the efficiency of muscle formation from cells of bone marrow origin in animal models has fallen far short of utility[57] in the mdx mouse or in the one DMD patient who has carried a long-term bone marrow transplant from his normal father.[58]

However, hope has again been revived by the identification of the talented class of stem cell derived from small blood vessels whose combined expression of mesodermal stem cells and endothelial markers has emerged in the coined epithet 'Meso-angioblast'.[48] Such cells have been used to partially correct the dystrophy associated with the α-sarcoglycan deficient mouse model of LGMD 2D Limb-Girdle dystrophy[49] (see Chapter 10). On injection into the femoral artery, these cells impact in the first downstream microvascular bed, mainly in the limb muscles in this instance, and emigrate into regions of muscle degeneration. This is the only example so far of repair of a substantial volume of muscle by cells delivered via the circulation. Because of their almost unlimited proliferative potential, it has proven possible to expand cells originating from the sarcoglycan deficient strain after genetic correction with a lentiviral vector to numbers large enough to treat entire limbs. This strategy falls short of the ideal of stem cell therapy in which a perpetual reservoir persists in for example the bone marrow, in that delivery of mesangioblasts is acute and would need to be repeated at intervals until the chronically degenerating diseased muscle is completely repopulated.

An important feature of any comprehensive cell-based therapy for muscular dystrophies is the replacement of the satellite cell compartment of the recipient muscle by the grafted cells. This has been shown to occur when for skeletal myoblasts are grafted[59,60] and bone marrow derived cells have also been shown to enter the satellite cell niche[61,62] but the myogenic capacity of these cells has not been reported.

Lessons from Past Mistakes

Shortly after the initial demonstrations that injected genetically normal myogenic cells were able to contribute to regeneration of muscle fibres of the dystrophic mdx mouse and to express the normal dystrophin gene within those muscle fibres, human trials of myoblast transfer were initiated, despite the expression of considerable unease by the majority of the scientific community.[63] None of the studies that took a rigorous approach to data collection demonstrated more than a trivial effect of myoblast transplantation on the dystrophin content of the recipient muscle.[64-70] The one centre to claim benefit from the therapy[71-73] failed to present its data in a manner that would permit rigorous evaluation.[74] In general, the problems previously identified in the animal studies, of inadequate dispersion of the injected cells from the graft site and of immune response against major and minor antigens, appear to be the major obstacles in man too. Jacques Tremblay's group have devoted considerable effort to schemes for overcoming these problems, and have demonstrated that the immune problem in particular is susceptible to modern immunosuppressive treatments, notably the use of FK506,[75] but have been able only marginally to improve the yield of muscle or the dispersion of myogenic cells from the injection site. A recent trial of myoblast transplantation has incorporated the improvements arising from these animal studies; preliminary results indicate a definite improvement in the proportion of dystrophin positive muscle fibres attributable to the donor cells but, at below 10%(Tremblay personal communication), still short of practical utility. At present, the only solution proposed to overcome these problems is to inject more cells at more closely spaced (<2mm) sites.

Outlook

Perhaps ironically, the most useful outcome to date from the early myoblast transplantation experiments in human subjects has been to engender a more cautious and thoughtful attitude to the application of gene and stem cell therapies to DMD. Fortunately, there were no severe adverse outcomes from the human trials, but the almost completely negative results of these studies did serve as a general lesson that seems to have moderated the unwarranted optimism that new discoveries commonly inspire in those involved in their development. It is, now quite clear that the effects of direct injection of myogenic cells into muscle is too local in effect to provide a useful restoration of function to more than a few selected muscles, and even this will require a significant increase in efficiency.

For body-wide treatment, cell based therapies, like any other genetic therapy, requires systemic delivery that can most plausibly be achieved via the blood vascular system. It is this realization that sustains a high degree of interest in the circulating stem cell phenomena. So far however, the demonstrated efficiency of delivery of these cells by the vascular route has been too low to be of practical use. Only the meso-angioblasts have showed any real promise in this respect, and in this instance the mechanism is not a targeted vascular delivery but impaction of the cells in the microvascular bed immediately downstream of the intra-arterial injection site. This loses the potential benefit of chronic delivery that might be expected of a bone marrow derived stem cell but certainly achieves a broader distribution than can be achieved with local intramuscular injection. In this respect, it is worth noting that it is this propensity for dispersed vascular delivery that is the most attractive feature of the bone marrow derived stem cell, while the plasticity of differentiative fate that is the most commonly lauded criterion of 'stem-cell-ness' is largely irrelevant; the only real fate in which we are interested is myogenesis, provided it is combined with the property of transendothelial migration. But, the potential for vascular delivery is not confined to stem cells, being clearly seen in the numerous inflammatory cells, especially macrophages that invade dystrophic lesions. It ought to be possible therefore to confer the mechanisms involved in macrophage homing to lesions onto committed myogenic cells,[76] and a recent publication suggests that it is indeed cells of this class that are responsible for the bone-marrow derived contribution to muscle regeneration.[77] Certainly it seems likely that a gene therapy alone will not provide a comprehensive therapy for DMD patients in whom there has already been serious disruption of the muscle architecture, and the best candidate for effective restoration of normal muscle structure is the normal cellular machinery of muscle repair.

References

1. Walton JN. Disorders of the volantary muscle. 2nd ed. London: J. & A. Churchill Ltd., 1969.
2. Webster C, Blau HM. Accelerated age-related decline in replicative life-span of Duchenne muscular dystrophy myoblasts: Implications for cell and gene therapy. Somat Cell Mol Genet 1990; 16:557-565.
3. Decary S, Mouly B, Hamida C et al. Replicative potential and telomere length in human skeletal muscle: Implications for satellite cell- mediated gene therapy. Hum Gene Ther 1997; 8:1429-1438.
4. Partridge TA. Invited Review: Myoblast transfer:A possible therapy for inherited myopathies? Muscle Nerve. 1991; 14:197-212.
5. Partridge TA, Morgan JE, Coulton GR et al. Conversion of mdx myofibres from dystrophin-negative to -positive by injection of normal myoblasts. Nature 1989; 337:176-179.
6. Karpati G. Dystrophin expression in mosaic skeletal muscle fibres. In: BA Kakulas FLM, ed. Pathogenesis and Therapy of Duchenne and Becker Muscular Dystrophy. New York: Raven Press, 1989:143-150.
7. Konigsberg IR. Cellular differentiation in colonies derived from single cell platings of freshly isolated chick embryo muscle cells. Proc Natl Acad Sci USA 1961; 47:1868-1872.
8. Mauro A. Satellite cell of skeletal muscle fibers. J Biophys Biochem 1961; 9:493-495.
9. Katz B. The terminations of the afferent nerve fibre in the muscle spindle of the frog. Phil Trans Roy Soc Lond [Biol] 1961; 243:221-240.
10. Moss FP, Leblond CP. Satellite cells as the source of nuclei in muscles of growing rats. Anat Rec 1971; 170:421-435.
11. Snow MH. Origin of regenerating myoblasts in mammalian skeletal muscle. In: Mauro A, ed. Muscle Regeneration. New York: Raven Press, 1979:91-100.
12. Reznik M. Origin of myoblasts during skeletal muscle regeneration. Electron microscopic observations. Lab Invest 1969; 20(4):353-563.
13. Lo DC, Allen F, Brockes JP. Reversal of muscle differentation during urodele limb regenereation. Proc Natl Acad Sci USA 1993; 90:7230-7234.
14. Velloso CP, Kumar A, Tanaka EM et al. Generation of mononucleate cells from post-mitotic myotubes proceeds in the absence of cell cycle progression. Differentiation 2000; 66:239-246.
15. Kinter CR, Brockes JP. Monoclonal antibodies identify blastemal celld derived from dedifferentiating muscle in newt limb regeneration. Nature 1984; 308:67-69.
16. Hay ED. Electron microscopic observations of muscle dedifferentiation in regenerating amblystoma limbs. Devel Biol 1959; 1:555-585.
17. Rosania GR, Chang Y-T, Perez O et al. Myoseverin, a microtubule-binding molecule with novel cellular effects. Nat Biotech 2000; 1-8.
18. Odelberg SJ, Kollhoff A, Keating MT. Dedifferentiation of mammalian myotubes induced by msx1. Cell 2000; 103:1099-1109.

19. Bateson RG, Woodrow DF, Sloper JC. Circulating cell as a source of myoblasts in regenerating injured mammalian skeletal muscle. Nature 1967; 213:1035-1036.
20. Grounds MD, McGeachie JK. Reutilisation of tritiated thymidine in studies of regenerating skeletal muscle. Cell Tiss Res 1987; 250:141-148.
21. Partridge TA, Grounds M, Sloper JC. Evidence of fusion between host and donor myoblasts in skeletal muscle grafts. Nature 1978; 273:306-308.
22. Grounds MD, Partridge TA. Isoenzyme studies of whole muscle grafts and movement of muscle precursor cells. Cell Tiss Res 1983; 230:677-688.
23. Morgan JE, Coulton GR, Partridge TA. Muscle precursor cells invade and repopulate freeze-killed skeletal muscles. J Musc Res Cell Motil 1987; 8:386–396.
24. Grounds MD. Skeletal muscle precursors do not arise from bone marrow cells. Cell Tiss Res 1983; 234:713-722.
25. Jockusch H, Voigt S. Migration of adult myogenic precursor cells as revealed by GFP/nLacZ labelling of mouse transplantation chimeras. J Cell Sci 2003; 116:1611-1616.
26. Ferrari G, Cusella-De Angelis G, Coletta M et al. Muscle regeneration by bone marrow-derived myogenic progenitors. Science 1998; 279:1528-1530.
27. Bittner RE, Schöfer C, Weipoltshammer K et al. Recruitment of bone-marrow-derived cells by skeletal and cardiac muscle in adult dystrophic mdx mice. Anat Embryol 1999; 199:391-396.
28. Gussoni E, Soneoka Y, Strickland CD et al. Dystrophin expression in the mdx mouse restored by stem cell transplantation. Nature 1999; 401:390-394.
29. Partridge TA, Sloper JC. A host contribution to the regeneration of muscle grafts. J Neurol Sci 1977; 33:425-435.
30. Peterson AC. Chimera mouse study shows absence of disease in genetically dystrophic muscle. Nature 1974; 248:561-564.
31. Law PK, Goodwin TG, Wang MG. Normal myoblast injections provide genetic treatment for murine dystrophy. Muscle Nerve 1988; 11:525-533.
32. Law PK, Goodwin TG, Li HJ. Histoincompatible myoblast injection improves muscle structure and function of dystrophic mice. Transplant Proc 1988; 20:1114-1119.
33. Law PK. Nondystrophic myoblast transplantation into dystrophic muscle. Muscle Nerve 1989; 337-338.
34. Vilquin JT, Kinoshita I, Roy R et al. Partial laminin $\alpha 2$ chain restoration in $\alpha 2$ chain-deficient dy/dy mouse by primary muscle cell culture transplantation. J Cell Biol 1996; 133:185-197.
35. Morgan JE, Watt DJ, Sloper JC Partridge TA. Partial correction of an inherited biochemical defect of skeletal muscle by grafts of normal muscle precursor cells. J Neurol Sci 1988; 86:137-147.
36. Karpati G, Pouliot Y, Zubrzycka Gaarn E et al. Dystrophin is expressed in mdx skeletal muscle fibers after normal myoblast implantation. Am J Pathol 1989; 135:27-32.
37. Rando TA, Blau HM. Primary mouse myoblast purification, characterization, and transplantation for cell-mediated gene therapy. J Cell Biol 1994; 125:1275-1287.
38. Rando TA, Pavlath G, Blau HM. The fate of myoblasts following transplantation into mature muscle. Expl Cell Res 1995; 220:383-389.
39. Huard J, Acsadi G, Jani A et al. Gene transfer into skeletal muscles by isogenic myoblasts. Hum Gene Ther 1994; 5:949-958.
40. Beauchamp JR, Morgan JE, Pagel CN et al. Dynamics of myoblast transplantation reveal a discrete minority of precursors with stem cell-like properties as the myogenic source. J Cell Biol 1999; 144:1113-1121.
41. Fan Y, Maley M, Beilharz M et al. Rapid death of injected myoblasts in myoblast transfer therapy. Muscle Nerve 1996; 19:853-860.
42. Morgan JE, Fletcher RM, Partridge TA. Yields of muscle from myogenic cells implanted into young and old mdx hosts. Muscle Nerve 1996; 19:132-139.
43. El Fahime E, Bouchentouf M, Benabdallah BF et al. Tubulyzine, a novel tri-substituted triazine, prevents the early cell death of transplanted myogenic cells and improves transplantation success. Biochem Cell Biol 2003; 81:81-90.
44. Hodgetts SI, Beilharz MW, Scalzo AA et al. Why do cultured transplanted myoblasts die in vivo? DNA quantification shows enhanced survival of donor male myoblasts in host mice depleted of CD4+ and CD8+ cells or Nk1.1+ cells. Cell Transplant 2000; 9:489-502.
45. Hodgetts SI, Grounds MD. Complement and myoblast transfer therapy: Donor myoblast survival is enhanced following depletion of host complement C3 using cobra venom factor, but not in the absence of C5. Immunol Cell Biol 2001; 79:231-239.
46. White JD, Davies M, Grounds MD. Leukaemia inhibitory factor increases myoblast replication and survival and affects extracellular matrix production: Combined in vivo and in vitro studies in post-natal skeletal muscle. Cell Tissue Res 2001; 306:129-141.

47. Qu-Petersen Z, Deasy B, Jankowski R et al. Identification of a novel population of muscle stem cells in mice: Potential for muscle regeneration. J Cell Biol 2002; 157:851-864.
48. Cossu G, Bianco P. Mesoangioblasts—vascular progenitors for extravascular mesodermal tissues. Curr Opin Genet Dev 2003; 13:537-542.
49. Sampaolesi M, Torrente Y, Innocenzi A et al. Cell therapy of alpha-sarcoglycan null dystrophic mice through intra-arterial delivery of mesoangioblasts. Science 2003; 301:487-492.
50. Morgan JE, Pagel CN, Sherratt T et al. Long-term persistence and migration of myogenic cells injected into preirradiated muscles of mdx mice. J Neurol Sci 1993; 115:191-200.
51. Morgan JE, Gross JG, Pagel CN et al. Myogenic cell proliferation and generation of a reversible tumorigenic phenotype are triggered by preirradiation of the recipient site. J Cell Biol 2002; 157:693-702.
52. Grounds M, Partridge TA, Sloper JC. The contribution of exogenous cells to regenerating skeletal muscle: An isoenzyme study of muscle allografts in mice. J Pathol 1980; 132:325-341.
53. Schultz E, Jaryszak DL, Gibson MC et al. Absence of exogenous satellite cell contribution to regeneration of frozen muscle. J Musc Res Cell Motil 1986; 7:361-367.
54. Ito H, Hallauer P, Hastings K et al. Prior culture with conconavalin A increases intramuscular migration of transplanted myoblast. Muscle Nerve 1998; 21:291-297.
55. Torrente Y, El Fahime E, Caron NJ et al. Intramuscular migration of myoblasts transplanted after muscle pretreatment with metalloproteinases. Cell Transplant 2000; 9:539-549.
56. Smythe GM, Grounds MD. Absence of MyoD increases donor myoblast migration into host muscle. Exp Cell Res 2001; 267:267-274.
57. Ferrari G, Stornaiuolo A, Mavilio F. Failure to correct murine muscular dystrophy. Nature 2001; 411:1014-1015.
58. Gussoni E, Bennett RR, Muskiewicz KR et al. Long-term persistence of donor nuclei in a Duchenne muscular dystrophy patient receiving bone marrow transplantation. J Clin Invest 2002; 110:807-814.
59. Blaveri K, Heslop L, Yu DS et al. Patterns of repair of dystrophic mouse muscle: Studies on isolated fibers. Dev Dyn 1999; 216(3):244-256.
60. Heslop L, Beauchamp JR, Tajbakhsh S et al. Transplanted primary neonatal myoblasts can give rise to functional satellite cells as identified using the Myf5nlacZl+ mouse. Gene Ther 2001; 8(10):778-783.
61. LaBarge MA, Blau HM. Biological progression from adult bone marrow to mononucleate muscle stem cell to multinucleate muscle fiber in response to injury. Cell Nov 15 2002; 111(4):589-601.
62. Dreyfus P, Chretien F, Chazaud B et al. Adult bone marrow-derived stem cells in muscle connective tissue and satellite cell niches. Am J Pathol. 2004; in press.
63. In: Griggs RC, Karpati G, eds. Myoblast transfer therapy. New York: Plenum Press, 1990.
64. Gussoni E, Pavlath GK, Lanctot AM et al. Normal dystrophin transcripts detected in Duchenne muscular dystrophy patients after myoblast transplantation. Nature 1992; 356:435-438.
65. Mendell JR, Kissel JT, Amato AA et al. Myoblast transfer in the treatment of Duchenne's muscular dystrophy. N Eng J Med 1995; 333:832-838.
66. Karpati G, Ajdukovic D, Arnold D et al. Myoblast transfer in Duchenne muscular dystrophy. Ann Neurol 1993; 34:8-17.
67. Huard J, Roy R, Bouchard JP et al. Human myoblast transplantation between immunohistocompatible donors and recipients produces immune reactions. Transplant Proc 1992; 24:3049-3051.
68. Huard J, Bouchard JP, R R et al. Human myoblast transplantation: Preliminary results of 4 cases. Muscle Nerve 1992; 15:550-560.
69. Tremblay JP, Malouin F, Roy R et al. Results of a triple blind clinical study of myoblast transplantations without immunosuppressive treatment in young boys with Duchenne muscular dystrophy. Cell Transplant 1993; 2:99-112.
70. Huard J, Roy R, Guerette B et al. Human myoblast transplantation in immunodeficient and immunosupressed mice: Evidence of rejection. Muscle Nerve 1994; 17:224-234.
71. Law PK, Goodwin TG, Fang Q et al. Human gene therapy with myoblast transfer. Transplant Proc 1997; 29:2234-2237.
72. Law PK, Goodwin TG, Fang Q et al. First human myoblast transfer therapy continues to show dystrophin after 6 years. Cell Transplant 1997; 6:95-100.
73. Law PK. Myoblast transfer therapy. Lancet 1993; 341:247.
74. Partridge TA. Letters to the editor—Is myoblast transplantation effective? Nat Med 1998; 4:1208.
75. Kinoshita I, Vilquin JT, Guerette B et al. Immunosuppression with FK 506 insures good success of myoblast transplantation in MDX mice. Transplant Proc 1994; 26:3518.
76. Parrish EP, Cifuentes-Diaz C, Li ZL et al. Targeting widespread sites of damage in dystrophic muscle: Engrafted macrophages as potential shuttles. Gene Ther 1996; 3:13-20.
77. Camargo FD, Green R, Capetenaki Y et al. Single hematopoietic stem cells generate skeletal muscle through myeloid intermediates. Nat Med 2003; 9:1520-1527.

Index

9 780367 446376